"Fazale Rana blew my mind with the ubiquitous extent to which chemistry and biochemical systems must be exquisitely fine-tuned to make any kind of physical life possible."

—Hugh Ross
Founder and President
Reasons to Believe

"Astronomers and astrophysicists have long marveled at the way the universe appears to be finely tuned to support life. Cells are equally complex (famed cell biologist Bruce Alberts described them as 'self-replicating robots') and they are the building blocks of all life. Could it be that the molecules that make up cells are equally fine-tuned? In this book, biochemist Fazale Rana makes a compelling case that not only do the molecules that make up cells possess the exact physical and chemical properties necessary for life as we know it to exist, but they are the only molecules with precisely these properties. This book provides a fascinating review of the biochemical properties of DNA, RNA, and other biomolecules, as well as studies demonstrating that alternative molecules would simply not be able to sustain life as we know it. Throughout this lesson in biochemistry, Rana weaves a captivating narrative of the development of his personal life's soundtrack, which includes everything from Eddie Money and Jeff Buckley to Bob Dylan and Led Zeppelin. This book serves as both a graduate-level course in biochemistry and a lesson in music history."

—Sharon R. Bloch
Professor of Science
Lincoln Christian University

"It's well-known that the laws of nature, their constants, and the forces that make up the universe have to all be just-right for the universe and for life to exist. This is called the anthropic principle, and it points to the origin of the universe requiring a mind and a purpose. In *Fit for a Purpose*, Rana describes how this principle extends to the very chemistry of the molecules necessary for life; molecules such as amino acids, nucleosides, carbohydrates, and lipids have just-right unique chemical properties that allow them to be used for the formation of proteins, nucleic acids (DNA and RNA), polysaccharides, and membranes and then used to form finely tuned biochemical systems. Rana further discusses the implication that this molecular fine-tuning supports the conclusion that a mind and a purpose are behind the origin of life; that the molecules of life are uniquely 'fit' for the 'purpose' of life. This book is both entertaining and very informative. I highly recommend it."

—Russell W. Carlson
Emeritus Professor of Biochemistry and Molecular Biology
and of the Complex Carbohydrate Research Center
University of Georgia

"When we hear words like logic, design, sophistication, elegance, and optimality, we naturally think of humanly created systems. Yet we can observe these special qualities in many parts of the natural world down to the most fundamental level. I see this in my own field of chemistry as well as in many other fields. But what about biochemistry? Does it hold there down to the level of the fundamental components of the cell? Fazale Rana takes us on a masterful tour of the inner workings of the cell, demonstrating that they have an uncanny resemblance to a well-designed factory."

—John Millam
Theoretical Chemist, PhD
Rice University

"*Fit for a Purpose* engages the reader with a fascinating question: Does the complex beauty of biochemistry amaze us because it is so improbable? Or does it amaze us because it hints at something more? To use a language metaphor, Fazale unpacks the letters, words, sentences, and paragraphs of biochemistry and uncovers a universal grammar and style guide along the way. I highly recommend it for science enthusiasts of all stripes."

—Darren Williams
Professor of Physical Chemistry
Sam Houston State University

"Fazale Rana probably has the broadest and deepest knowledge of biology and biochemistry of any scientist I know. With a doctorate in chemistry and biochemistry, a thorough knowledge of Scripture, and extensive experience, he is uniquely qualified to raise the question: Does the anthropic principle apply to biology and the design of life? Starting with an explanation of the anthropic principle and its origins in cosmology, Dr. Rana extends this principle to a number of areas of chemistry and biochemistry in his extensively referenced book. In this context he explains certain aspects of chemistry, protein and nucleic acid structure and synthesis, cell membranes, and energy-producing metabolic pathways. Using a balanced analysis, he proposes that the anthropic principle does apply to the chemistry and biochemistry of these topics, with which I agree. Further, he discusses the scientific and metaphysical implications of such biochemical anthropic principles in relationship to a revitalized Watchmaker argument, creationism, theistic evolution, and the origin of life."

—John C. Drach
Professor Emeritus of Medical Chemistry
University of Michigan

FAZALE R. RANA

FIT FOR A PURPOSE

Does the Anthropic Principle Include Biochemistry?

Covina, CA

Cover design: 789, Inc.
Interior layout: Christine Talley
Figure design: Heather Lanz

Names: Rana, Fazale, 1963-, author.
Title: Fit for a purpose : does the anthropic principle include biochemistry? / by Fazale R. Rana.
Description: Includes bibliographical references and index. | Covina, CA: RTB Press, 2021.
Identifiers: ISBN: 978-1-956112-00-9
Subjects: LCSH Biochemistry--Religious aspects--Christianity. | Anthropic principle. | Intelligent design (Teleology) | Life--Origin. | BISAC RELIGION / Christian Theology / Apologetics | RELIGION / Religion & Science | SCIENCE / Life Sciences / Cell Biology | SCIENCE / Life Sciences / Biochemistry | SCIENCE / Life Sciences / Genetics & Genomics
Classification: LCC BL255 .R36 2021| DDC 215/.7--dc23

Printed in the United States of America

First edition

1 2 3 4 5 6 7 8 9 10 / 25 24 23 22 21

For more information about Reasons to Believe, contact (855) REASONS / (855) 732-7667 or visit reasons.org.

Dedication

To Amy Rana

My precious angel
I am so glad that you were the one to show me
how weak was the foundation I was standing upon.

Contents

List of Figures and Tables

Figures

Tables

Acknowledgments

A number of people worked hard and sacrificed to make this book possible. First, I want to say thank you to my wife, Amy Rana. I'm grateful for your love, encouragement, and patience when this book project took "priority" over family matters.

I especially want to acknowledge the editorial team at Reasons to Believe who dedicated themselves to this book as if it were their own. Much thanks to Sandra Dimas, Joe Aguirre, Brett Tarbell, Helena Heredia, Ruth Alvarado, Maureen Moser, Jocelyn King, and freelance editor Linda Hucks for your expert editorial guidance. Your efforts transformed our manuscript into a book that I'm proud to call my own. Thanks to Charley Bell and Richard Silva from 789, Inc., for designing the book cover. Thanks to Heather Lanz for designing figures found in the book's interior. Also, thanks to Christine Talley for designing the interior layout of the book.

I'm indebted to the scholar team at Reasons to Believe. Thanks to Hugh Ross, David Rogstad, Jeffrey Zweerink, Kenneth Samples, Krista Bontrager, and George Haraksin for many insightful conversations in the hallway and during lunch over the years. These discussions helped to directly and indirectly shape the contents of this book. Thanks to Drs. Russell Carlson and Sharon Bloch, and Richard Deem for reviewing the manuscript. Your valuable input helped improve the technical accuracy and clarity of the book.

I am also grateful to each member of the Reasons to Believe team. You have supported me with your friendship and encouragement through the years. I am honored to call you friends. I am also grateful to Hugh and Kathy Ross for their friendship and guidance. Your influence has had a profound and lasting influence on my life.

To our ever-faithful Reasons to Believe donors: Dan and Katy Atwood, Ron and Bett Behrens, Ken and Tracy Camacho, Jamie and Maria Campbell, Mark and Valerie Durham, Rodney and Pam Emery, Otis Graf, Kirby and Karen Hansen, Mike and Marj Harman, Roger and Stefanie Joe, Mathew John, Matt and Janet Jones, Hal and Edie Kirman, Steve and Liz Klein, Francisco and Martha Larzabal, David La Pointe, Perry Lanaro, George and Valerie Leiva, Martin and Cynthia Levine, Steve and Kara Loupe, Moriah Martinez, Helen Masuda, Clark and Alice McKinley, Gary Parks, Steve and Eileen Rogstad, David Singh, Brant and Laura Ullman, Harold and Beverly Van Vuren, Darren and Jennifer Williams, John and Shera Williamson, Colleen Wingenbach, Josh and Kerri Wolcott, and John Yue. I owe the completion of this book to your dedicated partnership and believe God will use your generosity to impact more lives for his kingdom.

Introduction

I know. It's a morbid thought.

I can't help myself. This horrifying question just pops into my head practically every time I board an airplane. What would I do if this plane were about to go down during the middle of the flight?

Throughout my years of travel, I mulled over this scenario, and I decided that if the flight I am on is ill-fated, then during my last few minutes in the air I want to put on my headphones and play Lynyrd Skynyrd's signature song "Free Bird" as the plane drops from the sky.

Of course, first, I would say a prayer, thanking God for the incredible life I've experienced. And I would ask him to watch over my wife and children and, now, grandchildren. But then, the last thing that I would want to hear before I die is Allen Collins's legendary guitar solo.

From my morose perspective, "Free Bird" is the perfect song to listen to as one's life draws to a close. This song is about freedom and, as a Christian, I believe there is nothing more liberating than the freedom I'll experience when my earthly existence comes to an end and I am in the Lord's presence.

Fly away, free bird!

Historically, this song has become connected to the tragic 1977 plane crash that took the life of Lynyrd Skynyrd's lead singer Ronnie Van Zant, guitar player Steve Gaines, and his sister Cassie Gaines, one of Lynyrd Skynyrd's backup singers.

Before the plane crash, when Lynyrd Skynyrd performed this song in concert, they would often dedicate it to the memory of rock legend Duane Allman. In somewhat of a dark, poetic turn, this song has now become a tribute to Ronnie Van Zant after the crash. In some respects, there is an uncanny, prophetic feel to "Free Bird."

Perhaps this partly explains why this song uniquely captures some of my thoughts and feelings when I board an airplane and think about how my life on Earth might one day come to an end. It almost feels as if "Free Bird" was written and recorded just for me in those moments when I sit on a plane waiting for takeoff.

• • • • • • • • • • • •

Music is an integral part of our lives—and that seems to always have been the case. As musician and neuroscientist Daniel Levitin points out in his book *This Is Your Brain on Music,* music is an unusual human activity because of its antiquity and pervasiveness.[1]

Musical instruments are among the oldest artifacts from the archaeological record associated with modern humans. Archaeologists have found musical instruments in sites located in northern Africa, Europe, and Asia—places occupied by some of the earliest modern humans.[2] Typically, these instruments were created from the long bones of birds and functioned as whistles and flutes. In some cases, percussion instruments have also been recovered.

Music stems from our capacity for symbolism and our ability to manipulate symbols by combining them and recombining them in endless permutations. A growing number of anthropologists think that our capacity for music appeared well before we began to migrate around the world. In fact, we have evidence for symbolic expression that extends back to the time when modern humans first appear on Earth.[3]

Music not only traces back to the time of our origins, it is also everywhere today, all around us. As Levitin points out:

> Whenever humans come together for any reason, music is there: weddings, funerals, graduation from college, men marching off to war, stadium sporting events, a night on the town, prayer, a romantic dinner, mothers rocking their infants to sleep, and college students studying with music as a background.[4]

Music serves an important function for many of the most significant events in our lives and helps to make the most mundane parts of our daily routines enjoyable, even meaningful, at times. Music is a language, of sorts. It is a means to communicate information. However, it is so much more. And for that reason, perhaps, it isn't that surprising that music takes center stage in our lives. Music can inspire us. It can boost our confidence. It can give us hope. Music deeply connects to our emotions. It provides the means for us to express how we feel when words alone won't do. Music gives us chills. It moves us to tears. It comforts us. It allows us to express gratitude and joy. Music brings back memories of happier times and comforts us as we face difficulties and disappointments.

Many people view our capacity for language as one of the defining features of our species. A growing number of scientific studies indicate that our capacity for language is hardwired into our brains.[5] It also appears that our brains are hardwired for music.[6] Music played by people from disparate cultures all over the world shares universal features. And when people without any musical training write "songs," their compositions reflect these universal qualities. When people from all walks of life hear a new piece of music for the first time, the same areas of their brains display the same patterns of synchronized activity.[7] Researchers can predict if someone will like a song the first time they hear it based on the strength of certain neural connections. These connections seem to be governed by past musical experiences.

It goes without saying: music isn't key to our survival in the same way that food and sex are. But when we hear a musical piece we enjoy, it activates the same areas of the brain that food and sex activate. When we hear a song we like, the neurotransmitter dopamine floods our brain. Dopamine neurons connect to the area of the brain called the nucleus accumbens—nicknamed the pleasure center. The nucleus accumbens is also activated during sex, eating, and when we engage in addictive behavior, such as gambling. Our response to music also involves other brain areas connected to the nucleus accumbens, such as the amygdala (involved in processing emotions), the hippocampus (involved in learning and memory), and the ventromedial frontal cortex (involved in decision-making).[8] In other words, our response to music involves enhanced activity in several areas of our brain and increased connectivity between these areas.

Another area of the brain involved in our response to music is the medial prefrontal cortex, a part of the brain like the hippocampus that is also involved in memory. We have all had the experience of hearing a song and suddenly becoming overwhelmed with a deluge of memories, past experiences, and feelings we associate with that song. It's this part of the brain that seems to be responsible for evoking vivid memories of past events and emotions when we hear a familiar song.[9] Petr Janata, a psychology professor who works at the Center for Mind and Brain at UC Davis describes the phenomenon this way:

> What seems to happen is that a piece of familiar music serves as a soundtrack for a mental movie that starts playing in our head. It calls back memories of a particular person or place, and you might all of a sudden see that person's face in your mind's eye.[10]

Perhaps this phenomenon explains why we like to hear the same song over and over again.[11] There are songs we never tire of hearing because they help us recover a certain feeling. They give us a chance to relive some of our most important and meaningful experiences.

Songs also harbor meaning for us. Their sonic qualities and lyrics seem to perfectly capture the way we feel, the way we think. Songs connect us to other people with shared experiences and shared feelings. In this sense, songs become part of our identity. They become a reflection of who we are. According to Kenneth Aigen, the director of the music therapy program at New York University, "Each time we re-experience our favorite music, we're sort of reinforcing our sense of who we are, where we belong, what we value."[12]

• • • • • • • • • • • •

I find it interesting how quickly we embrace the songs written by other people. Often, a song is a personal expression by the songwriter, describing his or her experiences, feelings, and emotional reactions to life's circumstances. However, because of our shared humanity, a song written by someone we don't even know somehow resonates with us. It seems to perfectly capture how we feel and think when we experience the very thing that moved the songwriter to write and record a song.

At the memorial service for my friend John Harrelson (a musician and musicologist), one of John's lifelong friends recounted how the songs John wrote and recorded became the soundtrack to his life. At that moment, I realized I had a soundtrack for my life. In fact, most of us do.

The soundtrack to my life—which is still being assembled—consists of songs connected to some of the most significant and meaningful parts of my life. But the soundtrack to my life isn't made up of songs that happened to be playing when these events took place. It features songs that perfectly captured the moment—songs that seemed to be written for that particular instant of my life. I might say that the songs that make up my life's soundtrack seem to be fit for a purpose. It's almost as if the songwriter had me in mind when composing those songs, so ideal is the fit.

• • • • • • • • • • • •

The concept of fit for a purpose is usually associated with the business world. It refers to the ideal quality a product, service, or process should display based on its intended use. Usually there are many features of a product (or service or process) that need to perform well for a product to be considered overall fit for its intended purpose. Take a CD player, for example. (I know. Old school.) When buying a CD player, you don't want one that just plays CDs. You want one that gives you the best, most reliable sound quality possible. To get an optimal sound, the player should have a well-designed transport system that holds, spins, and reads the CDs. If you are an audiophile, you want a player that only reads CDs, instead of one that reads both CDs and DVDs. In this way the CD's transport system will have been optimized for audio quality, instead of a system designed to manage the trade-offs needed to deliver quality audio and video outputs. The CD player should have a superior digital-to-analog converter that transforms the digital data on the CD to analog output without distortion. The CD player needs a power supply that can deliver clean current. And, lastly, the CD player should be solidly built to minimize vibrations. In other words, it is the sum total of features, each one just-right, that makes a CD player fit for a purpose.

It wasn't in a business context that I first learned about the concept of fit for a purpose, but in a scientific one. Over the last several decades, astronomers and astrophysicists have discovered that a large collection of the fundamental constants, parameters, and characteristics that define the universe have to assume incredibly precise values for human life to even be possible in the universe. And, as it turns out, our universe is characterized by these exact values. This coincidence is remarkable. Our universe appears to be fit for a purpose—and it is not merely a single feature of our universe that makes it so. Just like a good CD player, it is a collection of features—constants, parameters, and characteristics—all with just-right values that make the universe fit for a purpose.

Another way to think about the universe: there is a soundtrack that plays along with it. But instead of the universe's soundtrack consisting of the perfect songs for the perfect occasions, the universe's soundtrack is made up of an ensemble of the just-right physical constants and parameters that make humanity's existence possible. And just as the music instinct seems to be hardwired into our brains, the fine-tuning of the universe seems to be hardwired into its very fabric.

Astronomers and astrophysicists have discovered this anthropic coincidence through counterfactual analyses. By systematically varying the numerical values that define the universe's constants and parameters they have shown, through theoretical calculations, that deviations from these values rob the universe of the capacity to harbor advanced human life. In some cases, all it takes is infinitesimally small deviations from these values to render the universe inhospitable to life. For example, as we will see in chapter one (pages 40–41), the ratio of the gravitational force constant to the electromagnetic force constant cannot differ from its value by any more than one part in 10^{38} without eliminating the possibility for life in the universe. To put it differently, it looks as if the fundamental constants and parameters of the universe have been exquisitely fine-tuned for life to be possible.

In some respects, the universe's constants and defining parameters can be likened to the old radio I used in high school. I had to adjust the radio's dial to find my favorite station (WKLC, 105.1 in St. Albans, West Virginia). If the dial was tweaked just a little to the right or just a little to the left, all I could hear was static. However, when I carefully positioned the dial to the precise spot, I was rewarded with the sweet sounds of my favorite rock tunes.

• • • • • • • • • • • •

The fine-tuning of the universe's dimensionless constants is referred to as the anthropic principle. This term was originally coined by physicist Brandon Carter in 1973.[13] At that time, Carter presented two conceptions of the anthropic principle: the weak and the strong versions. According to Carter, the weak anthropic principle denotes the observation that we live in a privileged location and time in the universe's history in which human life is possible. Human life simply wouldn't be possible in another location or at another time in the universe's history. As the name implies, the strong anthropic principle goes significantly further, referring to the observations that the fundamental constants, parameters, and characteristics of the universe *must* also be fine-tuned for life to be possible.

Since Carter's proposal, the scientific and metaphysical implications of the anthropic principle have been intensely debated. Some, like Brandon Carter, view the anthropic principle in tautological terms. Accordingly, they argue that if we didn't live at the right location and time in the universe's history or if the universe's constants weren't fine-tuned, then there wouldn't be observers in the universe to recognize that it wasn't fine-tuned. Differently stated, the anthropic principle is merely the consequence of a selection effect and it doesn't imply anything special about the universe or our place in the cosmos. Admittedly, it is hard to argue with this reasoning.

However, there are others who offer a different interpretation of the anthropic principle. They see the fine-tuning of the universe's dimensionless constants and parameters as a signature for a Creator's handiwork. Just like it took my hand to intentionally guide the dial on my radio to the precise location so that I could listen to WKLC, many people view the fine-tuning of the universe's constants and parameters as evidence that a Creator must have intended the universe to be the way it is—so that human life could be possible. In this framework, the fine-tuning of the universe's constants and parameters indicates that the universe was designed.

For those who interpret the anthropic principle as evidence for design, the case for a Creator runs much deeper than merely the design of the universe's constants. They see the anthropic principle as connoting meaning and purpose. That is, the universe seems to be fit for a purpose.

• • • • • • • • • • • •

Brandon Carter's conception of the strong anthropic principle has widespread acceptance among astronomers and astrophysicists, regardless of their worldview. Few of them dispute the idea that the universe's fundamental constants, parameters, and characteristics—features that define the universe—are fine-tuned in such a way that life is possible. Likewise, the idea that the universe is fit for a purpose—the advent of conscious, sapient beings—seems largely noncontroversial. What has become controversial is its interpretation. The focus of the debate centers on the metaphysical implications of the universe's fine-tuning and design. Specifically, does the anthropic principle point to the existence of a transcendent Creator who brought the universe into existence so that advanced sentient beings like us could exist?

One possible way to resolve the interpretive impasse may be to look outside of cosmology and physics to other arenas of science: chemistry, biochemistry, and biology. Up to this point in time, most people have regarded the anthropic principle to fall exclusively within the domain of astronomy and astrophysics.

But if a theistic interpretation of the anthropic principle is valid, it becomes reasonable to think that anthropic coincidences would also be observed in chemistry, biochemistry, and biology. It seems to me that if a Creator, indeed, intentionally designed the universe to be biofriendly, he wouldn't have limited those design features solely to the universe's physical constants and parameters. He would have made every aspect of the creation biofriendly. He would have designed all of creation to be fit for a purpose—namely humanity's advent.

These predictions bring us to the central question of this book. *Does the anthropic principle extend beyond physics and cosmology to chemistry and biochemistry?*

I first came to appreciate the possibility that anthropic coincidences could be found in the realms of chemistry, biochemistry, and biology nearly twenty years ago with the publication of Michael Denton's classic *Nature's Destiny*.[14] In this work, Denton presents a far-ranging case that the laws of the universe are structured in such a way that life—if it exists anywhere in the universe—must be precisely the way life appears on Earth. Life must exist in an aqueous environment and be formed from carbon compounds. Denton argues that life must be built from amino acids, proteins, nucleic acids, and DNA. Most significant to Denton's argument is the claim that the laws of physics, chemistry, and biology seemingly point to the inevitability of creatures like us—*Homo sapiens*. Denton concludes that, *in toto*, the laws of nature seem to be rigged to make the universe and Earth's biosphere precisely the way they need to be for human life to be possible.

About a decade later I came across an academic volume entitled *Fitness of the Cosmos for Life*, which also explored the prospects of the anthropic principle extending into the arena of biochemistry.[15] This book consists of a collection of essays written by scholars from a wide range of disciplines who took part in a symposium held at Harvard University in 2003 in honor of the famous physiologist Lawrence J. Henderson. The year 2003 marked the ninetieth anniversary of the publication of Henderson's work *The Fitness of the Environment*. Published in 1913, this volume may well represent the first modern-day presentation of the reasoning that undergirds the anthropic principle. Instead of focusing on astronomy and astrophysics, Henderson focused on the design of the chemical environment and the remarkable observation that this environment seems to be ideally suited to make life possible. In other words, if the chemical environment of Earth were any different, life would not be possible. Life couldn't have even evolved.

Over the course of the last decade, I have given the anthropic principle a lot of thought. I am now convinced that anthropic coincidences can be found throughout nature. I am of the opinion that the anthropic principle isn't merely confined to physics and cosmology. It encompasses chemistry, biochemistry, and biology. In this work, my hope is to present evidence from chemistry and biochemistry in support of this thesis. In doing so, I will use the insights of Michael Denton and the scientists who contributed to *Fitness of the Cosmos for Life* as a starting point. But I will also include my own ideas based on a decade of research about this question. During this process, I have discovered a growing number of insights into the structure and function of biochemical and biological systems that can be recruited as evidence for an extended anthropic principle. Interestingly, in many of these studies, the question of the fitness for purpose wasn't under consideration by the investigators. Yet, unbeknownst to them, their work has generated an understanding about biochemical systems that rightly serve as examples of anthropic coincidences.

•••••••••••••

This book addresses the broader question: does the anthropic principle apply to chemistry and biochemistry? My focus will be primarily on the prospects of a biochemical anthropic principle for the simple reason that I am a biochemist. Specifically, I am going to argue that:

- Instead of being shaped exclusively by historically contingent evolutionary processes, which produce happenstance outcomes, the molecular systems that define terrestrial biochemistry seem to be largely specified and dictated by the laws of physics and chemistry. To put it another way, the structure and function of biochemical systems appear to be prearranged by the laws of nature, instead of generated by natural selection.
- The properties of biochemical systems are precisely those that are needed for life.
- As a corollary, there don't appear to be alternate biochemistries. Though, admittedly, it is difficult to eliminate all conceivable biochemistries, it does appear as if terrestrial biochemistry is universal biochemistry.

This book is organized into four parts. The first part serves as an introduction to the anthropic principle. In chapter 1, I present an overview of the cosmological anthropic principle, describing some of the most salient features of the universe that must be fine-tuned for life to be possible. The content of this chapter is largely derived from the classic work *The Creator and the Cosmos* by Hugh Ross and *A Fortunate Universe* by Geraint Lewis and Luke Barnes. For those unfamiliar with the anthropic principle, this chapter will provide the necessary orientation to have a context for the rest of the book. I also use this chapter to highlight the methodology and mode of reasoning used by astronomers and astrophysicists who have concluded that the universe appears to be fit for a purpose. This will set the stage for a discussion in chapters 2 and 3 about the methodology I will use to assess whether the anthropic principle extends to chemistry and biochemistry.

Part two discusses the ideas of physiologist Lawrence J. Henderson—the father of anthropic reasoning. Even though Brandon Carter is often credited with the anthropic principle (he did coin the term, after all), it was Henderson, in 1913, who first articulated the idea that the environment displays features that make life possible. The purpose of chapter 2 is to give Henderson his proper due. In many respects, chapter 2 sets the stage for the remainder of the book. The approach I use to argue for anthropic coincidences in chemistry and biochemistry are modeled after Henderson's approach in *The Fitness of the Environment* (1913) and *The Order of Nature* (1917).

Chapter 3 makes the case that the anthropic principle extends to chemical systems. I include this chapter in the first part of the book because, in effect, much of the material presented in this chapter was already enlisted by Henderson when he made his case that the chemical environment is ideally suited for life. In 1913, when Henderson first made his case, the modern-day understanding of chemical bonding was yet to be developed. So, instead of merely recounting and summarizing Henderson's insights, I will use chapter 3 as an opportunity to update and extend Henderson's argument, making use of our current understanding of chemical bonding and molecular structure.

The third part of the book seeks to determine if there is a biochemical anthropic principle. In chapters 4 through 8, I will take a detailed look at the anthropic coincidences associated with structural features of proteins, DNA, and the lipid components of cell membranes, along with the key cellular process of DNA replication, protein synthesis, and intermediary metabolism. I chose to focus on these systems because they are universal—found in virtually every living organism—and they form the core biochemical systems. In fact, some origin-of-life investigators think these biochemical systems comprise the non-negotiable systems minimally required for life to exist.

Even though I am of the mindset that the anthropic principle extends into biochemistry, I intend to do my best to fairly and objectively present the scientific evidence, discussing studies that count for and against the existence of a biochemical anthropic principle, before presenting my conclusion. In other words, my intent isn't to argue for the existence of the biochemical anthropic principle, it is to determine if it exists or not.

The final chapter of the book explores the philosophical and metaphysical implications of the extended anthropic principle. In this part of the book, I present varying interpretations of the anthropic principle and make the case that a theistic understanding of the fine-tuning of the universe is the preferred understanding. I also make the argument that the anthropic coincidences observed in biochemistry and biology make better sense in a creation model / intelligent design framework than in one that employs some form of theistic evolution.

Does the anthropic principle apply to chemistry and biochemistry? Though my real interest is in the metaphysical implications of this idea, the answer to this question resonates and reverberates beyond philosophy and theology. It impacts attempts on the part of the scientific community to define life. It holds implications for how we understand the origin and chemical nature of life. It carries implications for astrobiology and the quest to discover life throughout our solar system and beyond. It is my hope that this book will contribute insight into these interesting and critical scientific questions.

So, let's tune up our guitars and rosin up our bows and get to it.

PART 1

The Anthropic Principle

Chapter 1

The Cosmological Anthropic Principle

When I was in high school, I *really* wanted to be cool. (I still want to be cool today, but now I'm more realistic about those prospects.)

To be considered one of the cool kids, you had to like the "right" musical artists. It was important your friends knew your album collection consisted of LPs recorded by the trendy musical acts of the day. And, more importantly, you wanted to be seen listening to their music—loudly.

And then there were other performers who were considered so uncool you wouldn't be caught dead listening to them. Yet I secretly liked some of these "uncool" musical acts. I even bought their records and listened to them—when I was sure no one was around. One of those uncool musical performers was Eddie Money. In fact, writing for *Rolling Stone* magazine on the day Money died (September 13, 2019), journalist David Browne described him as "the Patron Saint of Rock Uncool."[1]

Eddie Money was the stage name of Edward Joseph Mahoney (1949–2019). Money burst onto the musical scene in 1977 with an eponymous album released by Columbia Records. Three songs from this album charted: "Baby Hold On," "Two Tickets to Paradise," and "You've Really Got a Hold on Me." All three of these tunes ultimately became staples for the playlists of classic rock stations.

Money's easily recognizable voice was hoarse and gravelly. He wrote and recorded accessible, formulaic, radio-ready rock tunes. The lyrics of his songs often seemed uninspired and generic. They were cheesy, even cliché. Prone to excess, Money was a caricature of a rock star. As much as I liked his music, Money made me cringe at times. He always seemed a little clumsy and awkward, out-of-step with the rock scene. Yet there was something endearing about Eddie Money. Money never held back. He always seemed so sincere, as if he believed every word he sang. That made him relatable. Browne describes Money's appeal this way:

> He stood in for every person who was all sputtering emotions, bereft of the polished or articulate gene. . . . Money made you root for him, especially since so many of his songs amounted to confessions about how much he'd screwed up in one way or another.[2]

Maybe this is the reason I liked Eddie Money and his music. I identified with him. As a teenager (and still today) I was bereft of the polished gene. And I messed up plenty of times. Money's music adeptly captured how I felt about the things happening in my young life—even if his lyrics were a bit cliché. Given the success of Money's forty-five-year musical career, I suspect I wasn't the only one who secretly bought and listened to his music. His music really had a hold on me.

My favorite Eddie Money song, "Wanna Be a Rock 'n' Roll Star," appears on his debut album. As you might gather from the title, this song is a rocker. I liked the song for that reason, but the enduring appeal of this track is its inspiring message. The lyrics speak about Money's singular focus and unending drive to become what he felt he was born to be—a rock 'n' roll star. It is a song about grit and determination, about never giving up until you have accomplished your life's mission. As a high school student, I took this message to heart and carried this ideal with me through college, graduate school, my two post docs, and my career in research and development. I still feel inspired today when I put on this album—seeking a little musical guilty pleasure—and I hear this song. I wasn't born to be a rock 'n' roll star. But I was born to be a scientist. And I worked hard to fulfill that call, never giving up when things became difficult, and always pushing through with the end in view. I am going to continue in that vein until I fulfill my life's mission.

This song also resonated with me when I was a teenager for another reason. My stern, no-nonsense Indian father despised my passion for music. He loathed the rock music that I loved. To him it was noise—a cacophony. He thought the hours I spent listening to music were a total waste of time—time that would be better spent on my studies. Oh, the arguments we had! As a young man, there was no way I was going to let my father determine the way I spent my time or dictate the music I listened to.

In the 1970s, the spats between my father and me about my musical taste and listening habits weren't unique to our family. These types of arguments took place in households all over America. It was part of the times. I connected with Eddie Money as he sang about his own experience when he was in high school, arguing with his parents about the music he wanted to hear.

My mother would jump and shout,
she'd say what's that noise about
Finally turn that dial back to her station
But I knew right then I'd turn that dial back again

My father and I had similar arguments about the positioning and repositioning of the dial on the radios located around our house. No matter how many times my father changed the station, I would sneak the dial back to 105.1 FM. It's funny how positioning a radio dial a few quarter inches to the left or a few quarter inches to the right can make all the difference in the world.

• • • • • • • • • • • •

As mentioned in the introduction, the positioning of the dial on the old radio I listened to in high school serves as a useful metaphor that helps us understand the anthropic principle. It also gives us insight into the fundamental structure and design of the universe. (See page 25.) Astronomers and astrophysicists know that for life to even be conceivable, the "dials" corresponding to the universe's constants, parameters, and characteristics must be positioned precisely. If these dials deviate a little to the right or left, all that results is "static" in the universe and, therefore, life isn't possible. To put it plainly, the numerical values that define the universe must assume exacting values if life is to exist. Remarkably, our universe is characterized by these exact values. The fine-tuning of the universe's constants is called the anthropic principle. In the same way that Eddie Money saw rock 'n' roll as his life's purpose and I see science as mine, these anthropic coincidences make it appear as if there is a purpose to the universe—and that purpose is the advent of sapient beings.

For the most part, the anthropic principle holds widespread acceptance among astronomers and astrophysicists. Few scientists question the fine-tuning of the numerical quantities that define the universe.

The existence of the anthropic principle at the cosmic scale prompts the following questions:

- Do anthropic coincidences occur in chemistry and biochemistry?
- How pervasive are anthropic coincidences?

The central theme of this book addresses these two interrelated questions. Before we begin this query, it is worthwhile to examine the cosmological anthropic principle in detail. How did this idea originate? What are some of the most salient examples of fine-tuning in the universe? And how is this fine-tuning determined and measured?

In this chapter, an overview of the most significant anthropic coincidences discovered by astronomers and physicists will be presented. For some this will be a review. For others it will be an introduction to the anthropic principle. Most importantly this chapter will illustrate the methodology and mode of reasoning used by astronomers and astrophysicists to establish the anthropic principle and the universe's fitness for life. This illustration will serve as an important reference point for the approach I will take to determine if anthropic coincidences occur in the chemical and biochemical arenas.

The Genesis of the Anthropic Principle

Theoretical astrophysicist Brandon Carter formally introduced the scientific community to the anthropic principle in 1973 at a technical conference called the Krakow Symposium. This gathering was held to celebrate the 500th anniversary of Copernicus's birth.[3] Carter formulated the anthropic principle as a rejoinder to the Copernican principle, which posits that humans don't occupy a special place in the universe. Carter argued that the Copernican principle is not valid. Instead, humanity lives at the just-right location and the just-right time in the universe's history. According to Carter, we also live in a universe with the just-right physical constants. Along these lines, Carter constructed two versions of the anthropic principle: the weak anthropic principle and the strong anthropic principle. (Since Carter's time, physicists and astronomers have abbreviated these two versions of the anthropic principle as WAP and SAP, respectively). The WAP states that for life to exist it must reside at the *just-right space-time location* in the universe's history. The SAP states that for life to be possible, the universe must be structured around the *just-right physical constants.*

Carter wasn't the first physicist to recognize anthropic coincidences in cosmology, but he was the first to codify these coincidences into a scientific framework. For example, in 1953, Fred Hoyle discovered that the resonances for the nuclear energy levels of helium, beryllium, carbon, and oxygen must adopt precise values for life to be possible. (See pages 89–91.) In 1961, Robert Dicke recognized the same thing about the universe as Carter—namely biological constraints dictate the time in the universe's history when life is possible. Life couldn't have existed earlier because the necessary chemical elements did not occur at high enough levels for life to be possible. Moreover, the early universe's metallicity was too low for rocky planets to form. On the other hand, if life attempted to appear in a window of time too late in the universe's history, then main sequence stars wouldn't exist, making it impossible for stable planetary systems to form. And without stable planetary systems, life-support planets wouldn't exist.

Another important milestone in the history of the anthropic principle came in 1986 with the publication of *The Anthropic Cosmological Principle* by John D. Barrow and Frank J. Tipler. In this work, these two physicists delineate a number of the anthropic coincidences that impact the universe at a cosmic scale. They also addressed theological and philosophical implications of these coincidences, which Brandon Carter eschewed. Carter adeptly sidestepped the metaphysical questions that obviously arise as a consequence of the universe's fine-tuning. Instead, he interpreted his versions of the WAP and SAP strictly in scientific terms. His interpretation centered around an observer-centric tautology. According to Carter, for observers to exist with the capability to discover anthropic coincidences, the universe's physical constants must be fine-tuned and the observers must reside at the just-right location in the universe's space-time. If not, then no observers would exist. In other words, if observers reside in the universe, then the WAP must be true by necessity. As a corollary, if the space-time location and values of the physical constants were anything other than what they are, or if the universe resided at any other time in its history, then there wouldn't be any observers present to study the universe and recognize that it wasn't just-right for life.

Barrow and Tipler reformulated Carter's weak and strong anthropic principles. They redefined the WAP by combining Carter's versions of the WAP and the SAP. For Barrow and Tipler, the WAP refers to the idea that the values of the universe's physical and cosmological constants are not all equally probable. Instead, these values are restricted to only those quantities that make carbon-based life possible in the universe, at the appropriate location, after the necessary time has transpired. In effect, with their version of the WAP, Barrow and Tipler acknowledged the obvious point made by Carter about the connection between observers and the values of the universe's physical constants. But Barrow and Tipler also recognized that the fine-tuning of the universe prompts certain metaphysical questions. In response to this recognition, they offered up a new version of the SAP that introduced an imperative not found in either of Carter's versions of the anthropic principle. They claimed that the universe's physical and cosmological constants *must* assume precise, exacting values so that carbon-based life *will* exist. Barrow and Tipler suggested that this imperative arises for one of three possible reasons:

- The universe is the work of an intelligent Agent, influenced by classical design arguments for God's existence.
- The universe is the product of an observer-created reality, influenced by concepts from quantum mechanics.
- A near-infinite ensemble of universes with different physical and cosmological constants exist by necessity, with our universe being the one that is fine-tuned for life, simply as an outworking of chance.

In effect, by introducing a revised version of the SAP, Barrow and Tipler confronted the obvious philosophical implications of the anthropic principle and forced cosmology to straddle science and metaphysics. In the process, they reintroduced teleology into physics and cosmology.

So, what does the anthropic principle mean for humanity? What does the anthropic principle tell us about our place in the cosmos and the meaning and purpose of the universe? I will tackle those questions in chapter nine. For now, I will stick to specific scientific questions: Which of the universe's features must be fine-tuned for life to be possible? And why must these constants, parameters, and characteristics assume such exacting values?

The Cosmological Anthropic Principle

It is beyond the scope of this book to present an exhaustive description of the *cosmological* anthropic principle. Those details can be found in several works (some classics) that present the cosmological anthropic principle from a variety of worldview and philosophical perspectives. I refer motivated readers who have interest in a more technical discussion of the anthropic principle to Barrow and Tipler's *The Anthropic Cosmological Principle* and Reinhard Breuer's *The Anthropic Principle.* Readers who want a lay level discussion should consult Geraint F. Lewis and Luke A. Barnes's recent work, *A Fortunate Universe.* Finally, for readers interested in how the anthropic principle intersects with Christian theology, I recommend Hugh Ross's classic *The Creator and the Cosmos*, and Ross's *Why the Universe Is the Way It Is*, also destined to be a classic.

For the most part, the astronomers and physicists who study the anthropic coincidences in cosmology today rely on the same methodology Carter used in the early 1970s, which is a theoretical, counterfactual approach. Generally speaking, when researchers carry out these counterfactual analyses, they want to mathematically determine the consequences of varying the numerical value of a single constant, while keeping all the other constants unchanged. The consequences that typically interest them have to do with:

- galaxy, star, and planetary formation;
- the production and distribution of the chemical elements needed for life (carbon, hydrogen, oxygen, and nitrogen); and
- the formation and stability of atoms.

Based on hundreds of these types of studies, astronomers and astrophysicists have learned that, for life to exist, our universe's dimensionality and expansion rate have to be fine-tuned. The same is true for the values of the force constants and the numerical properties that describe the universe's fundamental particles.

Fine-Tuned Dimensionality

We live in a universe that consists of three dimensions of space (height, width, and length) and one dimension of time, in which time flows in a single direction. It is easy to take our universe's dimensional makeup for granted. In fact, I suspect that most of us just assume that there is no other way for the universe to be. Yet, hypothetically, it is conceivable that the universe could have consisted of only a single spatial dimension, or perhaps two dimensions of space. It is even possible that our universe could have been constructed from four or more spatial dimensions. Because we are confined to the space-time manifold of our universe, it is virtually impossible for us to visualize what these types of universes would be like. But, mathematically, physicists can vary the universe's dimensionality at will. In doing so they ask the questions:

- Is there something special about the space-time structure of our universe?
- What would our universe be like in a configuration other than one dimension of time and three dimensions of space?
- Could carbon-based life exist in alternative universes with a different

space-time manifold than ours?

In doing so, these investigators learned that altering the number of space and time dimensions of the universe significantly impacts the laws of physics in such a way that the origin and existence of life becomes impossible.[4]

For example, if the universe's dimensions increased from three to four spatial dimensions, it would alter the properties of gravity. In a three-dimensional universe, the gravitational attraction between two objects decreases by the inverse square of the distance separating them. But in a universe with four spatial dimensions, the gravitational attraction decreases by the inverse distance cubed. As a consequence, in a four-dimensional universe, planetary orbitals would be unstable unless they were exactly circular. In this model, astronomers and physicists have learned (through theoretical assessment) that introducing even the slightest amount of ellipticity to its orbit would send a potential life-support planet into the star around which it revolves or cause the planet to be ejected from its solar system into the recesses of space.

Varying the number of spatial dimensions in the universe from three to four would also have a devastating effect on the prospects for life due to impact at the atomic level. Theoretical calculations indicate that atoms could not exist because they wouldn't have a ground state. Without a ground state, electrons orbiting the nucleus would spiral inward, colliding with it.

If the universe consisted of either one or two spatial dimensions, then, according to the principles of general relativity, gravitational fields in empty space would not exist. Consequently, stars could not form and life would be impossible. On top of that problem, in a universe with only two spatial dimensions, most complex life-forms would be impossible. Two-dimensional organisms couldn't be any more sophisticated than a "blob." For example, many three-dimensional organisms possess a tube-within-a-tube body plan, allowing them to ingest foodstuff. But, in a two-dimensional universe, this type of body plan would split organisms down the middle into two halves.

A universe without any time dimension would be absurd. And a universe with two dimensions of time would be so complex that it would be impossible for organisms to anticipate future occurrences, making the world so unpredictable, it becomes hard to imagine how complex life-forms could exist.

Fine-Tuned Expansion

One of the most important discoveries about our universe is that it is expanding, not static. This expansion hasn't been monotonic throughout the universe's history. Instead, the expansion dynamics of the universe have been complex. During the initial stage of the universe's history, the expansion rate decelerated. During the current stage, the universe's rate of expansion is speeding up over time. (Currently, astronomers debate about the precise time in which this transition took place in the universe's history.) Astronomers and physicists have learned that unless the constants and parameters that dictate the expansion dynamics of our universe assume precise values, life cannot exist. In fact, the constraints on the constants associated with dark energy and dark matter are so exacting, they represent the most extreme examples of fine-tuning yet discovered.

The expansion of the universe is controlled by two factors: the density of matter in the universe and dark energy.[5] Because of gravitational attraction, the matter in the universe serves as a brake, slowing down the universe's expansion rate. The more matter in the universe (and the greater the mass density), the stronger the braking effect. During the first stage of the universe's history, this gravitational braking dominated the universe's expansion dynamics. Astrophysicists have learned that the universe's mass density must assume exacting values for the universe to be friendly for life. If the mass density were too high, then the expansion of the universe would decelerate too quickly and proceed too slowly, resulting in a universe inhospitable for life for a variety of reasons, such as orbital instability and excessive radiation bathing potential life-support planets. If the mass density were too low the expansion rate wouldn't decelerate quickly enough and the expansion of the universe would proceed too rapidly. As a consequence, star formation would be impacted, making unlikely the existence of the types of stars needed to sustain life-support planets.

Once matter gained sufficient separation as a result of the universe's expansion, gravity's influence waned. In the second and current stage of the universe's history, dark energy's influence dominates, causing the universe's expansion rate to accelerate. Dark energy comprises about 70 percent of the universe. At this juncture, astrophysicists do not fully understand dark energy. Dark energy appears to be intrinsic to space, exerting an antigravity effect. As a result, dark energy drives the universe's accelerating expansion.[6] Astrophysicists associate the cosmological constant in Einstein's field equations of general relativity with dark energy. As it turns out, the cosmological constant has to be fine-tuned on the order of 1 part in 10^{120}. If not, the universe will either expand so rapidly that it will prevent the sufficient clumping of matter necessary for galaxy and star formation, or it will expand too slowly, causing all the matter in the universe to clump together, making the conditions of the universe inhospitable for life.

Fine-Tuned Constants

The universe is characterized by four forces: gravity, electromagnetism, and the strong and weak nuclear forces. Of course, gravity refers to the force of attraction between objects, with the strength of the attraction proportional to the mass of the objects and inversely proportional to the inverse square of the distance separating the objects. Gravity is the weakest of the four fundamental forces. Electromagnetism refers to the attraction (or repulsion) of charged objects, with objects of opposite charge attracting one another and objects of like charge repelling one another. Like gravity, the strength of attraction (or repulsion) depends on the magnitude of the charge and is inversely proportional to the square of the distance separating the charged entities. The strong nuclear force refers to the attraction between protons and/or neutrons in the nucleus of atoms. The strong nuclear force is extremely strong. It is two orders of magnitude stronger than electromagnetism and 10^{38} times stronger than gravity. The strong nuclear force rapidly diminishes with distance, becoming insignificant when the distance between protons and/or neutrons is greater than 2.5 femtometers. The weak nuclear force exerts its influence at subatomic distances, just like the strong nuclear force, and it mediates radioactive decay.

Astronomers and physicists have discovered that the constants that define these forces must be fine-tuned for life to be possible in our universe.[7] If the gravitational force constant were stronger, then stars would burn too quickly to provide solar systems with the necessary stability for life to exist. If the gravitational force constant were weaker, then stars would be unable to produce the elements needed for life. If the electromagnetic force constant were either too strong or too weak, it would compromise chemical bonding to such an extent that molecules would not be possible in our universe. If the strong nuclear force constant were 0.3 percent stronger, no hydrogen would exist in the universe and the atoms needed for life wouldn't exist because their nuclei would be unstable. If the strong nuclear force constant were 2 percent weaker, hydrogen would be the only chemical element in the universe. If the weak nuclear force were stronger, only heavy elements would exist in the universe. If it were weaker, only light elements would exist in the universe. Either way, the universe would lack the chemical diversity needed for life.

Additionally, the ratio of the electromagnetic to gravitational force constants also needs to be fine-tuned for life to exist. If this ratio becomes larger than it is, only massive stars would exist in the universe. Stellar burning would be too erratic for life to exist. And if this ratio is smaller, only small stars would exist. This scenario would preclude the production of the heavy elements needed for life. For life to exist, this ratio can't vary by more than 1 part in 10^{38}. As such, it represents one of the most extreme examples of fine-tuning discovered.

Fine-Tuned Particles

The universe is comprised of a large number of subatomic particles. Some are familiar, such as protons, neutrons, and electrons. Others, such as quarks and leptons, are less-known to the average person on the street. The numerical parameters that define and describe the properties of these subatomic particles, too, must be fine-tuned for life to be possible.[8] Physical life can't exist in a universe without matter. Fortunately, we live in a universe in which matter does exist, thanks to a minuscule imbalance in the numbers of particles of matter to the number of particles of antimatter, the instant immediately after the universe began. Astronomers and physicists believe that during the earliest stages of the universe's history there were 1 billion and 1 particles of matter for every 1 billion particles of antimatter. Because matter and antimatter particles annihilate each other, this slight excess of 1 part in 10^9 resulted in a universe with the matter requisite for life.

The ratio of the number of protons to electrons in the universe also has to be precisely balanced for life to exist. If there were a slight excess in the number of either one of these particles, electromagnetism would be the dominant force in the universe, not gravity. The far-reaching ramifications would make it impossible for galaxies, stars, and planets to exist.

The masses of protons, neutrons, and electrons also have to be fine-tuned for a universe to be friendly for life. If the masses of protons and electrons don't assume exacting values, then the capacity for atoms to form chemical bonds would be negatively affected. The capacity for chemical bonding between atoms would also be compromised if the charge on protons and electrons wasn't precisely matched.

The masses of neutrons and protons also have to be carefully selected for life to exist. Neutrons are slightly more massive than protons. The mass of a neutron impacts its decay rate. If neutron mass were greater, then neutrons would decay so rapidly that there wouldn't be enough of these particles in the universe to form the heavy atoms needed for life. If neutron mass were lighter than it is, then neutrons would decay too slowly, leading to an overabundance of neutrons in the universe. This would cause stars to collapse into highly dense neutron stars, rendering the universe inhospitable for life. The decay rate of protons also has to be precisely adjusted for life to be possible. If protons decayed more rapidly than they do, the released radiation would preclude life's origin and existence. If the proton decay rate were slower, it would leave the universe with insufficient mass for life to be possible.

The Fitness of the Universe for Life

This brief overview represents a small sampling of the anthropic coincidences that astronomers and physicists have uncovered through theoretical counterfactual analyses. Many of the features included in this sample are fundamental numerical quantities that define and describe our universe. It is provocative to consider how many of the universe's constants, parameters, and characteristics have to simultaneously assume precise, exacting values for life to be possible. Though this survey has been limited, it should sufficiently illustrate why many scientists accept the anthropic principle, first proposed by Carter and further elaborated by Barrow and Tipler.

Even though this trifecta of physicists is credited with the anthropic principle, the first modern-day scientist to articulate this idea was Lawrence J. Henderson, a physiologist from Harvard University. In his classic work *The Fitness of the Environment* (1913), and later in the 1917 sequel *The Order of Nature*, Henderson argued that the environment displays features that make life possible.

In the next chapter, we will examine Henderson's ideas on the environment's fitness, in part, to give Henderson his proper place in the annals of science. More importantly, the next chapter sets the stage for the remainder of the book. Henderson's work is a bridge from the cosmological anthropic principle to the chemical and biochemical anthropic principles, and it provides us with the methodological framework we will need to determine if the anthropic principle exists in chemistry and biochemistry.

As we will see, Henderson's ideas were remarkably prescient.

PART 2

Fitness of the Environment for Life

Chapter 2

The Father of the Anthropic Principle

Occasionally, I find myself in a weird mood, where I feel sorrow and melancholy intermingled with joy and hope.

I don't think I experienced this combination of moods much when I was younger. But now that I am older, it isn't unusual for me to feel both sadness and hope at the same time. As someone who isn't all that in tune with my emotions, I honestly can't say why that is.

When I feel this way, I often catch myself humming—or singing, to the horror of those around me—the song "Hallelujah," written and first recorded by the Canadian singer and songwriter Leonard Cohen (1934–2016) for his 1984 album *Various Positions*.

This song has been described as the quintessential secular hymn, in part because it incorporates elements of rock and gospel music (though Cohen was Jewish) and refers to biblical themes and imagery. In some respects, the song seems to be about the conflict that resides deep within us between pursuing our deepest desires as human beings and seeking after wisdom, being never truly satisfied, forever disappointed.

When Cohen released this song as a single it achieved little commercial success. But the song gained prominence when it was covered by Welsh musical artist John Cale (one of the founding members of the Velvet Underground) in 1991. It was John Cale's version of the song that was used in the film *Shrek* (though Rufus Wainwright's version appears on the film's soundtrack).

Since Cale recorded his version of "Hallelujah" the song has been covered by a number of recording artists. For me and many others, Jeff Buckley (1966–1997) recorded the essential version of the song for his 1994 album *Grace*. Buckley's recording was released as a single in 2007, ten years after his death, and is arguably the most acclaimed version of the song. In fact, *Rolling Stone* included it in the list of the top 500 songs of all time.[1]

In his version, Buckley uses his incredible voice to careen between triumph and sorrow, between magnificence and agony. Maybe this is the reason that I am drawn to this song whenever I experience sorrow and joy at the same time.

"Hallelujah" is an example of a song in which the cover version(s) became more popular than the original. Other songs in this list include "Hound Dog" (originally recorded by Big Mama Thornton, but made more popular by Elvis Presley),"Respect" (originally recorded by Otis Redding, but made more popular by Aretha Franklin), and "Hurt" (originally recorded by Nine Inch Nails but made more popular by Johnny Cash).

I'm sure you could add some of your favorite covers to the list.

•••••••••••••

Sometimes scientific ideas can be like the cover versions of great songs. They take on a life of their own. And there are times when the people who originated these ideas aren't the ones whose names become associated with them. So is the case for the anthropic principle.

As pointed out in the previous chapter, the term anthropic principle was coined by physicist Brandon Carter in 1973 and further articulated by physicists Frank Tipler and John Barrow in 1986. This is why most people who are familiar with the anthropic principle associate these three scientists with the idea that the universe is designed to be fit for life.

But the recognition that anthropic coincidences exist throughout nature didn't originate with Carter. Though these ideas were explored perforce as part of the work in natural theology in the late 1700s and early 1800s, they were abandoned because of the influence of the evolutionary paradigm, ushered in by Charles Darwin's 1859 work *On the Origin of Species*. In this work, Darwin argued that the teleology of the natural theologians (such as William Paley) no longer belonged in biology because the remarkable designs so characteristic of biological systems could be explained by natural selection. And with this claim, design and purpose were stripped from biology.

Yet, nearly five decades later, Harvard physiologist Lawrence J. Henderson sought to bring design and purpose back into the fold of the life sciences and, in doing so, arguably presented the first modern example of anthropic reasoning. Working in the days prior to the rise of biochemistry as a discipline, Henderson wrote two books—*The Fitness of the Environment: An Inquiry into the Biological*

Significance of the Properties of Matter (1913) and *The Order of Nature: An Essay* (1917)—in which he proposed and defended the thesis that the environment is structured to have the necessary properties to make life possible. (Henderson's interest was the chemical environment, not the cosmic, planetary, or geological components of the environment. I will elaborate more on this distinction.) Henderson viewed his proposal as the reciprocal of Darwin's idea. Darwin, we know, argued that organisms evolve under the auspices of natural selection to become fit for their environment. But Henderson, recognizing that some environments could never harbor life of any kind, presented evidence that the environment we find ourselves in demonstrates a remarkable fitness that makes life possible in the first place. If it wasn't for the fitness of the environment, according to Henderson, life couldn't even originate, let alone evolve to adapt to its surroundings.

Arguably, he was the first to articulate concepts that we would recognize today as an integral part of the anthropic principle. But, like Cohen's recording of "Hallelujah," Henderson's conception of the anthropic principle is not the most acclaimed version. That distinction belongs to Carter, Tippler, and Barrow's expression of the idea.

Lawrence J. Henderson (1878–1942)

Formally trained as a medical doctor with an MD from Harvard Medical School (1902), Lawrence J. Henderson viewed himself as a physical chemist with interest in physiological systems.[2] (Today, Henderson would be considered a biochemist.) Before being appointed as a lecturer at Harvard Medical School in 1905, he spent three years receiving postdoctoral training in chemistry labs in Germany and the United States.

Over the course of his academic career at Harvard, Henderson, who was a member of the National Academy of Sciences in the US, pursued a wide range of interests beyond his work in physiology, including sociology and philosophy. But he is best known for his characterization of the acid-base buffering systems found in blood and other physiological fluids, such as the cytoplasm of the cell. His scientific magnum opus is his 1928 work *Blood: A Study in General Physiology.* This volume contains the insights he gleaned from a lifetime of work in this arena—insights still taught today as an integral part of medical school curricula.

Henderson discovered that the pH of blood and other biological fluids is regulated by a buffer system that primarily utilizes carbonic acid and salts derived from phosphoric acid. Carbonic acid forms when carbon dioxide dissolves in water. This reaction allows carbon dioxide to partition between the aqueous and gaseous phase. As a consequence of this property, the regulation of blood pH involves the complex interplay between the circulatory and respiratory systems, with the lungs, kidneys, and blood all playing a role. Danish chemist Karl Albert Hasselbalch made use of Henderson's discoveries to develop the famous Henderson-Hasselbalch equation used to calculate the pH of a buffer solution and interpret pH measurements of blood and other biological fluids.

Henderson's insights from studying biological buffering systems served as the inspiration for the ideas he presented in *The Fitness of the Environment*. In the preface, he writes:

> For it soon appeared that the key to the peculiar conditions of equilibrium between acids and bases in blood and protoplasm is to be found in such characteristics of phosphate solutions . . . and in the like behavior of similar solutions containing carbonic acid. When at length it became possible quantitatively to describe the chemical equilibria in such systems, it was at once clear that, of all known substances, phosphoric acid and carbonic acid possess the greatest power of automatic regulation of neutrality. . . . One does not like to accept a fact of such far-reaching importance as mere chance, and yet no other explanation was at hand. For after the briefest consideration, it was obvious that here, at least, natural selection could not be involved.[3]

The Fitness of the Environment

In other words, Henderson argued that the process of natural selection couldn't explain the use of carbonic and phosphoric acids as biological buffers. The use of these two acids in living systems wasn't a happenstance occurrence due to a mechanism characterized by chance events operated on by natural selection. These two substances are uniquely suited to serve as biological buffering agents. Biological systems must make use of them because no suitable alternatives exist. If phosphoric acid and carbonic acid didn't exist or didn't display their characteristic chemical properties, then life couldn't exist, let alone evolve. This idea presents far-reaching consequences. It means that natural selection doesn't provide the full explanation for life's design, history, and diversity.

Henderson was reticent to conclude that the distinctive and exceptional properties of carbonic and phosphoric acids were merely a lucky happenstance because he was aware of numerous other examples of chemical entities with uniquely useful and necessary properties for life. For Henderson, these additional examples constituted proof positive that the just-right properties of carbonic and phosphoric acids were more than mere chance. It indicated to him that the environment is fit for life. In fact, in *The Fitness of the Environment*, Henderson advanced the strong claim that the actual environment is the fittest possible one for life.[4]

Even though Henderson was aware of numerous chemical systems with the just-right properties for life, he opted against trying to present an exhaustive description. Instead, to evaluate and defend his hypothesis, Henderson chose to focus on the properties of water and carbon dioxide, two of the most prevalent materials on Earth's surface. To justify this approach, Henderson observed that water and carbon dioxide occur in high abundance in environments capable of sustaining life. Accordingly, if any reason exists to think that the environment is fit for life, it should be evident by considering these two materials alone. Henderson states:

> This restriction has been adopted in order to facilitate the logical discussion, and it should be borne in mind that other phenomena, dependent upon the properties of other substances, such as the above-mentioned characteristics of phosphate solutions, belong in the same category.[5]

Given his approach, Henderson spends significant space in *The Fitness of the Environment* discussing in detail a wide range of unusual, even unique, chemical and physical properties possessed by water and carbon dioxide that work together to create "great fitness for their biological role."[6]

Henderson also presents an extensive discussion about the uniqueness of chemical compounds made up of carbon, hydrogen, and oxygen, concluding that these elements form a vast number of compounds that uniquely possess the necessary structural complexity, chemical stability, and chemical reactivity to sustain life. Henderson concludes that "there are no other compounds which share more than a small part of the qualities of fitness of water and carbonic acid; no other elements which share those of carbon, hydrogen, and oxygen."[7]

Based on this insight, Henderson concluded that the environment is indeed fit for life and that the environment and the evolutionary process driven by natural selection worked "hand-in-glove" to produce the diversity of life on the planet. Henderson argued that natural selection and the environment's fitness formed a unit that resulted in the history and diversity of life found on Earth. Along these lines, the unique properties of materials (such as water, carbon dioxide, and organic compounds) meant that evolutionary processes were inevitably directed toward their use in living systems, constrained by the laws of nature, not by the influence of natural selection. For, if natural selection were the sole explanation for the use of these materials in living systems, then conceivable alternatives to water, carbon dioxide, and organic compounds would exist. But they don't.

What seems to have been most interesting to Henderson was the deeper meaning of the environment's fitness. He believed the anthropic coincidences couldn't be due to chance; they can't be "disavowed as gross contingency."[8] Instead Henderson viewed the fitness of the environment as endemic in the laws of nature and the universe's structure. As for the "why" and "how" questions, he proffered no insight, acknowledging that the environment's remarkable suitability for life is unaccounted for, with no genuine explanation at hand. Henderson felt it conceivable that some type of yet-to-be-discovered mechanistic explanation would make sense of the fitness of the environment.

Henderson also thought that the reciprocal interdependence between the environment's fitness and natural selection indicated a tendency or directional flow to life's evolutionary history that seemingly reflects a teleology or purpose to nature. Henderson states:

> We appear to be led to the assumption that the genetic or evolutionary processes, both cosmic and biological, when considered in certain aspects, constitute a single orderly development that yields results not only contingent, but resembling those which in human action we recognize as purposeful.[9]

At this point, Henderson recognized that two options exist to explain the apparent purpose of the universe: (1) "proof of supernatural purpose and design,"[10] or (2) some mechanistic explanation unrecognized by biologists.

In reaching the conclusion that there is a teleology to the universe, Henderson did everything he could to distance himself from the brand of teleology espoused by the natural theologians who lived and worked prior to Darwin. Henderson emphatically argued that this historical version of teleology was scientifically sterile and dangerous to the scientific enterprise. According to Henderson, discussions about this type of teleology, though of interest to philosophers and theologians, contribute nothing to the scientific endeavor.

For this reason, Henderson sought to account for the environment's fitness through a teleology that was mechanistic in nature.

Henderson proposed that the fitness of the environment arose as some type of "tendency" built into the structure of matter, energy, space, and time that worked in parallel with natural selection, but was completely independent of natural selection's influence, passively directing evolution in a specific inexorable direction that seemed to be predetermined. Henderson explained his interpretation of the environment's fitness this way:

> Our new teleology cannot have originated in or through mechanism, but it is a necessary and preestablished associate of mechanism. Matter and energy have an original property, assuredly not by chance, which organizes the universe in space and time.[11]

Henderson admitted that his new teleology had clear metaphysical implications, but ones he felt would be palatable to scientists.

In short, for Henderson, the fitness of the environment meant that we live in a biocentric universe. He hoped that insight would stimulate further research into the features of the environment that made the universe welcoming to life.

The Order of Nature

Henderson followed up *The Fitness of the Environment* with a much more philosophical work, *The Order of Nature*. In this volume, Henderson reprised the scientific case for the environment's suitability, primarily focusing on the unique chemical properties of compounds made from carbon, hydrogen, and oxygen (which include water and carbon dioxide). When Henderson left behind the scientific implications of the environment's fitness and turned his attention toward philosophical concerns, he did so tentatively and apologetically. He conceded that:

> The fact cannot be escaped that these considerations [the unique and necessary properties of water, carbonic acid, and compounds formed from carbon, hydrogen, and oxygen] have a philosophical as well as a scientific bearing. I have, therefore, after much hesitation, ventured to sketch the development of thought upon the problem of teleology, and at length to

> confront the scientific conclusions with the results of philosophical thought, in order to finally attempt a reconciliation.
>
> I fear that this task has been accomplished with feeble strokes. It was not undertaken confidently, but in the sincere belief that when such questions are involved men of science can no longer shirk the responsibility of philosophical thought.[12]

Though a scientist, Henderson felt as if he had an obligation to explore the obvious philosophical implications from *The Fitness of the Environment*. Along these lines, Henderson pointed out that the teleological appearance of nature is evident and unavoidable. He noted that, over the centuries, philosophers have devoted substantial thought to ferreting out implications that arise from the appearance of design and purpose in nature. Henderson began his exploration by surveying and sampling the history of thought in this area. In fact, Henderson spent a sizable portion of *The Order of Nature* commenting on the ideas that he believed represented major shifts in thinking about the teleological features of nature.

Exploring the ideas of Aristotle, Bacon, Descartes, Leibniz, Hume, Kant, Herbert Spencer, and others, Henderson plotted a trajectory that showed teleology go from front and center in discussions about the natural world to its rejection as a viable concept, at least in the sciences. Henderson noted that with the rise of the scientific method in the seventeenth century, thinkers such as Bacon, Descartes, and Leibniz connected mechanisms to the teleology apparent in nature. As scientists began to discover law-like behavior in the natural world, mechanisms became the explanation for the order and organization of the world, with the origin of teleology corresponding to the origin of mechanisms. As Henderson saw it, during the eighteenth century, the ideas of Hume and Kant became the harbinger for the death of teleology, with both philosophers denying teleology as part of nature's construct. Though both recognized order and organization in nature, they argued that the appearance of teleology arises in and is merely a projection of the human mind. And, of course, Darwin's discovery of natural selection and Herbert Spencer's view that evolution is a law-like process meant that the appearance of design and purpose, whether in the inanimate or animate realms of nature, arise through mechanism alone.

However, as Henderson points out, the discovery of the environment's fitness resurrects the idea of teleology in nature and presents a new way to account for the teleological appearance of nature. According to Henderson, an appeal to law can explain, in part, the orderliness and organization of nature. But a full account also requires consideration of the properties of matter. It is at this point Henderson revisited and expanded upon the ideas first presented in *The Fitness of the Environment*—namely, that the properties of compounds formed from carbon, hydrogen, and oxygen form a unique ensemble of properties that make the environment the fittest possible, making the existence and evolution of life possible. Once again, Henderson argued that these exceptional and unique properties of the environment cannot be an accident or the outcome of chance. Instead, they point to a deeper reality about the natural world. Henderson concluded:

> The properties of the universal elements antedate or are logically prior to those restricted aspects of evolution. . . . We are obliged to regard this collocation of properties as in some intelligible sense a preparation. . . . The properties of these elements must for the present be regarded as possessing a teleological character.[13]

Ruling out contingency and also recognizing that the endemic property of matter isn't due to cause and effect or some type of mechanistic explanation, Henderson speculated that the properties of matter may well reflect a final cause, echoing the ideas of Aristotle. Henderson reluctantly concluded:

> The whole problem of the teleological significance of our scientific investigation reduces to the simple but infinitely difficult question whether a final cause is to be postulated.[14]

For Henderson, the scientific insight about the fitness of the environment brings us face to face with the idea of *design*. And, whether we like it or not, Henderson concludes, we have to describe the environment's remarkable suitability for life as teleological because no other term fits.

Response to Henderson's Ideas

So, how were Henderson's ideas received? Did he achieve what he had hoped? To address these questions, Harvard University historian of science Everett Mendelsohn surveyed reviews of *The Fitness for the Environment* and *The Order of Nature* written by Henderson's contemporaries and published in scientific journals and other types of periodicals.[15] Henderson's scientific ideas encountered a mixed reception. Some found his idea of the environment's fitness for life insightful and groundbreaking. Others weren't impressed at all, marveling instead at Henderson's surprise that the environment is fit for life. These naysayers felt that his insights were self-evident to most biologists. In fact, these critics maintained that natural selection would be expected to produce organisms that would be so well adapted to the environment that it would appear as if the environment is, indeed, ideally suited to sustain life. Only those organisms that evolved to accommodate the environment would survive. The others would have simply died off.

While response to Henderson's scientific ideas ranged from excitement to ambivalence, the reaction to his new type of teleology was uniformly negative. Some saw his version of teleology as having little scientific value. Others saw it as an echo of earlier ideas associated with natural theology—ideas that had long been dispensed with by Darwin's theory, with the mechanism of natural selection leaving no place for purpose and design in biology, let alone any other scientific arenas. So, even though Henderson painstakingly tried to put a separation between his version of teleology and those of the natural theologians of the late 1700s and early 1800s, he seemed to be misunderstood by many of his contemporaries in precisely the way he hoped he wouldn't be.

As physicist John Barrow—one of the originators of the anthropic principle in cosmology—points out:

> One suspects that the lack of contemporary interest in Henderson's work might have been because such ideas were closely associated with the teleological design arguments of the past, which had been discredited by the emergence of natural selection as a better explanation for biological fine-tuning and, to a far lesser extent, by the philosophical objections of Hume and Kant.[16]

In fact, later in life Henderson regretted putting forth his new version of teleology and the metaphysical musings presented in both *The Fitness of the Environment* and *The Order of Nature*. He even referred to his own ideas as meaningless and reflective of an immaturity in thought.[17]

Though Henderson hoped his ideas on the environment's fitness for life would stimulate a new arena of scientific investigation, they seemed to have been largely ignored by his contemporaries—but not so today. Remarkably, Henderson's long-forgotten ideas are getting a new hearing, thanks in part to interest in questions related to biochemical and biological fine-tuning that arise because of the prominence and respectability of the anthropic principle in cosmology. Today, a growing number of researchers appreciate Henderson's insights as they seek to address such questions as:

- How is it possible for life to exist in the universe?
- Is the universe biocentric?
- Is the universe fit for life?

Remarkably, nearly sixty years before Carter introduced the scientific community to the anthropic principle in cosmology, Henderson had already laid its foundations, arguing for a teleological conception of nature. Henderson also recognized the philosophical and theological implications of the environment's fitness and courageously sought to tackle these ideas—albeit from a materialistic, naturalistic framework—enduring the ridicule of many of his contemporaries.

Chapter 3

The Chemical Anthropic Principle

For many people (including me), Bob Dylan (Robert Zimmerman) is the quintessential American singer-songwriter. Dylan has sold over 100 million records and received numerous awards, including the Presidential Medal of Freedom and the 2016 Nobel Prize in Literature.

As an iconoclast, controversy has surrounded Dylan's career. He first appeared on the music scene in the early 1960s, recording two folk albums that defied the conventions of pop music. He appealed to the countercultural movement of that time, but this appeal didn't last long. Dylan alienated his fanbase in 1965 when he used electrical instruments for the first time on his third record *Bringing It All Back Home*. Going electric provoked anger and hostility among Dylan's most devoted enthusiasts, who felt disaffected when he plugged in his guitar, leaving behind folk music for a more rock-oriented sound.

In many respects, Dylan's act of defiance in 1965 set a trajectory of a career-long love-hate relationship with his fans. Perhaps the most controversial move of Dylan's career took place in the late 1970s and early 1980s when he became an evangelical Christian. Dylan claimed that in a hotel room in Tucson, Arizona, he had a vision of Jesus Christ, and through Christianity he found answers to some of his most pressing needs. Following his conversion, Dylan recorded three albums of contemporary gospel music: *Slow Train Coming*, *Saved*, and *Shot of Love*. Dylan's embrace of the Christian faith was unpopular with many fans and fellow musicians. But, ironically, his conversion attracted newfound interest, causing his popularity to soar among Christians.

Though not a Christian at the time, I bought *Slow Train Coming* shortly after its release in 1979. It quickly became one of my all-time favorite rock albums and one of my favorite Dylan releases. The guitar work on the album stands out to me, thanks to the contributions of Mark Knopfler (founder and lead guitarist for Dire Straits).

People who are familiar with Dylan's work would readily recognize two songs of note on the album: "Gotta Serve Somebody" and "Slow Train Coming." But my favorite track is "Precious Angel," a composition written and sung to the woman who helped bring Dylan to Christianity.

This song took on newfound meaning for me about seven years later, when I converted to Christianity. I was an agnostic when I began my PhD program at Ohio University in the summer of 1985. By the next summer, I had become a Christian. The elegant, ingenious designs of biochemical systems coupled with the inadequacies of evolutionary explanations for the origin of life convinced me that a Creator must be responsible for the origin and fundamental design of life. This realization made me open to the gospel, though not right away.

Growing up in a Muslim home, I had little interest in or knowledge about Christianity—not until Amy (my wife-to-be at the time) began sharing her rediscovered faith with me. As a result, I began reading from Matthew's Gospel. And like Dylan, I had a religious experience one night in the chemistry lab after everyone had left for the day. While reading through the Sermon on the Mount, I experienced the presence of a person in the lab with me, after which I had the overwhelming sense that Jesus was who Christians understand him to be.

A few weeks later I informed Amy's pastor, Johnny Withrow, of my decision to embrace the Christian faith. He asked if I would recount what had happened to me with the congregation. As I did so, the lyrics of Dylan's song "Precious Angel" came flooding to my mind, and I realized that when I sang along to that song all those years ago, I was, in reality, addressing Amy.

Precious angel, under the sun,
How was I to know you'd be the one
To show me I was blinded, to show me I was gone

Almost 40 years after Dylan's conversion, another iconoclast in the musical world also converted to Christianity: rapper and singer-songwriter Kanye West. The similarities between Dylan and West are striking. One of the most widely acclaimed musicians in recent years, West has sold over 140 million albums. As with Dylan, controversy has surrounded West's musical career. Both Dylan and West continuously push the bounds of music—and cultural acceptability. Kanye's proclamation of faith, like Dylan's, has been met with a mixture of skepticism and excitement, accolades and criticism.

Though I am no fan of hip-hop, I find his 2019 *Jesus Is King* album a refreshingly eclectic mix of sounds combining hip-hop, elements of gospel music, and choral arrangements. One of my favorite tracks on the album is "Water," a song about the cleansing of sin and salvation offered to each of us through Jesus Christ.

Your love's water
Pure as water
We are water
Jesus, flow through us

In a short period of time this song has come to mean quite a lot to me. It reminds me of the cleansing love of Christ I experience every day. I am also drawn to this particular composition on the album because of my training and work as a biochemist. There are few chemical compounds as interesting as water. Though its chemical composition and structure are simple (H_2O), its physical and chemical properties are remarkably complex—and absolutely essential for life.

• • • • • • • • • • • •

West uses water to symbolize cleansing, as is often the case in music and literature. In "Water," West also uses this life-giving liquid as a metaphor for what he understands to be the source of our spiritual vitality—an apt metaphor, indeed. Water is a vital, indispensable source for our physical existence. All life depends on water. In fact, for many in the scientific community, the presence of water equates to life. In NASA's quest to discover life on Mars, Europa, and other bodies in our solar system, the mandate is to "follow the water." For where water exists, the greatest chance also exists for life to be discovered.

With the 1913 publication of *The Fitness of the Environment*, Lawrence Henderson proved to be, perhaps, one of the first scientists to fully appreciate the intimate connection between water and life. Water is the centerpiece of Henderson's argument for the environment's suitability for life. As mentioned in the previous chapter, Henderson delineated several special properties possessed by water that contribute significantly to the fitness of the environment for life. He argued that, if not for water's unusual suite of properties, life could never have emerged, let alone evolve to adapt to the environment.

Henderson's insights are truly remarkable because in 1913 scientists lacked a modern understanding of atomic structure and the nature of chemical bonding. This understanding had to wait for the advent of quantum mechanics and its application to chemical systems.

The Universe's Chemical Environment and Its Fitness for Life

This chapter will delve into some of the technical details that led Henderson to his perspective on the environment's fitness for life. In doing so, I will ask two interrelated questions:

- Do Henderson's insights still stand today?
- Does the anthropic principle extend beyond the confines of physics to include the chemical nature of the universe, as Henderson argued?

My focus in this chapter will center around the remarkable properties of water. Instead of merely recounting and summarizing Henderson's insights, I will revisit his argument, making use of our current understanding of the molecular structure of water and what it teaches us about water's vital role for life. Water's structural features fully explain the chemical and physical properties of this unusual liquid—properties that have led a number of scientists, not just Henderson, to regard this material as uniquely suitable to support life. As astronomer Owen Gingerich points out:

> In the years following the publication of Henderson's book, insights into atomic structure made . . . the unusual properties of water more understandable, without in any way diminishing Henderson's arguments or the awe that accompanies appreciation of this fine-tuning of our environment.[1]

Is Gingerich correct? I think so, but we need to objectively explore the question: Do our modern-day insights add to or detract from Henderson's argument for the fitness of the chemical environment and, hence, the existence of a chemical anthropic principle?

In *The Fitness of the Environment,* Henderson also describes the properties of carbon dioxide and the versatility of carbon-containing compounds. Henderson argued that both also contribute to the chemical environment's suitability for life. In my attempt to determine if the anthropic principle extends into the realm of chemistry, again, I will go beyond merely summarizing

Henderson's argument. Instead, I will seek to determine the impact our modern understanding of the molecular structures of carbon dioxide and carbon-containing compounds have on the chemical environment's fitness, if any.

Since Henderson's time, we have developed other insights from chemistry that could potentially add to the case for the environment's fitness and the chemical anthropic principle. Just like water and carbon dioxide, oxygen appears to possess the just-right set of properties that make it ideally suited for life, contributing to the environment's fitness.

We now have a good understanding of how chemical elements form during stellar burning and are subsequently distributed throughout the universe. The processes for the formation of carbon, oxygen, and nitrogen depend on the exquisite fine-tuning of nuclear energy levels. Traditionally, this insight has been used by physicists and cosmologists to argue for a cosmic-level anthropic principle. So, can this example of fine-tuning also be appropriated to the case for a chemical anthropic principle?

Finally, we have also come to learn about the abundance of elements in the universe. How do these quantities relate to abundances of elements needed for life? Do these insights form a set of anthropic coincidences that supports Henderson's thesis about the teleology of the universe embodied in the chemical environment?

Let's begin our query at the same place Henderson began his investigation of the chemical environment's fitness: with water.

Water

For those of us who live in the developed world, our day-to-day experiences make it easy to take water for granted. All we have to do is turn on the faucet and out comes clean, potable water. Water is everywhere, all around us. For that reason, water serves as the prototype for what a liquid should be. But, as I learned during the first few weeks of my organic chemistry course in college, few liquids behave like water. This all-too-common, all-too-familiar liquid is, in fact, one of the most unusual, anomalous materials that exist. And the unusual, odd properties of water seem, precisely, to be the very properties required for life to even be possible. This is how biologist Simon Conway Morris and physicist Ard Louis describe water:

> Colorless, transparent, and tasteless, the substance we call water is ubiquitous and commonplace. Arguably, it is also the strangest liquid in the universe with many peculiar

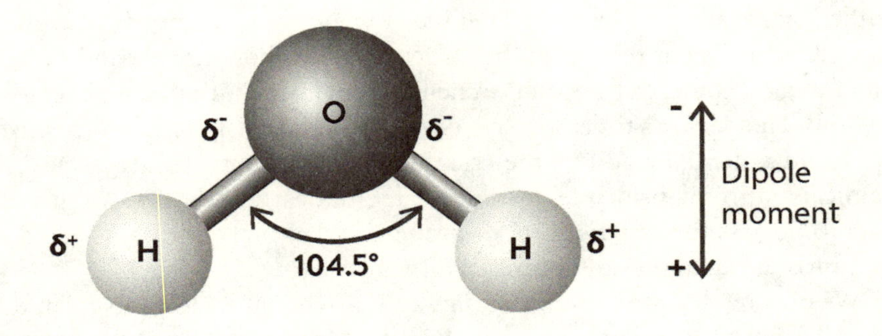

Figure 3.1: The Structure of Water

> counterintuitive properties that, it is widely proposed, are central to the existence of life. . . . Water is a "strange and eccentric" liquid. The anomalies of water, unsurprisingly, have been recruited by those who see an intriguing, if not suspicious, fitness to purpose, so far as life is concerned.[2]

So, what makes water so strange and eccentric? To understand its chemical and physical properties, we need to first understand its chemical structure. (For readers familiar with the structural and chemical properties of water, feel free to skip ahead to "Water's Unusual Properties.")

The Structure of Water

Water consists of two atoms of hydrogen bound to a centrally located oxygen atom. Instead of adopting a linear molecular geometry, water forms a bent structure with an H-O-H bond angle of 104.5°.

Water forms a bent geometry because of the unshared electron pairs of the oxygen atom. The two oxygen-hydrogen covalent bonds (formed by the two electrons shared between the oxygen and each of the hydrogen atoms bonded to the oxygen) and the unshared electron pairs orient in three-dimensional space to form a structure with a tetrahedral geometry. As a consequence of this molecular geometry, when considering only the arrangement of atoms, water appears to form a bent structure. But to truly appreciate water's structural properties and their impact on its physical and chemical properties, we really need to think about water as a tetrahedral molecule.

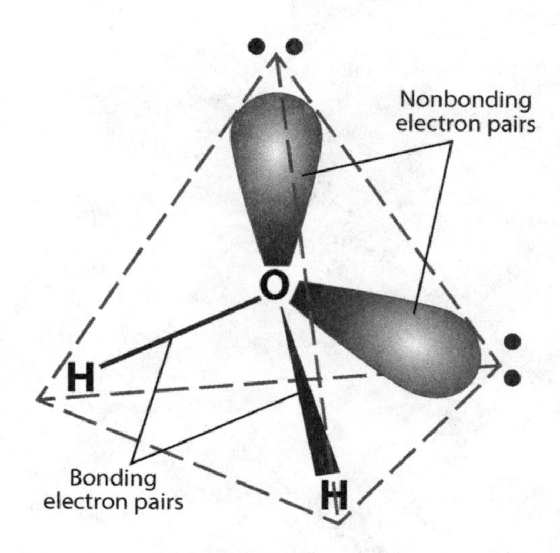

Figure 3.2: The Tetrahedral Geometry of Water

The unshared electron pairs are also responsible (at least in part) for another important structural feature of water: its charge polarity. To put it another way, though neutral in charge, some regions of the water molecule bear a partial negative charge and other portions of the molecule bear a partial positive charge.

The charge separation in the water molecule stems from two factors. The first has to do with the nature of the chemical bonds between oxygen and the two hydrogen atoms. Each bond forms when oxygen and the hydrogen atoms share two electrons, but the oxygen and hydrogen atoms don't share the electrons equally. Because oxygen has a greater electronegativity (the tendency of an atom to attract electrons to itself in a chemical bond) than hydrogen, the electrons spend more time in closer proximity to oxygen than hydrogen. This difference renders the hydrogen atoms with a partial positive charge and the oxygen atom with a partial negative charge. The second factor contributing to the polarity of the water molecule involves the unshared electron pairs of the oxygen atom, which imparts to this portion of the water molecule a partial negative charge. The tetrahedral geometry of the water molecule orients the partial positive and partial negative charges away from each other at opposite corners of the tetrahedron, creating polarity (charge separation) in the molecule.

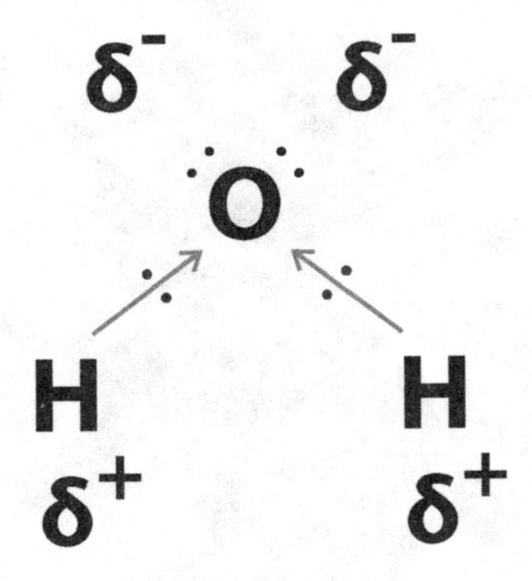

Figure 3.3: The Polarity of the Water Molecule

Hydrogen Bonding

Because of their polarity, water molecules interact with one another through a special type of interaction called hydrogen bonding. Hydrogen bonds occur between the partially positive hydrogen atoms of one water molecule and one of the partially negative unshared electron pairs of another water molecule. Each water molecule can interact with four other water molecules to form a hydrogen-bonded network of water molecules.

Water is not unique in its capacity to form hydrogen bonds. Any molecule with a hydrogen atom directly bonded to oxygen, nitrogen, or fluorine can form hydrogen bonds. But the impact of hydrogen bonding in water is more far reaching than in any other molecule.

The capacity to form hydrogen bonds dictates the long-range arrangement of water molecules in both the liquid and solid forms. In the gas phase (steam), negligible hydrogen bond interactions take place between the water molecules. On the other hand, in ice (the solid state of water), each water molecule interacts with four other water molecules to form an extensive, ongoing lattice of molecules. Hydrogen bond interactions generate an open network of water molecules. This open network explains the lower density of ice compared to water in a liquid state. Thus, hydrogen bonding accounts for one of water's most well-known and unusual properties: ice floats in liquid water. For nearly

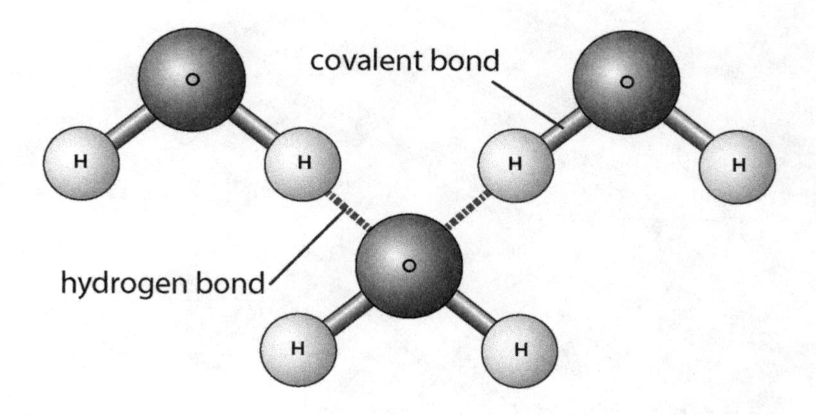

Figure 3.4: The Hydrogen Bond

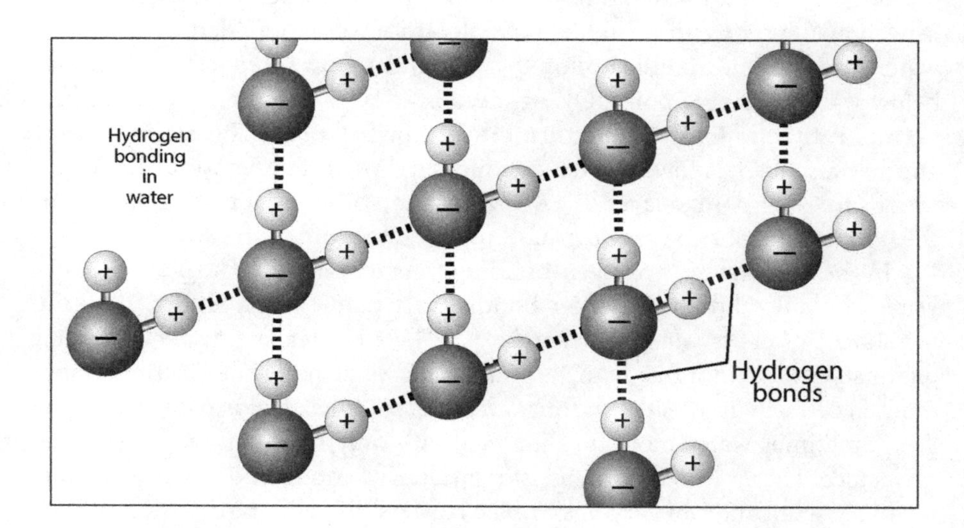

Figure 3.5: The Hydrogen-Bonding Network in Water

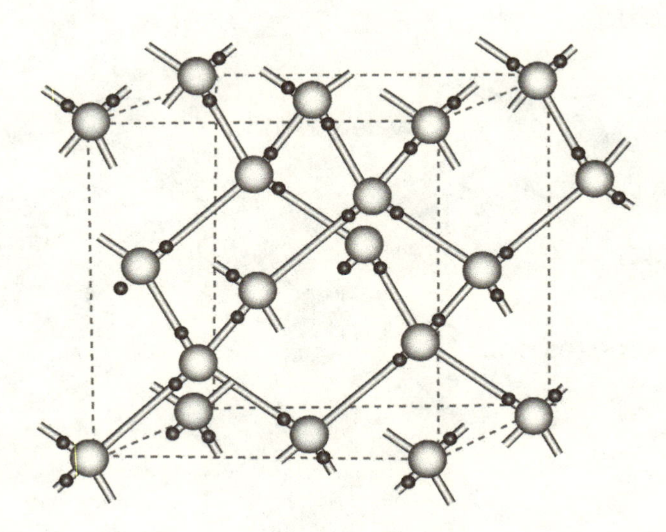

Figure 3.6: The Lattice Structure of Ice

every other known material, the solid state possesses a higher density than the liquid state and, therefore, sinks to the bottom of the liquid when the two phases coexist. Not so for water. If it weren't for this odd property, it is hard to conceive how life on Earth would be possible. If ice were more dense than liquid water, it would settle on the bottom of lakes and oceans when temperatures fall below water's freezing point. Over time, all the lakes and oceans would freeze permanently, and Earth would turn into a snowball planet. Because ice floats, and because ice has low thermal conductivity, when temperatures fall below water's freezing point a layer of ice forms on top of bodies of water, insulating the liquid beneath from the surrounding freezing temperatures.

In the liquid state, hydrogen-bonding interactions still take place between water molecules, but the hydrogen bonding is dynamic and sporadic. In the liquid state, clusters of water molecules exist. These clusters are dispersed among regions of water in a disorganized state. The water molecules in the clusters don't necessarily interact with four other water molecules, as with the case of ice. Sometimes water molecules interact with only two or three other water molecules. These clusters form, exist temporarily, and then dissociate. They're relatively small and can vary in size. The clusters' size and the duration of their existence varies inversely with temperature. At higher temperatures, clusters tend to be smaller and shorter-lived than at lower temperatures.

With this understanding in place, let's turn our attention to some of water's unusual physical and chemical properties. As we will discover, water's capacity to form hydrogen bonds is the chief reason water displays such anomalous physicochemical properties.

Water's Unusual Properties

It would take little effort to compile an extensive list of water's unusual properties that play a necessary role in creating a chemical milieu suitable for life. Some properties exert their influence on a planetary scale, such as the lower density of ice compared to liquid water. Other properties exert their influence at the organismal level. And still others exert their primary influence in the molecular arena. It is beyond the scope of this book to detail all of these properties. (A number of scientists have already done this.[3]) So, in the interest of brevity, I am going to survey those properties germane to the biochemical anthropic principle. My discussion won't be exhaustive, either. I am merely going to present a survey, limiting the discussion to those properties that I deem to be most salient for the structure and function of biochemical systems. Even though I am presenting a truncated list of water's unusual properties, I think it is extensive enough for us to address questions about the suitability of the chemical environment for the existence of living organisms and determine if a chemical version of the anthropic principle exists.

Water's anomalous thermal properties. Because of water's capacity to form hydrogen bonds, it has an unusually high boiling point and melting point. At Earth's atmospheric pressure water boils at 100°C and melts at 0°C. If water molecules didn't interact via hydrogen bonding, then, based on the trends of the boiling and melting points of other hydrides, water would be predicted to boil at -100°C (at Earth's atmospheric pressure). As a point of reference, hydrogen sulfide (H_2S), which has a molecular geometry similar to water, boils at -60°C.

The high melting and boiling points of water force this material to adopt a liquid phase within the just-right temperature range to render water a suitable matrix for living organisms. As a rule of thumb, the rate of chemical reactions doubles for each 10°C increase in temperature. And the converse is also true. The rate of chemical reactions becomes halved for every 10°C decrease in temperature.

At -100°C (the predicted boiling point of water if it didn't form hydrogen bonds) it is too cold for most chemical reactions to proceed. Yet at 0°C, (the freezing point of water at atmospheric pressure) most chemical processes readily occur. Water's boiling point of 100°C is also fortuitous. At high temperatures, chemical reactions proceed rather quickly, which would be desirable for living systems. But if temperatures exceed 100°C it leads to chemical instability for many biomolecules. For example, at temperatures above 100°C proteins readily denature, making it impossible for these critical biomolecules to adopt stable three-dimensional structures that are crucial for their biochemical roles. Denaturation occurs under these temperature conditions because delicate noncovalent intermolecular interactions become disrupted. These interactions play a role in stabilizing the higher-order structures necessary for proteins to adopt functional three-dimensional structures.

The separation of water's boiling and melting points by 100 degrees (at atmospheric pressure) is significant. It ensures water remains liquid over a fairly broad temperature span, making life possible under a wide range of environmental conditions. Water also has the highest heat of evaporation of any known substance, thanks to its capacity for hydrogen bonding. (The heat of evaporation is the energy needed to convert a unit amount of material from a liquid to vapor at its boiling point.) Because of its high heat of evaporation, water has a tendency to remain in the liquid phase, even at its boiling point.

Water also has an unusually high heat capacity. This quantity refers to the amount of energy needed to elevate the temperature of a unit amount of a substance by 1°C. Water has the second highest heat capacity of any known substance, again, thanks to its capacity to form hydrogen bonds. Its high heat capacity also helps water retain a liquid state and helps keep the temperature of water stable amid environmental heat fluctuations. Temperature stability is critical for life because of the close connection between temperature and chemical reactivity and the stability of the higher-order structures of proteins and other types of biomolecules.

Water's high thermal conductance keeps cells from "cooking" themselves. Thermal conductance refers to the capacity of a substance to conduct heat away from its source. Biochemical processes generate heat. If the heat can't radiate quickly away from the cell, it will lead to an elevation in the local temperature of the cellular environment. And this elevation in temperature would compromise biomolecule stability. But because water can quickly and efficiently dissipate heat away from its source—again, thanks to its capacity to form hydrogen bonds—it stabilizes the temperature of the local cellular environment.

Water's unusual properties as a solvent. Because of water's polarity and its capacity to form hydrogen bonds, a wide range of chemical materials readily dissolve in water. For this reason, water is often called the universal solvent,

The Liquid Phase is a Requirement for Life

Living organisms can exist only in water's liquid phase. In the gas phase, the solvent and solute molecules are so far apart from each other, it severely limits intermolecular interactions. Additionally, chemical reactions (which require molecules to collide with each other) in the gas phase are a relatively rare occurrence. Also, noncovalent associations among molecules would be too infrequent to form molecular complexes, which are critical for a number of life processes, including the formation of higher-order biomolecular structures and other types of subcellular structures. If the matrix were to exist in the solid state of ice, then molecular mobility would be highly restricted. This lack of mobility would prevent the chemical reactions and other dynamic processes necessary for life from occurring. In the liquid phase, molecular density and molecular mobility are both high enough to allow molecules to interact with each other, creating a highly dynamic environment where chemical reactions and other noncovalent interactions occur at a high enough frequency to sustain life.

though this label is a misnomer because there are many materials that won't dissolve in water. Still, one would be hard-pressed to find a solvent that is more versatile than water.

Water's capability to dissolve a wide range of materials creates a liquid milieu inside and outside the cell that allows for sufficient chemical diversity to support the multitude of molecular structures and chemical activities needed for life. If water lacked versatility as a solvent, the resulting physicochemical restrictions would most likely preclude life altogether.

Water's polarity allows it to dissolve ionic materials, which possess positive and negative charges. Plus, its capacity to form hydrogen bonds allows water molecules to form intermolecular, noncovalent associations with hydrophilic materials, especially those also capable of forming hydrogen bonds. On the other hand, water lacks the capacity to solubilize nonpolar (oily) chemical compounds. Water's inability to dissolve nonpolar materials is also related to its capacity to form hydrogen bonds. As it turns out, this inability contributes to water's unusual and just-right properties for life by giving rise to a phenomenon known as the hydrophobic effect.[4] The hydrophobic effect describes the aggregation of water-insoluble, "oil-like" materials in aqueous systems. This aggregation occurs not because of attraction taking place among the aggregates, but because of repulsion between water and the oil-like materials. The aggregation process segregates the oil-like compounds from water molecules. If water were to dissolve nonpolar materials, then individual water molecules would have to surround each molecule of the nonpolar substance. But water does not interact with the nonpolar molecules. And because the water molecules can't interact with the nonpolar solute, they are forced to adopt an icelike structure, increasing the ordered structure of the molecules while reducing the entropy of the system. This reduction in entropy is thermodynamically unfavorable. However, the segregation and aggregation of the nonpolar materials allows the water molecules to remain in a disordered state that keeps the system's entropy at a relatively high level.

Proteins, RNA, DNA, and cell membrane components possess nonpolar, oil-like regions as part of their molecular makeup. To maintain water in the maximum disordered state, the nonpolar regions of these biomolecules segregate away from the water and aggregate. This sequesters them from contact with water. If not for this hydrophobic effect, the stable higher-order, three-dimensional structures of proteins and RNA molecules, along with DNA's double helix and cellular membranes, would not form, and life simply wouldn't be possible. Water is one of the few materials that manifest the hydrophobic effect and, with respect to other materials that do, water displays a significantly stronger hydrophobic effect.

Water's unusual capacity to participate in biochemical processes. Not only does water serve as a solvent system that creates the just-right milieu for life, it also plays an active role in the chemical processes that support the cell's biochemical operations. Because of water's capacity for hydrogen bonding, it undergoes a chemical process called self-ionization. This reaction makes possible the acid-base chemistry that is central to so many biochemical processes.

The chemical properties of water also make possible hydrolysis reactions, in which water reacts with molecules (such as polysaccharides, triglycerides, proteins, nucleic acids, etc.) breaking them down into constituent parts. This process consumes the water molecule, with portions of the molecule becoming incorporated into the breakdown products. Hydrolytic reactions play a central role in cellular metabolism, particularly those metabolic pathways which liberate chemical energy needed to power biochemical operations.

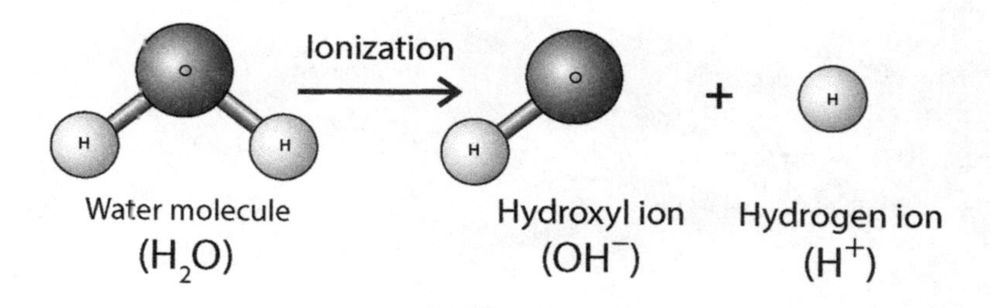

Figure 3.7: Self-Ionization Reaction of Water

Water also takes part in reduction and oxidation reactions, known as redox reactions. These chemical processes are indispensable to cellular metabolism. For example, during cellular respiration oxygen serves as the final electron acceptor in the electron transport chain, generating water as the end product of this sequence of redox reactions. (For more details, see chapter 8.) Water also serves as a source of electrons during photosynthesis, generating oxygen as the product.

Water can also serve as a ligand that binds to metal centers in proteins and, in doing so, participates in several key biochemical processes. Typically, it is the unshared electron pairs on the oxygen atom that allow water to bind to protein metal centers. Water also actively contributes to the higher-order, three-dimensional structure of biomolecules, such as proteins, by forming a hydration shell around the protein's exterior surface. This shell contributes to the protein's structure and function and is rightly understood to be part of the biomolecule's structural features.

Because of water's polarity and hydrogen-bonding capacities, charged materials can readily move through aqueous systems. Water's charge conductance makes a whole host of electrochemical processes possible inside the cell. Because of its hydrogen-bonding capacity, water also possesses the ability to make use of quantum tunneling to transport protons through protein channels embedded in cell membranes.

Based on this brief survey, it becomes apparent that water displays an impressive set of fortuitous properties that significantly contribute to the fitness of the chemical environment for life. It is also evident that these just-right properties arise, in large measure, from water's hydrogen-bonding capacity.

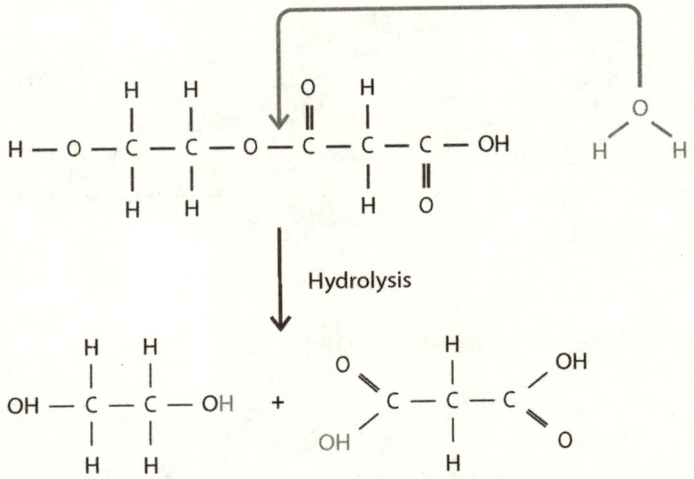

Figure 3.8: Hydrolysis Reaction

But it isn't just the existence of hydrogen bonding in water that is critical. It is also the strength of the hydrogen bonds formed by water. In fact, chemist Martin Chaplin's study of varying the strength of the hydrogen bond on water's chemical and physical properties has shown that the hydrogen bond strength must be fine-tuned for water to have its life-giving properties.[5]

Fine-Tuning of Water's Chemical Properties

Through a counterfactual analysis, Chaplin demonstrated that if the hydrogen bond strength were *weaker*:

- *It would lower the melting and boiling points of water.* This lowering would require life to exist at lower temperatures. This would be an impediment to life, because as temperatures lower, so do the rates of chemical reactions. Decreasing hydrogen bond strength in water would also mean that hydrogen bond strength would decrease in proteins and DNA as well, destabilizing the higher-order, three-dimensional structures of these biomolecules.
- *It would compromise water's ability to solubilize hydrophilic and ionic materials.* As a consequence, water would not serve as a suitable matrix for life, because it wouldn't permit the requisite chemical diversity.

Figure 3.9: Hydration Shell around a Protein

- *It would lead to a loss of water's hydrophobic effect.* This loss would prevent proteins and RNA from forming stable three-dimensional structures, compromise the formation of the DNA double helix, and prevent cell membranes from forming.
- *It would reduce water's ability to self-ionize.* This loss of self-ionization would alter the acid-base chemistry necessary for life.

Chaplin has also demonstrated that if the hydrogen bond strength were *stronger*:

- *It would raise the melting and boiling points of water.* This temperature increase would force liquid water into a temperature regime that would destabilize the higher-order, three-dimensional structures of biomolecules, such as proteins and nucleic acids.
- *It would compromise water's ability to solubilize hydrophilic and ionic materials.* Stronger hydrogen bonds would prevent water molecules from dissociating from the clusters they form. This reduced dissociation would keep water molecules from surrounding and interacting with hydrophilic solute molecules. As a consequence, water wouldn't have the capacity to support the chemical diversity necessary for life.
- *It would reduce water's ability to self-ionize.* This effect may seem

> counterintuitive. Increased hydrogen bond strength would lead to enhanced self-ionization, but because of the strength of the hydrogen bond network among water molecules, the resulting hydrogen and hydroxide ions couldn't efficiently diffuse away from the site of the reaction. As a result, the self-ionization reaction would have more opportunity to reverse itself and regenerate the water molecule. The net effect of the more efficient reformation reaction would result in an overall reduction in self-ionization and an alteration of the acid-base chemistry necessary for life. As a consequence, the hydrolysis reactions described would not be able to occur.

Clearly, water's just-right hydrogen-bonding capabilities result in a special set of just-right properties that unduly contribute to the environment's fitness for life. Is water the only material with these properties or are there other materials that could serve as a suitable matrix for life?

Is Water the Only Solvent Suitable for Life?

Instead of carrying out counterfactual analysis using theoretical techniques, we can address this question experimentally by performing actual comparisons of the physical and chemical properties of water with other potential solvents. Along these lines, it is quite commonplace for origin-of-life researchers and life scientists to speculate about nonaqueous materials that could serve as a life-support matrix. While a number of proposed alternatives seem intriguing, in that these materials possess some properties that would serve to sustain life, ultimately, they all fall short for a variety of reasons. While we can't decisively rule out the possibility that some of these materials could form a chemical environment that could support life—perhaps different than life as we currently know it—the evidence seems to argue against such a conclusion.

Let's consider a sampling of this evidence for some of the most commonly proposed alternatives to water.

Deuterium oxide (D_2O). The optimal qualities of water for life become no more apparent than when considering the effect of replacing the hydrogen atoms in water with deuterium, an isotope of hydrogen. (D_2O is known as heavy water.) This change to the chemical composition of water seems to be trivial. In reality, it would have a profound effect on the environment's capacity to support life. Replacing hydrogen with deuterium in the water molecule would be deleterious for life. In fact, life on Earth (at least, as we know it) could not exist in a deuterium oxide matrix. Replacing hydrogen with deuterium alters the mass of water, which in turn alters the rate of every chemical process that involves water. These altered rates would pervasively disrupt metabolic processes to such an extent that life couldn't exist.

Ammonia (NH_3). Like water, ammonia is abundant in the universe. And like water, ammonia adopts a tetrahedral geometry, with three hydrogens bound to a centrally located nitrogen atom. Unlike the oxygen atom found in water, the nitrogen atom has only one unshared electron pair. Because of the electronegativity of the nitrogen atom and its unshared electron pair, ammonia has the capacity to form hydrogen bonds.

Ammonia also undergoes a self-ionization reaction, making possible acid-base chemistry in an ammonia matrix. Ammonia also dissolves a wide range of organic materials and even some metals.

But the hydrogen bonds in ammonia are not nearly as strong as they are in water. One reason for this difference in strength relates to the lower electronegativity of nitrogen compared to oxygen. This difference reduces the magnitude of the partial positive charge of the hydrogen atom in the three chemical bonds between nitrogen and hydrogen. Also, because nitrogen has a single unshared electron pair (instead of water's two), the clustering of ammonia molecules in the liquid matrix isn't as extensive as it is in water. The reduced strength of the hydrogen bond results in a lower melting point (-77°C) and boiling point (-33°C) for ammonia compared to water. As a result, the rate of chemical reactions in liquid ammonia would be too slow to support the biochemical processes necessary for life.

Another problem with the reduced strength of hydrogen bonds in ammonia compared to water is the loss of the hydrophobic effect. This loss compromises the ability of cell membranes to assemble and proteins to adopt higher-order, three-dimensional structures. Also, the reduced strength of the hydrogen bond interactions, combined with ammonia's molecular structure, impacts the self-ionization of ammonia, skewing the equilibrium so that this molecule is more likely to accept a hydrogen ion than give one up.

Hydrogen sulfide (H_2S). In terms of its chemical structure, hydrogen sulfide has the closest similarity to water compared to any other chemical compound. Yet it forms extremely weak hydrogen bonds compared to water. Hydrogen sulfide is not nearly as good a solvent as water. Nor does it display a hydrophobic effect. Hydrogen sulfide has a boiling point of -60°C and remains liquid over a narrow temperature range. None of these features make this material suitable as a matrix for life.

Hydrogen fluoride (HF). Like water, hydrogen fluoride is a polar molecule

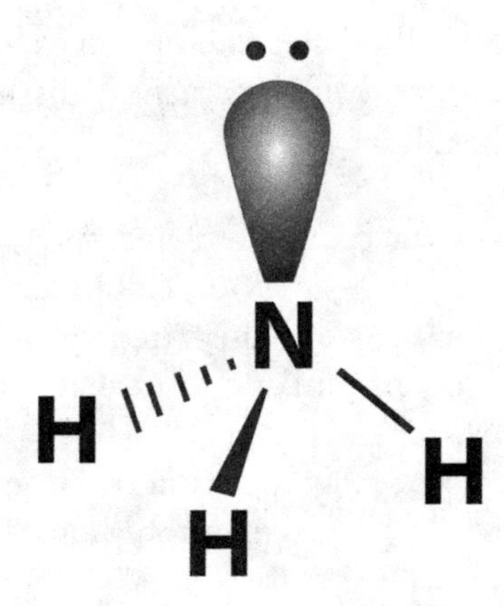

Figure 3.10: The Molecular Structure of Ammonia

that can dissolve a wide range of ionic and polar compounds. It has a melting point of -84°C and a boiling point of 19.5°C and is also capable of hydrogen bonding.

It is unlikely that hydrogen fluoride could serve as a solvent for life because it is a cosmically rare substance. Oxygen occurs at five times the level of fluorine in the cosmos and three times the level of fluorine in Earth's crust. Also, hydrogen fluoride is such a strong acid that it is chemically destructive to organic materials.

Methane (CH_4). Methane is one of the most abundant materials in the cosmos. Methane is a liquid at extremely cold temperatures (the melting point is -182°C and the boiling point is -162°C). It is a nonpolar compound that can't dissolve hydrophilic and ionic materials. It doesn't display the hydrophobic effect and it is chemically inert. None of these features bode well for methane as a matrix for living systems.

This survey—albeit brief and incomplete—makes it clear that there appears to be no other solvent like water. Experimental comparisons of water's properties and the properties of other potential candidates as a life-support matrix justify this conclusion. Simon Conway Morris and Ard Louis sum it up this way:

> Although many fluids are reasonable candidates in one respect or another, there is also the sense that no other fluid possesses

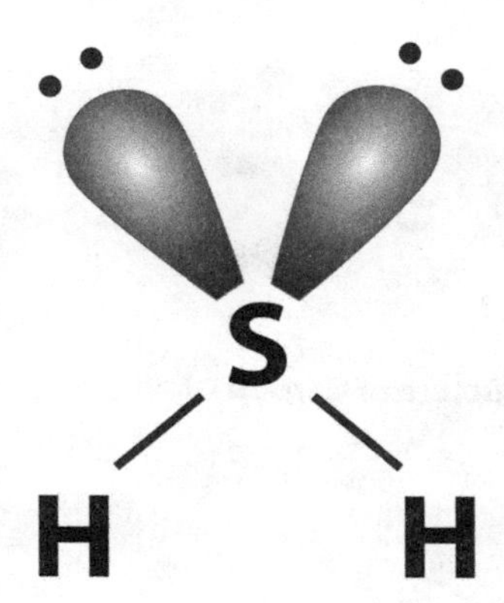

Figure 3.11: The Molecular Structure of Hydrogen Sulfide

> *all* the properties that make water so biologically versatile. Not only does it provide a broad and flexible canvas for life to paint its full multifaceted tableaux, it is itself part of the palette.[6]

Water is not the only substance with the just-right properties needed for life. In *The Fitness of the Environment*, Henderson devotes significant space to the properties of carbon dioxide, arguing that this compound also possesses a unique set of just-right properties that contribute to the chemical environment's suitability for life. So, what is it about carbon dioxide that makes it necessary for life?

Carbon Dioxide

Because of the threat of climate change, most people think of carbon dioxide (CO_2) as a dangerous, life-threatening gas. Yet, life couldn't exist on Earth if not for this odorless, colorless gas. In fact, carbon dioxide has a unique set of just-right properties that contribute to the suitability of the environment—when water serves as the matrix for life. In other words, the just-right properties of carbon dioxide work in conjunction with the just-right properties of water to help form an environment fit for life.

Through his study of the physiological properties of blood, Henderson was the first to recognize carbon dioxide's life-giving properties. So, have the intervening years of scientific advance affirmed or undermined Henderson's views

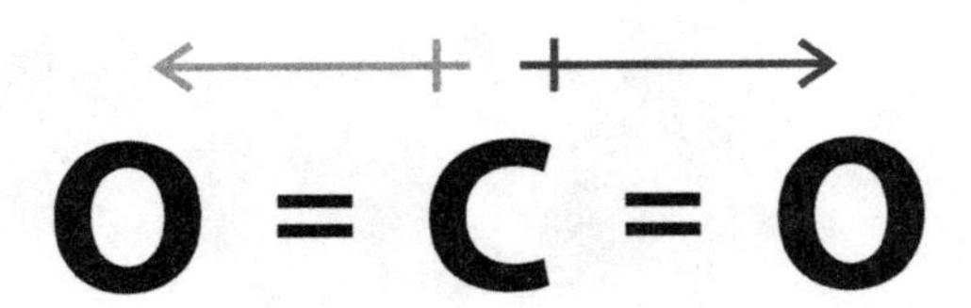

Figure 3.12: Chemical Structure of Carbon Dioxide

on carbon dioxide?

To address this question, it is important to review our modern-day understanding of the chemical structure of CO_2. Carbon dioxide consists of two oxygen atoms bonded to a central carbon atom. Because carbon has a valence of four (meaning it must form four chemical bonds when it combines with other atoms to generate a stable molecule), the two oxygen atoms form double bonds with the carbon atom. Both oxygen atoms have two pairs of unshared electrons.

Unlike water, which is a bent molecule, carbon dioxide adopts a linear geometry. This configuration provides the molecule's electrons, shared and unshared, the greatest possible separation in space, minimizing charge repulsion in the molecule. Because oxygen displays a greater electronegativity than carbon, the electrons in the double bonds have a closer association with the oxygen atoms than with the carbon atoms. The net result of the difference in electronegativities of the carbon and oxygen atoms creates a partial negative charge on each of the oxygen atoms and a partial positive charge on the carbon atom. But because the direction of the carbon-oxygen bonds is diametrically opposite, the polarities of these two bonds cancel each other, rendering carbon dioxide a nonpolar molecule.

What fortuitous properties of carbon dioxide lead to an environment that can uniquely harbor life? As with the previous discussion about the unusual and advantageous properties of water, I will limit exploration of carbon dioxide's beneficial properties to those with the most bearing on biochemical systems.[7] Toward this end, one of the most important properties of carbon dioxide is its solubility in both air and water. Compared to carbon monoxide (CO)—a close chemical analog to CO_2—carbon dioxide's solubility in water is an order of magnitude greater than the solubility of carbon monoxide. Two

Carbon Dioxide (CO_2)

$$\overset{-}{O} = \overset{+}{C} = \overset{-}{O}$$

Figure 3.13: Bond Polarity in Carbon Dioxide

physicochemical properties account for carbon dioxide's solubility in water. The first relates to its molecular structure. Even though carbon dioxide is a nonpolar molecule, it is soluble in water because of the polarity of the carbon-oxygen bonds. Because of the partial negative and positive charges on the oxygen and carbon atoms, respectively, water molecules can surround and interact with the carbon dioxide molecules in solution. The second relates to the chemical reactivity of carbon dioxide. When solubilized, carbon dioxide reacts with water. The partial positive charge on the carbon atom plays a role in carbon dioxide's reactivity toward water. This reaction yields carbonic acid (H_2CO_3). Because of Le Chatelier's principle (a system in equilibrium will dynamically change when conditions impacting the equilibrium are altered in such a way as to reestablish the equilibrium), the reaction of carbon dioxide with water "pulls" this carbon dioxide from the surrounding atmosphere into water.

As noted, carbon monoxide is much less soluble in water than carbon dioxide. At first glance, this doesn't make sense because it would seem that this molecule would be equally water soluble. However, the chemical bonding in carbon monoxide is unusual, rendering the oxygen atom with a partial positive charge and the carbon atom with a partial negative charge. This internal charge separation offsets the charge separation that results from the electronegativity differences of the two atoms, which renders the carbon atom with a partial positive charge and oxygen with a partial negative charge. The net effect: carbon monoxide is nonpolar. The offsetting differences in partial charges on the carbon and oxygen atoms also make carbon monoxide unreactive toward water.

Conversely, carbon dioxide's solubility in air and water impacts a number

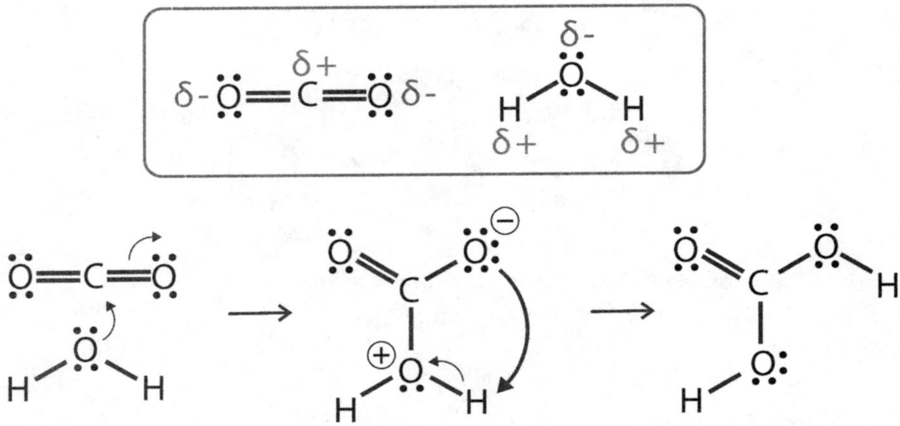

Figure 3.14: Reaction of Carbon Dioxide with Water

of critical biochemical processes. Because carbon dioxide is soluble in both air and water, it can freely exchange between both mediums. Carbon dioxide's solubility in air leads to its global distribution throughout the atmosphere. This distribution pattern makes carbon dioxide readily available at any location on the planet's surface to take part in photosynthesis. Because carbon dioxide is water soluble and can freely exchange between air and water, it can dissolve into intercellular and intracellular fluids, becoming available for the carbon fixation reactions of photosynthesis—a biochemical process that forms the foundation of many ecosystems. Unlike carbon monoxide, which readily oxidizes, carbon dioxide is chemically inert. Yet, as noted, because the carbon atom in CO_2 bears a partial positive charge, under the right circumstances this molecule can react with negatively charged functional groups to add a carbon atom to organic compounds. This reactivity makes it possible for carbon fixation reactions to take place in photosynthesis and other types of autotrophic pathways.

The dual air and water solubility of carbon dioxide plays another vital role in making life possible. It allows organisms, whether single-celled or multicellular, to efficiently eliminate metabolic wastes. The breakdown product of fuel molecules (such as sugars and fats) is carbon dioxide. These fuel molecules exist in a chemically reduced state, which means that energy needed to power cellular operations becomes liberated when chemical bonds in these molecules are broken. This breakdown process fragments the molecules into smaller and smaller pieces, eventually yielding a molecule of carbon dioxide (as the waste product) for every carbon atom in the fuel molecule. If carbon dioxide were like most oxides, it would be a solid material. If that were the case, waste products would accumulate in the cell and foul up cellular operations—unless each cell (and the organism as a whole) expended significant amounts of energy to remove the waste products from the cell and organism. An impractical energy expenditure such as this would perhaps make it impossible for life to exist. The same energy demand would exist if carbon dioxide was water insoluble, because even in the gaseous form carbon dioxide would accumulate in the cell. Fortunately, carbon dioxide is a gaseous material that *is* soluble in water. This allows the waste products that result from the breakdown of fuel molecules to readily diffuse away from the cell without any energy expenditure whatsoever.

Carbon dioxide's chemical reactivity with water is fortuitous for another reason. It establishes a buffering system known as the bicarbonate buffering system. Many multicellular organisms use this buffering system to maintain a stable pH in intercellular fluids, such as blood. (Stable pH is maintained inside cells with a phosphate buffering system.) It is provocative that the carbon dioxide that diffuses away from cells as a waste product, in turn, plays a role in maintaining the pH of the blood. In mammals, once carbon dioxide leaves the cell, it reacts with water to form carbonic acid. (This reaction is facilitated by the enzyme carbonic anhydrase in red blood cells circulating through the bloodstream.) Carbonic acid dissociates to form a hydrogen ion (acid) and the bicarbonate ion (HCO_3^-), which is called a conjugate base.

If the blood becomes too acidic, then the bicarbonate ion can react with hydrogen ions to reform carbonic acid. If the blood becomes too alkaline, then the bicarbonate ion can further dissociate into the carbonate ion (CO_3^{-2}) and a hydrogen ion, which then reacts with the hydroxide ion (OH^-) to form water. This buffering system helps keep the pH of blood near 7.4. In mammals, the lungs convert the bicarbonate back into carbonic acid, which then reacts to form water and carbon dioxide. This gas is then expelled from the body by the lungs.

Buffering of the blood is critical because the structure and function of many proteins depend on pH. Even small deviations from physiological pH compromises the function of these biomolecules. A stable physiological pH is also necessary for glucose (the primary fuel molecule used by mammals) to be transported through the blood to various tissues in the body. Glucose becomes chemically unstable under alkaline conditions. If pH varied in the blood, it

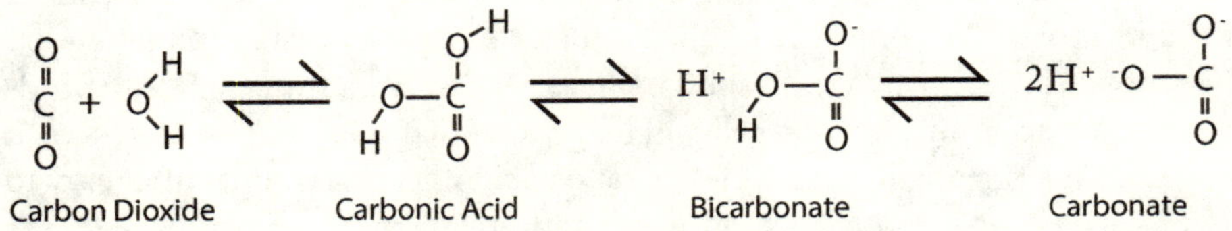

Carbon Dioxide | Carbonic Acid | Bicarbonate | Carbonate

Figure 3.15: The Bicarbonate Buffering System

would compromise the delivery of glucose to tissues.

The chemical reactivity of carbon dioxide with water to form the carbonic acid / bicarbonate system plays another vital role for life on our planet. This reaction helps some photosynthetic organisms, particularly microbes, use bicarbonate from the environment as a carbon source. This process involves carbonic anhydrase converting bicarbonate to carbon dioxide at the cell surface. Once formed, carbon dioxide can diffuse through the cell membrane into the cell interior. The reverse reaction can also be used by plants under low carbon dioxide conditions to sequester carbon dioxide. Carbonic anhydrase also facilitates this process, converting carbon dioxide to bicarbonate in the cell. Because cell membranes are impermeable to bicarbonate, it can be stored in the cell. And when carbon dioxide is needed, the bicarbonate can be reconverted to this gaseous molecule.

I hope this brief survey provides enough evidence to establish that carbon dioxide's unique and just-right properties contribute to the environment's suitability for life. As Henderson concludes in *The Fitness of the Environment*:

> There can be no doubt that the physical properties of carbon dioxide are less important to the living organism than are those of water. . . . However, the less conspicuous substance is not without its physical fitness.[8]

Carbon dioxide is not the only molecule in Earth's atmosphere that plays a role in the environment's suitability for life—so does oxygen.

Oxygen

If there's a gas that everyone associates with life, it's oxygen. We all know that without access to oxygen, many organisms (including human beings) will quickly die. As a chemist, life's dependence on oxygen is somewhat counter-intuitive to me—at least at first glance. Oxygen is a highly reactive, chemically destructive material. If anything, organisms should try to do whatever they can to avoid contact with oxygen. In fact, there are organisms that eschew oxygen (anaerobes), and others have biochemical systems in place that help to mitigate the potential damage oxygen can cause. Interestingly, these mechanisms are often coupled to biochemical processes that help the cell to maximally extract energy from fuel molecules, such as sugars and fats. For this reason, these organisms depend on ready, uninterrupted access to oxygen. In effect, these processes require the very chemical properties of oxygen that make it so reactive and destructive.

The molecular structure of oxygen explains its chemical reactivity. As it turns out, oxygen, like water, is a weird molecule. These weird structural features lead to a set of fortuitous properties that contribute to the environment's fitness for life.

I already discussed the impact of oxygen's high electronegativity. The only chemical element with a higher electronegativity than oxygen is fluorine. Oxygen's high electronegativity plays a key role in the capacity of water to form hydrogen bonds. What are some of the other properties of oxygen that make this material uniquely suited for life? As with water and carbon dioxide, our discussion of oxygen's fitness will benefit from a description of our modern-day understanding of its chemical structure and physicochemical properties. Molecular oxygen (or more appropriately dioxygen) consists of two oxygen atoms bound together to form a linear molecule. Because the valence of oxygen is two, the two oxygen atoms are joined by a double bond, with each oxygen atom possessing two pairs of unshared electrons.

Unfortunately, this depiction of oxygen's molecular structure doesn't account for some of oxygen's physicochemical properties. In reality, oxygen is better depicted as two oxygen atoms joined by a single bond, with partial double-bond character. In this alternative structure, each oxygen atom possesses two unshared electron pairs and a single, isolated unshared electron. (Molecules with unshared electrons are called free radicals. Molecular oxygen is sometimes referred to as a diradical.)

Molecular oxygen's diradical structure explains its aggressive chemical reactivity. Because of the diradical structure, molecular oxygen is a powerful

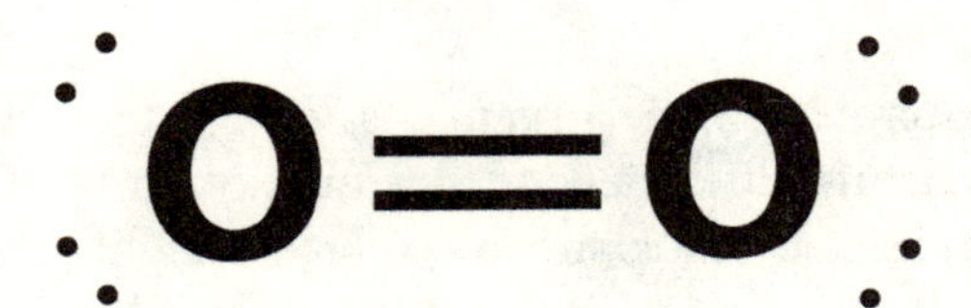

Figure 3.16: The Molecular Structure of Oxygen

oxidizing agent. Molecular oxygen reacts with a wide range of materials forming oxides—compounds consisting of a chemical element combined with oxygen. (Carbon dioxide is one of the oxides of carbon.) As a rule, during the process of oxidation, the near maximal amount of energy is liberated from the chemical bonds of the substance undergoing oxidation. This chemical property allows oxygen to extract, practically speaking, the maximum amount of energy from the breakdown of sugars and fats (two classes of molecules used as an energy source by cells) as they are transformed into carbon dioxide.

Oxygen plays another role in the breakdown of fuel molecules. It serves as the terminal electron acceptor in the electron transport chain (ETC), generating water during the process. (See chapter 8.) The ETC harvests energy needed to carry out the various biochemical operations that take place within the cell. For the most part, the ETC is comprised of a series of protein complexes, conceptually organized into a linear array. The first complex of the ETC receives chemically energetic electrons (ultimately derived from the breakdown of sugars and fats) and passes them along to the next complex in the ETC. Eventually, these electrons are handed off from complex to complex, until they reach the terminal part of the ETC. When shuttled from one complex to the other, the electrons give up some of their energy. This released energy is captured and ultimately used to produce compounds, such as ATP (adenosine triphosphate), which serve as energy currency inside the cell. One of the final steps carried out by the ETC is the conversion of molecular oxygen into water, with oxygen receiving the de-energized electrons.

Because it is such a powerful oxidizing agent, oxygen is the ideal molecule to serve as the terminal electron acceptor in the ETC. Other conceivable materials such as the halogens (fluorine, chlorine, etc.) and nitrogen lack oxygen's physicochemical properties. For example, nitrogen is not as soluble in

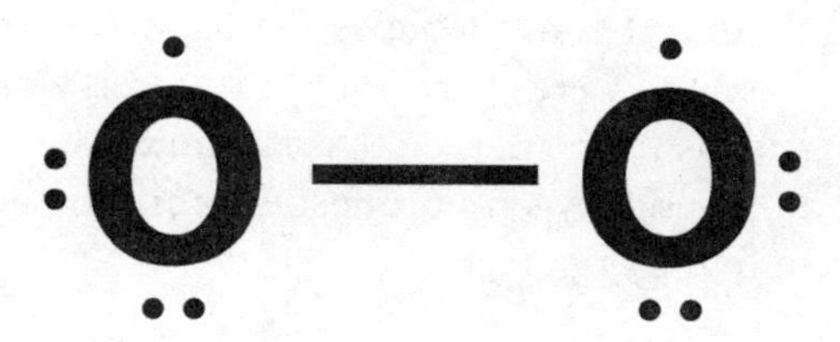

Figure 3.17: A More Accurate Depiction of the Molecular Structure of Oxygen

water and not as strong an oxidizing agent. As biochemists Ali Naqui, Britton Chance, and Enrique Cadenas conclude:

> It may be interesting to note why O_2 has been "chosen" by nature to act as the terminal oxidant of the respiratory chain. . . . The kinetic reactivity of the halogens makes them unsuitable as biological oxidants, and nitrogen is too poor an oxidizing agent. Oxygen is thus the only element in the most appropriate physical state, with a satisfactory solubility in water with desirable combinations of kinetic and thermodynamic properties.[9]

Despite molecular oxygen's high level of chemical reactivity, it oddly becomes chemically inert, relatively speaking, under the precise set of conditions required for life to exist: moderate temperatures (which correspond to the temperature range of living organisms) and in aqueous solutions. This unusual and just-right chemical behavior stems from oxygen's diradical structure and the constraints it places on the reaction mechanism of this molecule. Interestingly, the cell overcomes this constraint through the use of enzymes with metal centers. These metals have the just-right electron configuration that makes them compatible with oxygen's diradical structure. If it weren't for oxygen's chemical inertness, life couldn't exist in an environment with appreciable atmospheric oxygen. The organic materials that make up living organisms would spontaneously combust.

Oxygen also displays just-right solubility properties. Like carbon dioxide, it is soluble in air and water. This dual solubility allows the oxygen generated as a by-product of photosynthesis to diffuse away from plant cells and tissues into the environment. It also allows oxygen to dissolve in intercellular and intracellular fluids so that it is available as the terminal electron acceptor of the ETC. However, if oxygen were more soluble, its levels would build up in the fluids of living organisms to the degree that it would cause wanton oxidative damage to cells and tissues. Oxygen's properties prove to be critically right for essential functions in living organisms.

Uniqueness of Carbon-Containing Compounds

In *The Fitness of the Environment*, Henderson described another fortuitous aspect of nature that makes life possible: the chemical properties of carbon and the compounds it forms. Since Henderson's classic work, it has become commonplace for origin-of-life researchers and life scientists to speculate about the existence of alternative forms of life—life based on another chemical element, such as silicon, instead of carbon. In fact, in some quarters of the scientific community it is a foregone conclusion that life as we *don't* know it exists elsewhere in our solar system and beyond. And they believe we will one day discover it. Yet there are others in the scientific community who would affirm Henderson's original conclusion. So, does Henderson's insight about organic compounds still hold? Is carbon the only chemical element that can form a wide range of stable compounds with the chemical diversity and complexity necessary to support life? Or could other chemical elements form the basis for life? The best place to begin this query is with the evidence that led Henderson, and many other scientists since his time, to think that carbon is uniquely suited for life.

The Chemistry of Carbon-Containing Compounds

Chemical complexity is one of the requirements for life. Even the simplest cells require a diverse collection of chemical compounds to carry out the basic activities needed for life, such as:

- Taking up nutrients from the environment
- Converting nutrients into energy to power life's operations
- Turning nutrients into biomolecules that perform life's activities
- Using nutrients to generate molecules that assemble to form subcellular and cellular structures
- Growing and reproducing
- Repairing damaged biomolecules and subcellular and cellular structures
- Breaking down damaged or unneeded cell components
- Eliminating waste

No one disputes the fact that carbon displays the chemical properties to form compounds with sufficient diversity and complexity to support these activities. Thanks to our modern-day understanding of chemical bonding, we understand why carbon possesses this capability.

- Carbon has a valence of four. This property means carbon forms compounds with four chemical bonds.
- Carbon can form single, double, or triple bonds. This bonding capacity means carbon can form a diverse array of chemical compounds with differing physical properties and chemical reactivities.
- Carbon forms strong bonds with itself. The capacity of carbon to form strong carbon-carbon bonds means that it can form compounds that consist of long, linear, and branched chains, as well as compounds with single and fused ring structures.
- Carbon forms strong bonds with hydrogen, oxygen, nitrogen, and sulfur. These atoms can be incorporated into the linear, branched, and ringed structures formed by carbon. Carbon's ability to bond with heteroatoms (atoms that are not carbon or hydrogen) generates an enormous range of diverse compounds with a wide range of chemical and physical properties.
- Carbon forms aromatic bonds. These bonds have special properties that make compounds unusually stable. It also imparts compounds with special chemical and physical properties. Aromaticity in a molecule also allows for special types of intermolecular interactions.
- Carbon compounds are metastable. Generally speaking, organic carbon compounds display a balance of chemical stability and reactivity.

Without question, this set of properties makes carbon ideally suited for life. The diversity and complexity of organic compounds are bewildering. In fact, upwards of tens of millions of organic compounds exist or could conceivably exist. Carbon does indeed contribute to the environment's fitness for life. Still, is carbon unique? Is it possible for life to be based on a chemical element other than carbon?

The Chemistry of Silicon-Containing Compounds

According to its position in the periodic table of elements, the most likely alternate candidate would be silicon. This element appears right below carbon in the periodic table, and, for this reason, it shares many of carbon's useful chemical properties. Silicon has a valence of four. It forms silicon-silicon bonds. Silicon can form linear and branched chains and ringed compounds. On top of that, silicon-based compounds tend to be more thermally stable than carbon compounds. This allows silicon compounds to exist at higher temperatures, opening the possibility that silicon-based life could exist in more extreme environments than carbon-based life.

Yet our modern-day understanding of the chemical properties of silicon raises serious doubts that life could be based on silicon instead of carbon.

- Silicon-silicon bonds are much weaker than the corresponding carbon-carbon bonds.
- Silicon lacks the capacity to form double and triple bonds.
- Silicon-silicon bonds are more reactive than carbon-carbon bonds.
- The linear and ringed compounds formed by silicon are more constrained and less stable than the corresponding carbon-based compounds.
- Silicon-hydrogen bonds are unstable.
- Silicon-silicon bonds are highly susceptible to oxidation.

On top of these problems, silicon oxides are solid materials, unlike carbon dioxide which is a gas. This difference means that it would be profoundly difficult to eliminate waste from the breakdown of silicon-based fuel molecules.

The bottom line is that Henderson was correct, and his conclusion still stands. A comparison of the properties of carbon and silicon highlights the unique and exceptional properties of carbon and the compounds it forms—compounds with the just-right properties that make life possible.

The Abundances of Chemical Elements in the Universe

Up to this point, our survey demonstrates that carbon, oxygen, and hydrogen, and the compounds they form—such as water, carbon dioxide, molecular oxygen, and a suite of organic materials—display the just-right properties that make life possible. Restated differently, the chemical and physical properties of these substances render the environment fit for life. If, indeed, the universe has been structured to be biogenic, it is reasonable to think that the elemental abundances of the universe should reflect that fact. In other words, if the universe is indeed fit for life, then we should expect that the elements needed for life would be the most abundant.

Unlike in Henderson's day, astronomers now can measure the cosmic abundances of the elements. In terms of relative abundance, the top ten elements are hydrogen and helium (which formed shortly after the big bang), followed by oxygen, carbon, neon, iron, nitrogen, silicon, magnesium, and sulfur.

So, how does the cosmic abundance of elements compare to the abundance of chemical elements in living systems? Generally speaking, life is built around carbon, hydrogen, oxygen, nitrogen, phosphorus, and sulfur (CHONPS). Of course, other elements are critical for life to exist. As a specific point of reference, the most abundant elements in the human body are hydrogen, carbon, nitrogen, calcium, phosphorus, potassium, sulfur, chlorine, sodium, and magnesium. A visual inspection of figure 3.18 reveals a close correspondence between the cosmic abundances of elements and the elemental abundances in living organisms.

How should we interpret this intriguing correspondence? One possible explanation follows Brandon Carter's interpretation of the anthropic principle—of course the abundance of elements in living systems roughly mirrors cosmic abundances. Evolutionary processes took advantage of the available materials as life emerged and evolved.

To be fair, this explanation isn't unreasonable, but, from my perspective, it doesn't feel very satisfying. It overlooks the unusual—perhaps, even solely unique—properties of carbon and oxygen (two of the most abundant elements in the universe) that are indispensable and just-right for life. It is eerie to think that the two chemical elements most necessary for life also turn out to be among the most abundant in the universe. It's doubly eerie to recognize that the process that generates these two elements—the triple alpha process—must be exquisitely fine-tuned.

The Triple Alpha Process

During the early moments of the universe's history—the first three minutes—the elements hydrogen, helium, and lithium formed. The other chemical elements that naturally occur in the universe formed later in cosmic history. Given their importance for life, the formation of carbon and oxygen is of special interest. How do these chemical elements form? Why are they two of the most abundant elements in the universe?

These two elements form as a result of stellar burning through the triple alpha process. As the hydrogen fuel in stars burns, helium forms. Two helium nuclei (called alpha particles) combine to form a beryllium nucleus (^{8}Be). This particle is unstable and quickly decays back into the two helium nuclei. If, during its brief existence, the beryllium nucleus collides with another helium

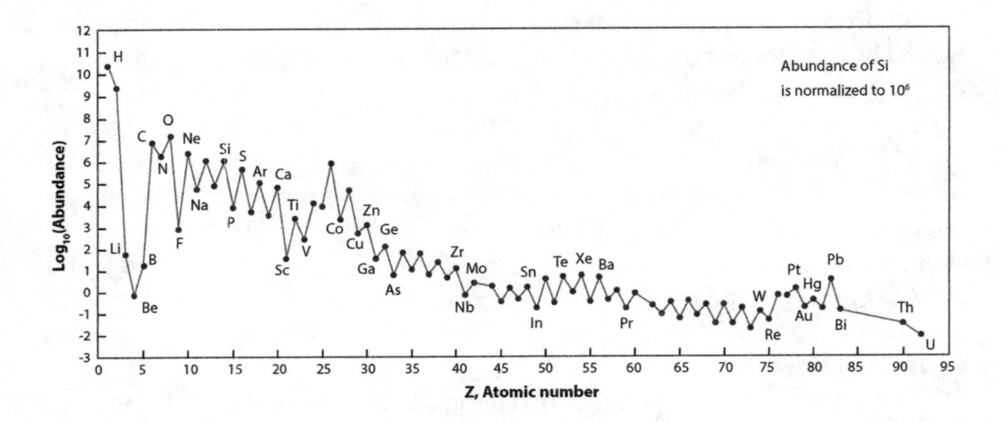

Figure 3.18: Cosmic Abundance of Elements

nucleus, carbon (^{12}C) forms. Unlike the beryllium nuclei, the carbon nuclei are stable. Some of the carbon nuclei will react with helium to form oxygen (^{16}O) nuclei.

In principle, the oxygen nuclei can fuse with additional helium nuclei to form successively heavier nuclei. However, this typically doesn't happen in stellar cores that support the triple alpha process because these additional fusion events require much higher temperatures and pressures. As a result, carbon and oxygen are the primary products of stellar nucleosynthesis, explaining their high abundance in the universe. These elements will be retained in the star until its lifetime comes to an end. At this point, these elements are released into space and eventually become incorporated into gaseous nebulae.

The formation of carbon and oxygen require the fine-tuning of several parameters. First, the energy of the two helium nuclei just happen to precisely match the energy of the beryllium nucleus in its ground state. If not for this close match in energy (called a resonance of the energy levels), the probability of forming the beryllium nucleus would be quite remote. The resonance in energy levels dramatically increases the likelihood of forming the beryllium nucleus. Even though the beryllium nucleus is unstable, it persists long enough to support the formation of carbon nuclei, thanks to the fact that its decay rate is four orders of magnitude slower than the scatter time of the helium nuclei after they collide with one another. The formation of carbon nuclei also requires another energy level resonance. In this case, the resonance is between the energy of the beryllium and helium nuclei and an excited state for the carbon nucleus. The kinetic energy associated with the collision of the beryllium and helium nuclei is sufficient to promote the resulting carbon nucleus to its excited state. Afterwards, the carbon nucleus returns to its ground state. While carbon nuclei can collide with helium to form oxygen, this process is relatively inefficient due to the fact that there isn't a resonance of energy levels with the carbon and helium nuclei and the energy levels of the oxygen nucleus. If such a resonance existed, carbon would be readily consumed, oxygen would readily form, and little to no carbon would exist in the universe. In fact, if the energy of the carbon nucleus were four percent lower, the formation of oxygen nuclei would be so probable that the process would proceed with a high degree of efficiency. As it stands, the production of oxygen is just efficient enough to make it highly abundant without interrupting carbon production.

In other words, a fortuitous set of circumstances is responsible for the relatively high abundances of carbon and oxygen in the universe, two chemical elements with a set of just-right properties that make the chemical environment of the universe fit for life.

The Fitness of the Chemical Environment and the Anthropic Principle

The goal of this chapter was to address the question: Does the anthropic principle extend to the chemistry of the universe? I began the inquiry predicting that it would, based on Henderson's work over a century ago. When Henderson argued that the chemical environment displays a fitness for life, he made his case without the benefit of our modern-day understanding of chemical bonding. With that understanding in place, Henderson's conclusion about the environment's suitability for life isn't diminished one bit. We now have deeper insight as to why water, carbon dioxide, molecular oxygen, and organic compounds possess the uniquely fortuitous properties that make life possible. We also have a good understanding as to why life must be carbon-based.

The last century has led to additional insights that reinforce Henderson's original conclusion. It is provocative that the two chemical elements most necessary for life, oxygen and carbon, are also among the most abundant chemical elements in the cosmos. And the triple alpha process that forms these two elements in the cores of stars relies on a finely tuned mechanism. If even one of several parameters deviated much beyond its actual value, carbon and oxygen would exist at relatively low levels in the universe, and life wouldn't be possible.

$$^{4}He + ^{4}He \rightleftharpoons ^{8}Be$$

$$^{4}He + ^{8}Be \rightleftharpoons ^{12}C^{*} \rightarrow ^{12}C + 2y$$

Figure 3.19: Triple Alpha Process

In this respect, the cosmological and chemical anthropic coincidences intertwine and are inseparable.

Clearly, the finely tuned fundamental parameters, constants, and characteristics of the universe discovered by astronomers and physicists dictate the properties of chemical systems, which make these systems just-right for life. In some respects, this insight is not unexpected. Many scientists, including me, largely adopt the view that the laws of physics explain the laws of chemistry (and the laws of physics and chemistry explain biochemistry). Keep in mind, however, that the formal description of the cosmological anthropic principle focuses primarily on the effects that the *dimensionless* constants of physics have on galaxy and star formation, the production of the chemical elements necessary for life, and atomic structure. The chemical anthropic principle can be considered an extension of this anthropic reasoning into the realm of chemistry, focusing on chemical properties that arise from the laws of physics, but manifest exclusively in the realm of chemistry.

This discussion brings us to my key question. Does the anthropic principle include biological systems? From my perspective as a biochemist, the place to begin that query is with the structure and function of key biochemical systems—life's most basic and foundational systems—looking for anthropic coincidences.

PART 3

The Biochemical Anthropic Principle

Chapter 4

Proteins

It was the first time I tried to learn all the lyrics to a song.

I was inspired to this feat by my third-grade music teacher, Mrs. Milliken. Our music lessons took place twice a week in our elementary school auditorium. We sat near the stage in the first couple of rows, next to an old piano, while Mrs. Milliken taught us the rudiments of music theory. Then we would sing a few songs together as she accompanied us on the piano. One of those songs was "Sixteen Tons," a composition made popular by Tennessee Ernie Ford in 1955. In retrospect, I can understand why Mrs. Milliken wanted us to learn this song. In the late-1960s it was still an important part of the American pop cultural milieu. The song also aptly spoke to the challenges that many of my classmates' families experienced living in West Virginia.

I liked the song enough that I sang it most mornings and afternoons as I walked to and from school. I was mesmerized by Tennessee Ernie Ford's deep resonant voice, and I tried to emulate it as I sang:

You load sixteen tons, what do you get?
Another day older and deeper in debt

Ford's recorded version of the song reached number one on the Billboard charts. Yet he didn't write the song, and he wasn't the first person to record it. Those accomplishments belong to Merle Travis, the famed country and western singer and guitar player from Rosewood, Kentucky. Travis wrote the song for his 1946 record, *Folk Songs of the Hills*. Like many of Travis's songs, "Sixteen Tons" describes the exploitation of coal miners and the brutal lives they lived. Growing up in Muhlenberg County in Kentucky, Travis had firsthand knowledge of the hard lives coal miners led. His family members worked in the mines.

Growing up in West Virginia, where many of my friends' fathers and family members worked in the mines, I knew what the life of a coal miner was like, even at a young age. I admired how dedicated these men were to their families and their willingness to work long and difficult hours, under dangerous conditions. But this commitment to hard work wasn't just true for the coal miners. It was true for practically everyone I knew. When I was growing up, hard work and sacrifice were a way of life for most people in West Virginia. (I'm sure the same is true today.) For many people I knew, it took everything they had to make it through life. But hard work wasn't just a necessity; it was also a source of pride. I learned that the people around me would put up with a lot of bad behavior, but laziness was the one vice few would overlook. If you wanted to earn people's respect, you had to be willing to work hard. I internalized that message.

My parents taught me the same lesson. Even though my parents weren't from West Virginia, they believed in the virtue of hard work. My mom grew up in the aftermath of the Great Depression and my father's family lost everything when the Partition of India occurred. The only way they were able to recover from these devastating experiences was through hard work—and education.

Both of my parents were highly educated. My father was a nuclear physicist and college professor. My mom used her degrees in physics and math to teach science and math in high school and junior high. Education was the number one priority in our home. My parents expected that I would sacrifice and work hard at my studies. I ended up synthesizing both messages and, as a result, I brought a blue-collar attitude to my schoolwork. I carried this mindset through undergraduate training, graduate studies, and postdoctoral work. To this day, I often envision myself putting on the proverbial hard hat each day as I head off to work at the office.

My West Virginia roots influenced my musical tastes quite a bit. Throughout my life, many of my favorite songs—those that really speak to me—involve the theme of hard work. Some of my favorites on the list include:

- Bachman-Turner Overdrive's "Takin' Care of Business"
- Merle Haggard's "Workin' Man Blues"
- Jim Croce's "Workin' at the Car Wash Blues"
- Lee Dorsey's "Working in the Coal Mine"

The song that tops my list is an obscure tune recorded by Black Oak Arkansas (BOA), "Sure Been Workin' Hard." Fronted by Jim "Dandy" Mangrum, this southern rock band was one of the top grossing live acts in the early- to mid-1970s. I loved their unique blend of rock, country, blues, and gospel.

Every time I hear (sing) "Sure Been Workin' Hard," it conjures up images of laborers laying railroad tracks, singing this song to distract themselves from the difficulty of their work. It is a song about someone who puts in days of never-ending hard work, whose only true reward at the end of the day is to return home to the love of his woman.

This song has become an anthem for me. Over the decades, I have probably sung the chorus to this song at least once or twice a day. I usually catch myself singing this song when I am in the midst of a grind at work. Just singing the chorus helps spur me on. It helps me keep my head down and continue my work. I find myself automatically singing the chorus just after I finish a hard day of work, as I walk to my car to head home. It just seems to be the perfect punctuation to the end of my workday as I head home to the warmth and love of my wife and family.

•••••••••••••

If any class of biomolecules emulates hardworking people, it would be proteins. In fact, biochemists sometimes refer to proteins as the workhorse molecules of life. Even the simplest, free-living microbes possess several thousand distinct types of protein molecules operating in concert to carry out life-essential processes and to form subcellular, cellular, and extracellular structures.[1] Some proteins reside in the cell's cytoplasm and others in the lumen of organelles. Some proteins combine with others to form larger structures, such as the cytoskeleton. Some proteins are associated with cell membranes by binding to their surfaces. Other proteins embed—either partially or entirely—within the membrane's interior. Proteins catalyze chemical reactions and play a central role in harvesting chemical energy. Other proteins comprise the cell's defense systems. Others store and transport molecules. Proteins play a role in replicating DNA. Proteins even make other proteins. In this capacity, a special array of proteins work together to make all the proteins the cell needs, including those proteins that are part of the array—a true chicken-and-egg system. This list of functions provides just a sampling of the key activities that proteins mediate. Talk about an exemplary molecular-scale work ethic. If not for the activity of these hardworking molecules, life wouldn't be able to make it.

Because of proteins' central importance to living systems, biochemists have a special interest in studying these biomolecules. Over the last several decades, life scientists have amassed an incredible amount of information about the structure and function of proteins. There is still much more to research, but what scientists have learned is provocative, considering our interest in trying to understand the extent and scope of the anthropic principle.

The goal of this portion of the book is to determine whether the anthropic principle applies to biochemical systems. As noted in the introduction, based on the pioneering work of a handful of biochemists, there is good reason to expect anthropic coincidences in the design and activities of biochemical systems. Given the critical contribution proteins make to living systems, these biomolecules serve as an ideal starting point in our quest to determine whether the anthropic principle includes biochemical systems. As biochemist Michael Denton writes, "If the cosmos is uniquely fit for life as it exists on Earth, then, given their primal significance, the proteins should be another uniquely fit ensemble of natural forms."[2] To put it another way, if the anthropic principle extends into the biochemical arena, it should be most evident in the structure and function of proteins. And, conversely, if we don't observe anthropic coincidences associated with protein biochemistry, then we wouldn't expect to see them elsewhere in the cell's chemical systems.

Before we roll up our sleeves and begin mining the data for any anthropic coincidences related to proteins, a brief review of protein structure is in order. (If you are already familiar with the basics of protein structure, please feel free to skip ahead to page 103.)

An Overview of Protein Structure

Proteins are massive, complex biomolecules made up of chain-like molecules, called polypeptides, which are folded into precise three-dimensional structures. One or more of the same or different folded polypeptides interact to form most protein molecules. The polypeptide's three-dimensional architecture is critical, determining the way it combines with other polypeptides to form a given protein. Consequently, the three-dimensional structure of the folded polypeptides and the precise nature of their interactions with one another determine the protein's function.

The cell's machinery assembles polypeptide chains by joining together smaller molecules called amino acids. These building-block compounds are characterized by having both an amino group and a carboxylic acid bound to a central carbon atom. Also bound to this carbon are a hydrogen atom and a

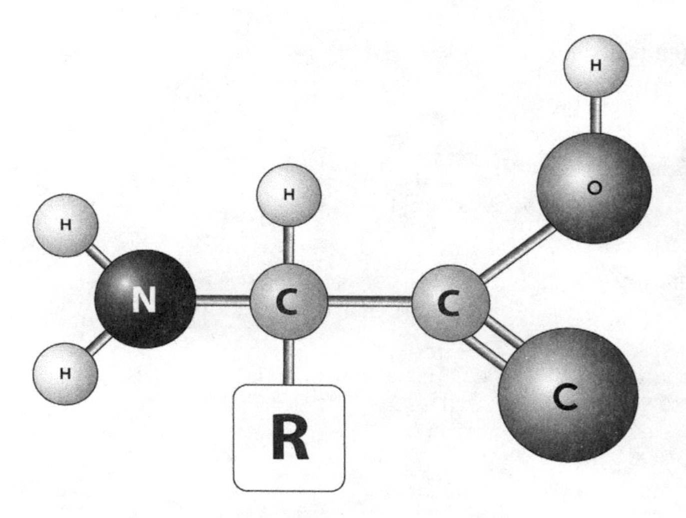

Figure 4.1: The Structure of a Typical Protein Amino Acid

substituent that biochemists call an R group.

The R group determines the amino acid's identity. For example, if the R group is hydrogen, then the amino acid is called glycine. If the R group is a methyl group, then the amino acid is called alanine. Close to 150 amino acids are found in proteins, but only 19 amino acids (plus 1 imino acid, called proline) are specified by the genetic code. Biochemists refer to these 20 as the canonical set.

Polypeptides form when the cell's machinery guides amino acids to react with each other to form a linear chain, with the amino group of one amino acid combining with the carboxylic acid of another to form an amide linkage. (Sometimes biochemists call this linkage a peptide bond.)

The repeating amide linkages along the amino acid chain form the protein's backbone. The amino acids' R groups extend from the backbone, creating a distinct physicochemical profile along the protein chain for each unique amino acid sequence. To first approximation, this unique physicochemical profile dictates the way the polypeptide chain folds into its higher-order structure and, hence, specifies the protein's function. Biochemists refer to the amino acid sequence that forms a specific polypeptide chain as the primary structure.

The secondary structure refers to the three-dimensional arrangement of the polypeptide chain's backbone and the stabilizing interactions between

Amino acid	Structure	Amino acid	Structure
Glycine	H–C(H)(NH$_3^+$)–COO$^-$	Isoleucine	CH$_3$–CH$_2$–CH(CH$_3$)–C(H)(NH$_3^+$)–COO$^-$
Serine	HO–CH$_2$–C(H)(NH$_3^+$)–COO$^-$	Proline	H$_2$C–C(H$_2$)–C(H)(COO$^-$)–N(H)(H)–CH$_2$ (ring)
Threonine	CH$_3$–CH(OH)–C(H)(NH$_3^+$)–COO$^-$	Phenylalanine	C$_6$H$_5$–CH$_2$–C(H)(NH$_3^+$)–COO$^-$
Cysteine	HS–CH$_2$–C(H)(NH$_3^+$)–COO$^-$	Tryptophan	(indole) C–CH$_2$–C(H)(NH$_3^+$)–COO$^-$; N(H)–CH
Tyrosine	HO–C$_6$H$_4$–CH$_2$–C(H)(NH$_3^+$)–COO$^-$	Methionine	CH$_3$–S–CH$_2$–CH$_2$–C(H)(NH$_3^+$)–COO$^-$
Asparagine	NH$_2$–C(=O)–CH$_2$–C(H)(NH$_3^+$)–COO$^-$	Aspartic Acid	$^-$O–C(=O)–CH$_2$–C(H)(NH$_3^+$)–COO$^-$
Glutamine	NH$_2$–C(=O)–CH$_2$–CH$_2$–C(H)(NH$_3^+$)–COO$^-$	Glutamic Acid	$^-$O–C(=O)–CH$_2$–CH$_2$–C(H)(NH$_3^+$)–COO$^-$
Alanine	CH$_3$–C(H)(NH$_3^+$)–COO$^-$	Lysine	H$_3$N$^+$–CH$_2$–CH$_2$–CH$_2$–CH$_2$–C(H)(NH$_3^+$)–COO$^-$
Valine	(CH$_3$)$_2$CH–C(H)(NH$_3^+$)–COO$^-$	Arginine	H$_2$N–C(=NH$_2^+$)–NH–CH$_2$–CH$_2$–CH$_2$–C(H)(NH$_3^+$)–COO$^-$
Leucine	(CH$_3$)$_2$CH–CH$_2$–C(H)(NH$_3^+$)–COO$^-$	Histidine (at pH 6.0)	HC=C–CH$_2$–C(H)(NH$_3^+$)–COO$^-$; HN$^+$–C(H)–NH (ring)

Figure 4.2: The Protein-Forming Amino Acids

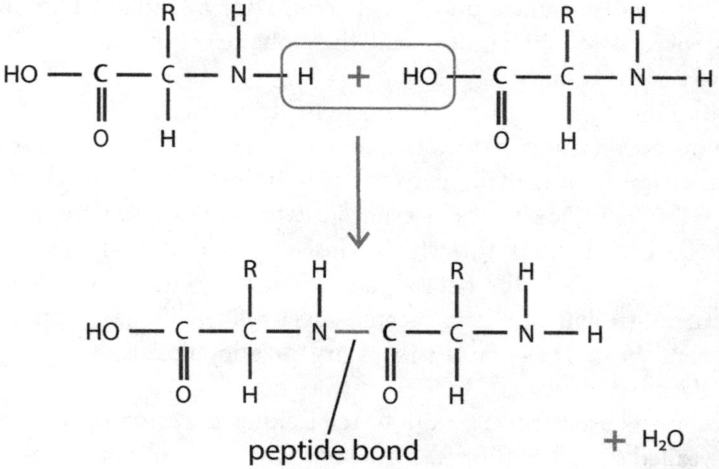

$$\mathrm{HO{-}C({=}O){-}C(R)(H){-}N(H){-}H + HO{-}C({=}O){-}C(R)(H){-}N(H){-}H \rightarrow HO{-}C({=}O){-}C(R)(H){-}N(H){-}C({=}O){-}C(R)(H){-}N(H){-}H + H_2O}$$

peptide bond

Figure 4.3: The Chemical Linkage between Amino Acids

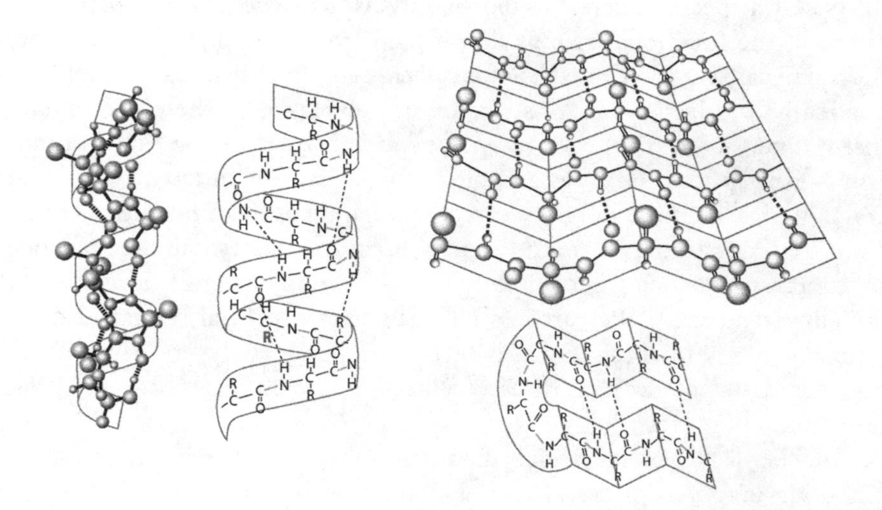

Figure 4.4: Protein Secondary Structure—Random Coil, Alpha Helix, and Beta Pleated Sheet

chemical groups that make up its backbone. Four of the most common secondary structures are the random coil, the alpha (α) helix, the beta (β) pleated sheet, and the (β) beta turn.

By studying protein structures, biochemists discovered that the same combination of secondary structures frequently interact to form structural motifs. These structural units contribute to the overall three-dimensional structure of the folded polypeptide chain. For example, two common structural motifs are (1) the helix-turn-helix motif, which consists of two alpha-helices connected by a beta turn, and (2) the beta-alpha-beta motif, which consists of the sequential arrangement of a beta-pleated sheet followed by an alpha-helix and another beta sheet. These motifs form organizational units referred to as supersecondary structures.

Biochemists use the term *motif* to refer either to recurring amino acid sequences (called sequence motifs) or to recurring three-dimensional structural features (structural motifs). This practice can be confusing because biochemists' reference to these motifs may or may not correspond to the structural motifs that arise from the specific combination of secondary structural elements defined as supersecondary structures.

Tertiary structure describes the three-dimensional structure of a polypeptide chain. Part of this description includes the location of each of its atoms in space. Of specific interest to biochemists is the structure and spatial orientation of the functional groups that extend from the polypeptide backbone. Biochemists have discovered compact, self-contained regions that fold independently, within the tertiary structure of polypeptides. These self-contained three-dimensional regions of the polypeptide's structure are called domains. Some polypeptides consist of a single compact domain, but many possess several domains. In effect, domains can be thought of as the fundamental units of a polypeptide's tertiary structure. Each domain possesses a unique biochemical function. Biochemists refer to the spatial arrangement of domains as the domain architecture. Researchers have discovered several thousand distinct domains. Many of these domains recur in different polypeptides, with each one's tertiary structure comprised of a mix-and-match combination of protein domains.

Biochemists have also identified a number of recurring domain combinations in polypeptides. These domain combinations are referred to as superdomains.

Sometimes biochemists use the term *fold* to refer to structural features within the three-dimensional architecture of a polypeptide. Folds can be found

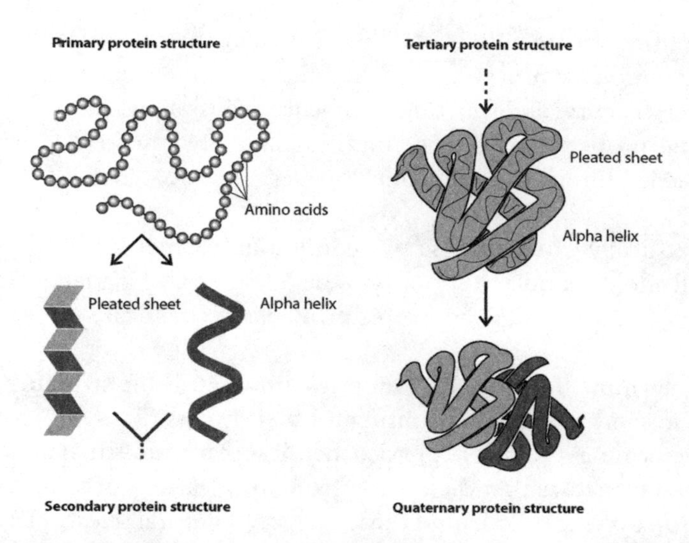

Figure 4.5: Protein Structure—Primary, Secondary, Tertiary, and Quaternary

in a range of proteins and serve as the basis for classifying proteins into groups or families. In other words, the terms *fold* and *motif* are somewhat ambiguous in their usage. On the other hand, the terms *secondary structure*, *supersecondary structure*, *domain*, *superdomain*, and *tertiary structure* are much more precise in their meaning.

Quaternary structure arises when individual polypeptide chains interact to form a complex. These individual polypeptide chains are called protein subunits. These complexes can consist of two or more of the identical subunits, or they can consist of combinations of different types of subunits.

Protein Primary and Secondary Structures and the Anthropic Principle

Let's tackle the central question of this chapter. Do we see evidence for anthropic coincidences in the structural features of proteins?

As I discussed in the introduction (page 25), if the biochemical anthropic principle exists, two conditions must be true:

- The molecular systems that define terrestrial biochemistry must be primarily specified and dictated by the laws of physics and chemistry, not by the forces of natural selection. In other words, if the biochemical anthropic principle exists, then I predict that historically contingent evolutionary processes have had limited influence—if any at all—on the structure and function of biochemical systems.

- The properties of biochemical systems are precisely those that are needed for life to be possible.

To specifically gauge if these two conditions are met for biochemical systems, I will adopt a similar approach to the one used by Henderson to demonstrate the application of the anthropic principle to chemical systems:

- Determine if a molecular rationale undergirds the structure of amino acids and proteins, accounting for their fitness for life.
- Determine if there are physicochemical constraints that arise from the laws of nature that dictate the structural features of amino acids and proteins (independent of the influence of natural selection), rendering them fit for life.
- Compare the structural features of amino acids and proteins with close chemical analogs to determine if the characteristics possessed by amino acids and proteins make them unusually fit for life.

Let me elaborate on these criteria. I think the second entry in this list is self-evident. If, indeed, biochemical systems have been prescribed by the laws of nature to have the just-right properties for life, then we expect that we would be able to identify those constraints and demonstrate that they lead to the precise set of features in biochemical systems necessary for life to exist. Along these lines, I also predict that a "logic" based on molecular principles would define biochemical systems if, indeed, these systems have been prescribed by the laws of nature to make life possible. In other words, there should be good reasons why biochemical systems are the way they are. In effect, the first and second criteria are interrelated. Because of the complexity of biochemical systems, the first criterion is much easier to satisfy than the second. At times, it can be challenging to determine if a particular aspect of biochemistry was predetermined by constraints arising out of the laws or was shaped by natural selection. Therefore, out of necessity, much of our focus will center on trying to understand why biochemical systems are the way they are.

The first criterion is important for another reason. If historically contingent evolutionary mechanisms are largely responsible for shaping biochemical systems, then we wouldn't necessarily expect a molecular rationale to undergird the structure and function of biochemical systems. To appreciate why that is the case, we need to think through the idea of historical contingency. This concept was elaborated on by evolutionary biologist Stephen Jay Gould (1941–2002).[3] According to Gould, the evolutionary process consists of an extended sequence of unpredictable, chance events. Altering any of these events would send evolution down a different trajectory. In other words, chance and history determine evolutionary outcomes. To help clarify this concept, Gould used the metaphor of "replaying life's tape." If we were to push the rewind button, erase life's history, and then let the tape run again, the results would be completely different each time because the evolutionary process will not proceed to a predictable endpoint. As a consequence, historically contingent evolutionary processes wouldn't necessarily be expected to produce biochemical systems that made sense. Instead, this type of evolutionary process would only be expected to produce what worked well enough for survival. That is, if biochemical systems were generated by historically contingent processes, they would be defined more so by their evolutionary history than any type of molecular logic.

The final criterion serves as a measure of the fine-tuning of biochemical systems. Instead of performing a counterfactual analysis based on theoretically altering the value of dimensionless constants (as is the case when working with the cosmological anthropic principle), we will assess the fine-tuning of biochemical systems by comparing them with both real and conceptual alternatives, asking the question: Are the biochemical systems found in nature uniquely suited for life? If the answer is yes, then we can be confident that these systems have been fine-tuned in such a way as to make life possible.

I would like to make one final point before we begin our investigation. Even though I am an old-earth creationist (who is skeptical about facets of the evolutionary paradigm), at this juncture, as I seek to determine if a biochemical anthropic principle exists, I am not pitting evolutionary explanations for the origin and design of biochemical systems against explanations that rely on the work of an intelligent Agent. Instead, I am seeking after the best scientific explanation for the origin and design of biochemical systems by pitting historical contingency against law and necessity. I am asking: Are these systems mere happenstance, the outworking of a historically contingent evolutionary process? Or have the structure and function of biochemical systems been dictated and constrained by the laws of nature independent of the influence of natural selection? While there are metaphysical implications connected to these two questions, I am setting those implications aside as I attempt to adjudicate between these two options based on the scientific evidence, independent of any worldview considerations.

With that clarification behind us, let's begin our quest by examining amino acids, the building blocks used by the cell's machinery to assemble polypeptide chains.

Why Amino Acids?

As noted, 20 chemically distinct amino acids comprise the proteins found in every organism on Earth. This set of amino acids is universal. Yet hundreds of amino acids exist in nature. In fact, 86 different amino acids have been found in the Murchison meteorite.[4] Many scientists believe that the organic fraction of meteorites, such as the Murchison, give insight into the chemical inventory and processes that took place on early Earth. In light of this finding, biochemists have long been interested in what, if anything, makes the specific set of 20 protein amino acids special. This curiosity leads to more precise questions:

- Why are proteins built primarily using the universal canonical set of amino acids? Could other amino acids have been used?
- Why are proteins built from amino acids at all? Why not build them from the chemically simpler hydroxy acids?
- Why are the amino acids in proteins alpha (α) amino acids? Why not beta (β) or gamma (γ) or delta (δ) amino acids?
- Why do all the amino acids in proteins have an α-hydrogen?
- Why are there no N-alkyl amino acids in proteins?

Many naturally occurring amino acids, such as those found in the Murchison meteorite, possess some of the structural features in this list. Shouldn't at least some of these alternative amino acids have made their way into proteins? At first glance, it seems conceivable that other amino acids could have been chosen to fulfill the role of some, if not all, of the canonical amino acids. Yet close examination of the properties of the 20 canonical amino acids indicates that these are the only ones with the just-right set of features that make them uniquely fit for life.[5]

Why not hydroxy acids? The closest chemical analogs to amino acids are hydroxy acids. Just as amino acids can be linked to make proteins, hydroxy acids can be linked together in a head-to-tail fashion to form molecular chains. However, proteins cannot be built from hydroxy acids because the bond that joins these molecules together consists of an ester linkage, not an amide linkage, as found in proteins. Ester linkages are not as chemically stable as amide linkages. (They are susceptible to hydrolysis under alkaline conditions.) Additionally, ester linkages do not form a planar bond between subunits. On the other hand, amide bonds are planar, which allows intrastrand hydrogen bonding. These bonds stabilize the higher-order protein structures that dictate the function of these biomolecules. (The importance of this property will be further elaborated.)

Why not diacids and diamines? Amide bonds could be generated if the subunit molecules used to make up the protein chains alternated between diacids and diamines. Diacids are compounds with two carboxylic acid functional groups. Diamines are compounds with two amino groups. Nylons, molecular analogs to proteins, form from diacids and diamines. However, the usage of these subunits introduces unnecessary chemical complexity into the system without any additional payoff. In other words, amino acids generate amide bonds in the simplest, most efficient way possible.

Why α-amino acids? Unlike alpha (α) amino acids, beta (β) or gamma (γ) or delta (δ) amino acids can't form orderly secondary structures, such as alpha (α) helices and beta (β) pleated sheets, which are critical for protein structure. N-alkyl amino acids are also unsuitable as protein subunits because they cannot participate in hydrogen bond interactions needed to form the secondary structures of proteins. Additionally, the α-amino acids must possess a hydrogen atom in the α position. This helps avoid steric hindrances when the protein folds into higher-order, three-dimensional structures.

This brief survey demonstrates that a molecular rationale undergirds the use of α-amino acids (with an R group and a hydrogen atom bound to the central carbon) to build proteins. Moreover, it appears as if α-amino acids uniquely display the just-right set of properties. The most reasonable structural analogs of α-amino acids simply do not have all the necessary physicochemical properties to form proteins with stable three-dimensional structures.

Why These R Groups?

The α-amino acids found in proteins have another important structural feature called an R group. The R group varies from amino acid to amino acid and contributes to each amino acid's distinct chemical and physical properties.

Biochemists have some understanding about these R groups' physicochemical utility. It turns out that, thanks to the R groups, the relatively small set of 20 amino acids possesses a wide range of chemical and physical properties, which impart a significantly diverse array of structural and functional possibilities to polypeptide chains.

A survey of the physicochemical diversity. Some R groups are hydrophobic. (Recall that these types of compounds avoid contact with water because of the hydrophobic effect. See pages 68–70.) These amino acids are often buried in the protein interior where they play a key role in stabilizing the protein's three-dimensional structure. It turns out that the R groups of several of the hydrophobic amino acids consist of branched alkyl chains. This property reduces their mobility within the protein interior and helps hold together the protein structure. This benefit does not exist for R groups consisting of linear, unbranched chains. The R group of glycine is a hydrogen atom. This, too, has steric benefits for protein structure, allowing the protein chain to make tight turns where glycine residue occurs.

Some R groups are hydrophilic. These amino acids have a diversity of useful physicochemical properties. The amino acids serine and threonine are the simplest possible amino acids possessing primary and secondary alcohols. Alcohols are useful for several chemical processes. They serve as attachment points for chemical groups like phosphates (which modify protein structure and function) and take part in numerous important chemical reactions. The amino acid cysteine is the simplest possible amino acid with a sulfhydryl R group. Thiols can take part in important chemical reactions as well. In addition, the R group of one cysteine residue in a protein can react with the R group of another cysteine to form a crosslink that helps stabilize protein three-dimensional structure. Aspartic acid and glutamic acid provide proteins with carboxylic acid groups, which give proteins the capacity for acid-base chemistry. When the carboxyl group is ionized, these amino acids become negatively charged and can interact with amino acid groups bearing a positive charge to form a salt bridge. These molecular-scale bridges help stabilize protein three-dimensional structure. Interestingly, the R group of glutamic acid is longer than that of aspartic acid. This difference allows for salt bridges of varying lengths to form within a protein's interior. Lysine, histidine, and arginine are the three positively charged amino acids. The R group of arginine bears a permanent positive charge. On the other hand, the positive charge of the lysine and histidine R groups depends on the specific chemical environment. In other words, the positive charge of these side groups can be "turned on and off." Lysine's positive charge stems from the presence of an amino group, while histidine's positive charge is a property of the imidazole functional group. Not only do these two R groups provide a source of positive charge, but they also have distinct chemical properties that add to the range of protein functional properties. For example, positively charged R groups help proteins bind to and interact with negatively charged biomolecules. As we will see in the next chapter, the DNA backbone possesses negatively charged residues. The proteins that bind to DNA bear positive charges thanks to R groups such as lysine and arginine.

Phenylalanine, tyrosine, and tryptophan all have R groups made up of aromatic hydrocarbons. Phenylalanine possesses the simplest possible R group structure that can contain a benzene ring. The benzene ring is hydrophobic and planar, making it structurally rigid. Additionally, the electron delocalization within the aromatic ring imparts phenylalanine with the capacity to take part in chemical interactions while typical hydrophobic R groups cannot. Likewise, tyrosine's phenolic R group gives this aromatic amino acid the wide range of interesting chemical properties possessed by phenol.

Technically, given the chemical nature of its R group, proline is not an amino acid. Rather it is an imino acid. Proline is an extremely rigid molecule. When proline is incorporated into a protein chain, its R group takes part in forming the chain's backbone. This forces the protein chain to make a sharp turn that proline's structural rigidity locks into place.

Of note are the functional R groups absent from the 20 protein-forming amino acids, particularly aldehydes and ketones. These two chemical functionalities take part in a wide range of interesting chemistry, which raises questions about their omission from the set of 20. As it turns out, these two functional groups are so reactive that their presence in a protein molecule would be disruptive. For example, both groups can form chemical crosslinks. This reaction could easily (and randomly) tether two protein molecules together. Unwanted crosslinks would form within a protein as well.

From this brief survey, it becomes evident that the R groups of the 20 canonical amino acids have a diverse and impressive array of physicochemical properties that contribute to protein structure and function. This insight reveals that a molecular rationale undergirds the selection of the 19 amino acids and 1 imino acid that constitute the canonical set. This leads to a follow-up question. Is the canonical set unique? Or, hypothetically, could other amino acids have been included in the canonical set instead?

An optimal set of amino acids? Several studies have addressed this question. For example, work by researchers from the University of Hawaii suggests that the canonical set of amino acids may well be optimal.[6] In other words, the canonical set may be uniquely suited to render protein molecules structurally and functionally fit for life. These investigators conducted a quantitative comparison of the range of chemical and physical properties possessed by the 20 protein-building amino acids versus random sets of amino acids that could have been selected from early Earth's hypothetical prebiotic soup. It turns out that the set of amino acids possesses properties that evenly and uniformly vary across a broad range of size, charge, and hydrophobicity. These researchers also demonstrate that the amino acids selected for proteins form a "highly unusual set of 20 amino acids; a maximum of 0.03% random sets outperformed the standard amino acid alphabet in two properties, while no single random set exhibited greater coverage in all three properties simultaneously."[7] As I have discussed, the canonical amino acid set's wide range of physicochemical properties make it possible for proteins to carry out critical chemical operations necessary for life.

With these previous studies as a backdrop, investigators from the Earth-Life Science Institute in Tokyo pursued a better understanding of the optimal nature of the universal set of amino acids used to build proteins. To do this they used a library of 1,913 amino acids (including the 20 amino acids that make up the canonical set) to construct random sets of amino acids. The researchers varied the set sizes from 3 to 20 amino acids and evaluated the performance of the random sets in terms of their capacity to support (1) the folding of protein chains into three-dimensional structures, (2) protein catalytic activity, and (3) protein solubility. They discovered that if a random set of amino acids included even a single amino acid from the canonical set, it dramatically outperformed random sets of the same size without any of the canonical amino acids. The researchers concluded that each of the 20 amino acids used to build proteins stands out, possessing highly unusual properties that make each one ideally suited for its biochemical role and confirming the results of previous studies.[8]

A team of German researchers demonstrated that quantum mechanical effects also contribute to the optimal properties of the universal set of 20 protein amino acids.[9] To do this, they examined the gap between the HOMO (highest occupied molecular orbital) and the LUMO (lowest unoccupied molecular orbital) for the protein amino acids. The HOMO-LUMO gap is one of the quantum mechanical determinants of chemical reactivity. More reactive molecules have smaller HOMO-LUMO gaps than molecules that are relatively nonreactive. The German biochemists discovered that the HOMO-LUMO gap was small for six of the 20 amino acids (histidine, phenylalanine, cysteine, methionine, tyrosine, and tryptophan) and, hence, these molecules display a high level of chemical activity. Interestingly, some biochemists think that these six amino acids are not necessary to build proteins. Previous studies have demonstrated that a wide range of foldable, functional proteins can be built from only 13 amino acids (glycine, alanine, valine, leucine, isoleucine, proline, serine,

propanoic acid + ethanol ⟶ ethyl propanoate + water

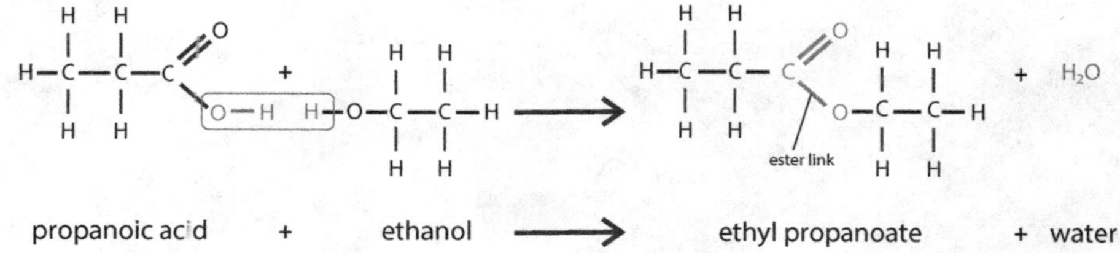

Figure 4.6: The Chemical Formation of an Ester Linkage

threonine, aspartic acid, glutamic acid, asparagine, lysine, and arginine). As it turns out, this subset of 13 amino acids has a relatively large HOMO-LUMO gap and, therefore, is relatively unreactive. This property makes them ideally suited for forming stable protein structures. So, why would the six highly reactive amino acids also be included in the canonical set? As it turns out, these amino acids readily react with the peroxy free radical, a highly corrosive chemical species that forms when oxygen is present in the atmosphere. The German biochemists believe that when these six amino acids reside on the surface of proteins, they help protect the proteins from oxidative damage.

This brief survey indicates no happenstance in the canonical set of amino acids found in proteins. It cannot be the outworking of a historically contingent evolutionary history that shaped the origin and early evolution of life, including its biochemical composition. If this scenario *were* the explanation, I would not anticipate the level of optimization displayed by the canonical amino acids. Instead, I would expect the canonical set to possess a suboptimal collection of properties. The optimal nature of the canonical set strongly suggests that the selection process that chose its constituent amino acids reflects an exquisite underlying molecular rationale. In addition to its optimal properties, comparisons with other types of amino acids and some of their close chemical analogs reveal the unique and just-right physicochemical properties possessed by the canonical set—properties that fortuitously make the canonical amino acids ideally fit for their role in constructing life's workhorse molecules.

This remarkable insight leads to questions about other characteristics of proteins: Is there anything special about the amide linkages that join amino acids together to form the primary structure of proteins? How does the amide

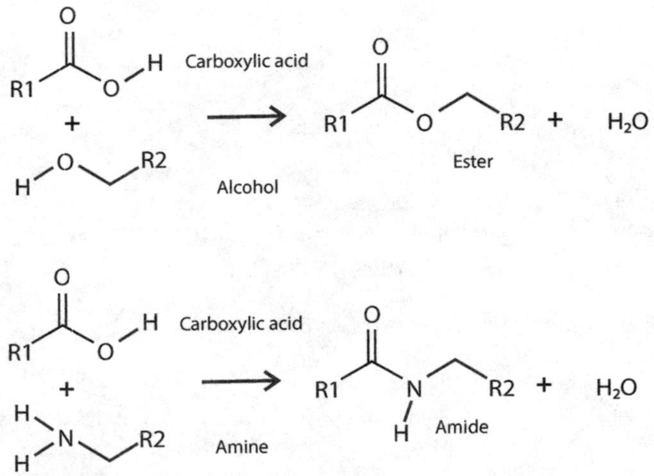

Figure 4.7: A Comparison of Amide and Ester Linkages

bond compare to other types of chemical bonds?

Why an Amide Linkage?

The closest chemical analog to an amide bond is an ester linkage, formed when a carboxylic acid reacts with an alcohol. Both amide and ester bonds form through a chemical process known as a condensation reaction.

At first glance, it seems conceivable that biochemical systems could have included a class of "workhorse" molecules consisting of polyesters built from α-hydroxy acids. As noted, this class of compounds is the closest chemical analog to the α-amino acids. Presumably, α-hydroxy acids would have been available on early Earth. In prebiotic experiments designed to simulate the conditions of primordial Earth—such as the famous spark discharge experiments that Stanley Miller carried out in the 1950s at the University of Chicago—α-hydroxy acids formed at levels comparable to amino acids. In fact, the reaction mechanism for the formation of α-hydroxy acids is simpler than the reaction pathway that generates α-amino acids. So, based on the abundance and the comparative ease of α-hydroxy acids' formation, it seems more reasonable to think that, if life arose through chemical evolution, α-hydroxy acids would have been selected instead of α-amino acids.

Though α-hydroxy acids are the closest chemical analogs of amino acids,

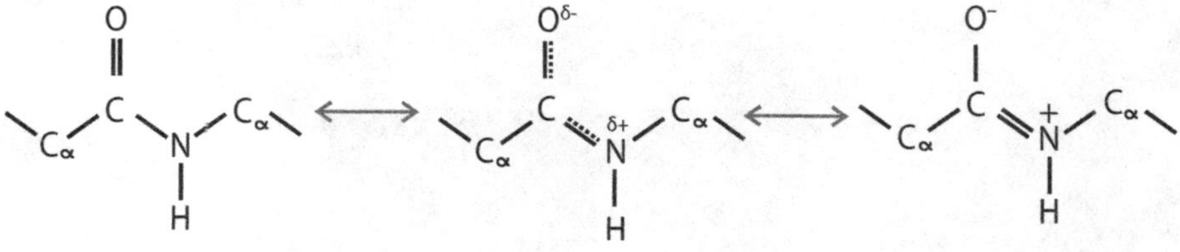

Figure 4.8: The Double-Bond Character of Amide Linkages

the polymers that result from each of these monomers (polyesters and polypeptides, respectively) display radically different physicochemical properties. Many of these differences become evident when the unique and unusual nature of the amide bond is compared to an ester linkage.

The polyester linkage consists of a single bond between the carbonyl carbon of a carboxylic acid and the oxygen atom of an alcohol. At first glance, it would appear that the amide linkage, too, should consist of a single bond between the carbonyl carbon of the carboxylic acid and the nitrogen atom of the amino group. But such is not the case. As it turns out, the amide linkage has partial *double*-bond character.

To first approximation, differences in the electronegativities of oxygen, nitrogen, and carbon help account for the differences in the bond order of the amide and ester linkages. Because oxygen has a greater electronegativity than carbon, it more closely attracts the electrons in the carbon-oxygen double bond of the amide carbonyl group. This excessive attraction creates a partial negative charge on the carbonyl oxygen and a partial positive charge on the carbonyl carbon. In turn, the partial positive charge on the carbon tends to "pull" the unshared electron pair of the nitrogen atom in the amide linkage into the bond between these two atoms, giving the amide linkage partial double-bond character.

Despite the structural similarities between the ester and amide linkages, the ester bond has negligible double-bond character, even though the same inequity in electron attraction exists in the ester carbonyl as it does in the amide carbonyl. The partial positive charge on the ester carbonyl can't pull either of the two unshared electron pairs of the oxygen atom into the ester bond, because

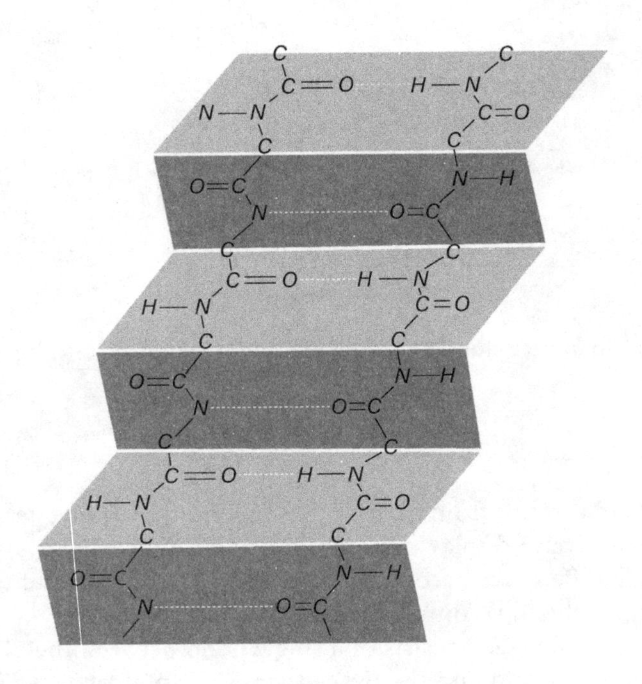

Figure 4.9: The Rigidity of the Protein Backbone

oxygen has a greater electronegativity than nitrogen. As a result, the oxygen atom hangs onto its unshared electron pairs more ardently than the nitrogen atom. The net outcome is that the ester linkage, in effect, only forms a single bond, while the amide linkage forms a linkage with double-bond character.

The double-bond character of the amide linkage has some important consequences for protein primary structure and the tendency of the protein chain to fold into higher-order, three-dimensional structures. In terms of protein primary structure, the double-bond character of amide linkages renders this bond more chemically stable than the ester linkage. Specifically, amide linkages are far less susceptible to hydrolytic cleavage than ester bonds (though amide bonds will undergo hydrolysis under highly alkaline conditions or highly acidic conditions). This greater degree of chemical stability ensures that polypeptide chains, once formed by the cell's machinery, are much more likely to stay intact than the corresponding polyesters.

The amide double bond also imparts a rigidity to the protein backbone

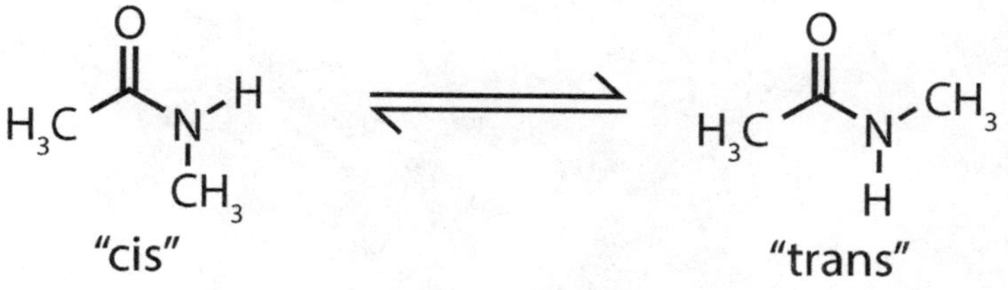

Figure 4.10: The *cis* and *trans* Geometries of the Amide Linkage

that the single bond of an ester linkage does not afford. When joined together via a single bond, two atoms (and the chemical groups connected to those atoms) can rotate freely around the bond, relative to one another. (Think of two boards hammered together by a single nail.) On the other hand, a double bond adjoining two atoms restricts rotation. (Again, think of two boards, this time secured together by two nails.) The free rotation around the ester single bond results in "floppy" polyester molecules that can't form stable higher-order, three-dimensional structures. The restricted rotation around the amide double bond imparts a stiffness to the backbone that helps to stabilize the three-dimensional structure of the polypeptide.

The partial double bond of the amide linkage makes it possible for proteins to adopt stable secondary structures. Because of its double-bond character, the amide linkage displays a planar molecular geometry. Not so for esters. This functional group adopts a tetrahedral geometry because of its single-bond character. The planar geometry forces the carbonyl oxygen and the hydrogen bound to the nitrogen atom of the amide bond to assume one of two distinct configurations: *cis* and *trans*. In the *trans* geometry, the carbonyl oxygen and the hydrogen bound to the nitrogen reside on opposite sides of the double bond. In the *cis* geometry, the carbonyl oxygen and the hydrogen bound to the nitrogen reside on the same side of the double bond.

The amide linkages in proteins all adopt the *trans* configuration. This geometry is dictated by steric effects. In the *cis* configuration, the R groups on adjacent amino acid residues would clash with one another. This undesired interaction does not occur when the peptide bond adopts a *trans* geometry. This situation is fortuitous because the *trans* configuration positions the carbonyl

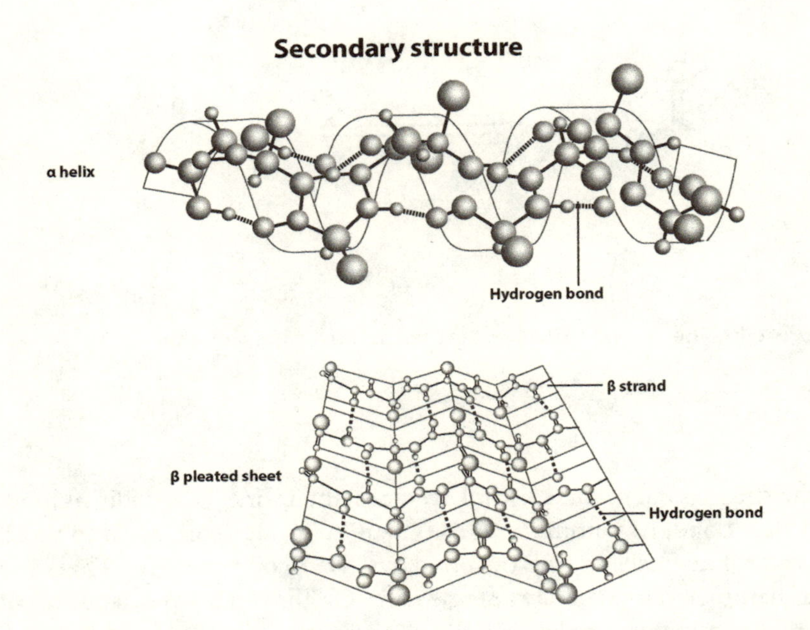

Figure 4.11: The Stabilizing Hydrogen-Bonding Interactions in Protein Secondary Structures

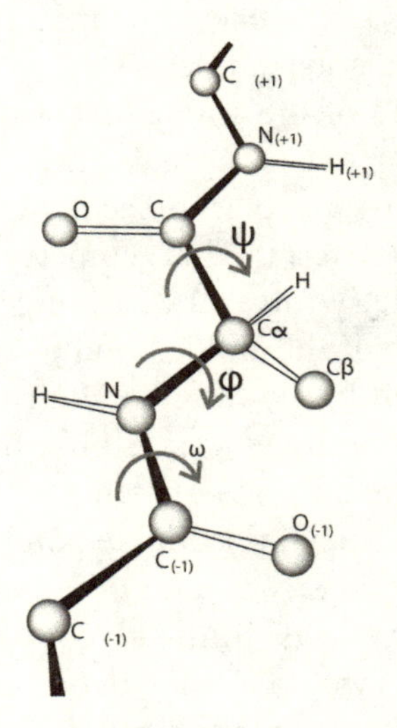

Figure 4.12: The Omega and Psi Bond Angles of the Polypeptide Backbone

oxygen and the hydrogen (bound to the nitrogen) in the right orientation in space to allow for the intrachain hydrogen bond interactions that stabilize protein secondary structures, such as the alpha-helix and beta-pleated sheet. If not for the amide bond's planar geometry, polypeptides would be unable to form ordered, regularly shaped secondary structures.

Another consequence of the planar geometry of the amide linkage is the restriction it imposes on the rotation of the two single bonds located on either side of the alpha carbon of each amino acid residue in the polypeptide backbone. Biochemists designate these angles as *omega* and *psi.* In principle, the chemical groups attached to these bonds should be able to rotate freely, adopting any angle between 0° and 360°. In reality, the rotation around these bonds is hindered because of the volume and spatial orientation of the chemical groups associated with the amide bond (called steric hindrance). Steric hindrance reflects the fact that two chemical groups can't occupy the same region of space at the same time. Thus, only certain values of omega and psi are allowed because of these steric constraints. The permitted values of omega and psi (allowed because of these steric constraints) are *precisely* those required for the protein backbone to adopt either an alpha-helix or a beta-pleated sheet.

The influence of the steric constraints on the omega and psi bond angles can be visualized in a Ramachandran plot. This diagram graphs allowed values of psi against allowed values of omega and identifies regions of permitted bond angles. The four permitted regions on the Ramachandran plot correspond to the alpha-helix and the beta sheet secondary structures, along with two relatively rare secondary structures, the left-handed helix and the polyproline secondary structure.

So, is there anything special about the amide linkage? Even a limited comparison of the properties of the amide and ester bonds and a brief survey of the physicochemical properties of the amide linkage highlight the unusual and remarkable nature of the chemical bond that forms the primary structure of proteins by joining together amino acids. It's a little uncanny that the amide bond possesses the just-right properties that make it possible for polypeptides to form stable secondary and tertiary structures. If not for these properties, the workhorse molecules that make life possible would not exist. Biophysicists Jayanth Banavar and Amos Maritan expertly capture the provocative implications of the anthropic coincidences associated with proteins' primary and secondary structures:

> One cannot but marvel at how several factors—the steric

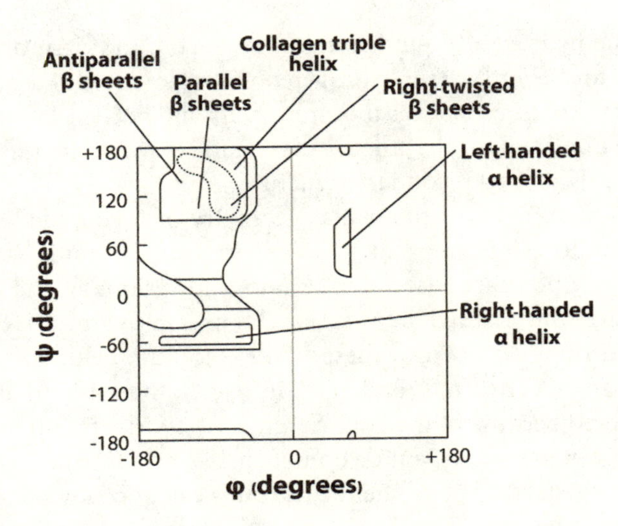

Figure 4.13: A Ramachandran Plot

> interactions; hydrogen bonds, which provide the scaffolding for protein structures; the constraints placed by quantum chemistry on the relative lengths of the hydrogen bonds and covalent bonds; the near planarity of the peptide bonds; and the key role played by water—all reinforce and conspire with one another to place proteins in this novel phase of matter.[10]

This remarkable insight propels us forward in our quest to determine if the anthropic principle applies to biochemical systems and leads to obvious questions: Are the anthropic coincidences associated with proteins limited to the 20 canonical amino acids and the amide linkage that forms the primary structure of proteins? Or do the higher-order, three-dimensional supersecondary and tertiary structures adopted by proteins also display their own set of anthropic coincidences?

Protein Folds and the Anthropic Principle

To address this question, it is helpful to understand a bit about protein space. This concept refers to all the possible sequences available for a polypeptide chain of a specified length of amino acids. For example, for a tripeptide comprised of 3 amino acids, the protein space consists of 8,000 possible sequences.

(There are 20 amino acids that form the canonical set. Therefore, 20^3 possible sequence combinations exist.) For a peptide twice that length, comprised of 6 amino acids, the protein space consists of 64,000,000 possible amino acid sequences. And for a polypeptide chain of 150 amino acids (which is the typical size of a polypeptide chain used to form proteins in nature), protein space consists of 20^{150} sequences, corresponding to approximately 1.43×10^{195} distinct polypeptide chains.

Clearly, protein space is vast and expands dramatically as the polypeptide chain length increases, but considering only amino acid sequences (which corresponds to a protein's primary structure) does not provide a true description of protein space. The higher-order, three-dimensional arrangements of atoms in space dictate a protein function, not the primary structure. So, to define protein space accurately we must consider a protein's primary, secondary, and tertiary structures. Biochemists have estimated that, at a minimum, a polypeptide chain of 150 amino acids can fold in 10^{68} possible ways. In other words, for a polypeptide chain of 150 amino acids, the best representation of protein space consists of approximately 1.43×10^{195} sequences that collectively adopt approximately $10^{13,260}$ possible three-dimensional structures.

Despite the vastness of protein space, biochemists have discovered that polypeptide chains can only assume a limited number of three-dimensional shapes. Biochemists estimate that the number of distinct folds falls between 3,000 to 10,000.[11] Even though 10^{68} folds exist, *in principle*, for each polypeptide chain of 150 amino acids, *in actuality*, only 10^4 to 10^5 folds appear to be physically possible. In other words, only a small portion of protein space can be occupied by folded proteins. In support of this observation, biochemists have discovered that different amino acid sequences adopt the same fold. (As a corollary to this discovery, biochemists also learned that the same fold can support a wide range of functions.)[12] Collectively, these two observations indicate that physicochemical constraints restrict the available regions of protein space that can be occupied by actual proteins.

We shouldn't be surprised to learn that physicochemical restrictions limit protein folding. We already discussed the influence the amide bond's planar geometry places on the omega and psi bond angles of the polypeptide backbone due to steric hindrance. These limitations mean that the polypeptide chain can adopt only certain conformations (spatial arrangements of atoms in molecules) that, fortuitously, lead to the formation of alpha-helical or beta-pleated sheet secondary structures.

Constraints of Supersecondary Structures

As already noted, alpha-helices and beta-pleated sheets interact and combine to form regularly recurring supersecondary structures, such as the helix-turn-helix motif and the beta-alpha-beta motif. Different proteins, with different three-dimensional structures, possess the same supersecondary structures. However, not every conceivable combination of secondary structures exists in proteins. By surveying protein structures, biochemists have observed that a limited number of supersecondary structural motifs actually exists. The two most prominent secondary structural elements found in proteins (alpha-helix and beta-pleated sheet) can only combine and interact with one another in specific ways. This limitation arises from steric effects and topological considerations.[13] The well-known protein biochemist Cyrus Chothia summarizes this insight this way:

> In most proteins the alpha-helices and beta-sheets pack together in one of a small number of ways. The connections between secondary structures obey a set of empirical topological rules. . . . These similarities arise from the intrinsic physical and chemical properties of proteins.[14]

Biochemists specializing in the study of protein structure currently seek to elaborate these topological rules, with an eye toward using this growing understanding to predict the three-dimensional structures of proteins, starting from amino acid sequences. They also hope these rules will aid in the *de novo* design of novel proteins. However, the laws of nature clearly put constraints on the supersecondary structural motifs of proteins. And yet these structural motifs are sufficient to form the wide range of functional proteins needed to support the complexity, richness, and diversity of Earth's life.

Constraints on Domain Architecture

As noted, biochemists know that proteins' tertiary structure is comprised of compact, self-contained regions that fold independently and possess unique biochemical functions (domains). While the tertiary structure of some proteins consists of a single domain, many proteins are made up of several domains that are spatially arranged to form a domain architecture. Among the several thousand distinct protein domains discovered, many recur in different proteins, with most tertiary structures consisting of a mix-and-match combination of domains.

In 2019, a research team from the University of Alabama at Birmingham and the National Institutes of Health (NIH) discovered evidence that a set of topological rules also dictate and limit the way domains interact when they form the tertiary structure of proteins.[15] These investigators carried out an n-gram analysis on 23 million protein domains found in the protein sets of 4,800 species representing all three domains of life (Bacteria, Archaea, and Eukarya). The algorithm used by the research team to study domain interactions is the same one used by text messaging apps to anticipate the next word the user will likely type. This algorithm determines the probability of a word being used based on the previous word (or words) entered by the user. (If the probability is based on a single word, it is called a unigram probability. If the calculation is based on the previous two words, it is called a bigram probability, and so on.)

The researchers treated adjacent pairs of protein domains in the tertiary structure of each protein in the sample set as a bigram. Surveying the proteins found in their data set of 4,800 species, they discovered that only 5 percent of all the conceivable domain combinations exist in actual proteins. This discovery means that in the same way that certain word combinations never occur in human languages because of the rules of grammar, certain domain combinations never occur in proteins. In other words, there appears to be a protein "grammar" that constrains the domain combinations in proteins. This insight implies that physicochemical constraints (which define protein grammar) dictate the naturally occurring range of protein tertiary structures, permitting only 5 percent of imaginable domain-domain interactions.[16] At this juncture, nobody knows what the grammatical rules for proteins may be, but knowing that they exist paves the way for future studies. It also generates hope that one day biochemists might understand them sufficiently so they can use them to predict protein structure from amino acid sequences.

Again, it is remarkable that the constraints of domain-domain interactions in proteins lead to tertiary structures that support the wide range of functional properties proteins require to sustain the complexity and diversity of Earth's life.

The Geometric Constraints on Protein Folding

Further evidence for steric and topological constraints on protein structures comes from a theoretical study carried out by biophysicists Jayanth Banavar and Amos Maritan. These researchers modeled the range and types of folds adopted by polypeptide chains by treating them as flexible tubes formed by linking together individual disks (representing amino acids). These researchers varied the hydrophobicity of each disk, generating tubes with varying hydrophobicity profiles along their length. Additionally, they set up their model so that each tube could interact with itself and fold into compact structures that would expel water from its interior.[17]

As part of their modeling work, the researchers defined a quantity they called a folding parameter (X). This parameter corresponds to the ratio of the tube diameter to the minimal distance required to separate adjacent regions along the tube before these regions can interact. If the tube diameter is large, then X becomes much greater than 1 ($X>>1$) and the tube is too "fat" to fold. If the tube diameter is small, then X becomes much smaller than 1 ($X<<1$) and a near-infinite number of folds results. On the other hand, if X is approximately 1 ($X{\sim}1$), the number of folds is limited to a few thousand folds. Banavar and Maritan discovered that the folds their hypothetical tubes adopted when $X{\sim}1$ are the same folds observed for the proteins found in nature. This result is nothing short of remarkable. As it turns out, the folding parameter for polypeptide chains is approximately 1. In other words, the geometry and the thermodynamic properties of polypeptide chains dictate how they fold into the three-dimensional structures that constitute protein tertiary structure.

To me, this finding is nothing short of astounding because it means that the proteins' amino acid sequence isn't the sole determinant of its three-dimensional structure and function, as biochemists have long thought. Instead, the hydrophobicity profile along the polypeptide chain (which is influenced by amino acid sequence) and its chain geometry determine protein three-dimensional structure. Differing amino acid sequences can yield nearly the same hydrophobicity profile, which means that different amino acid sequences can adopt the same fold. Banavar and Maritan summarize their results this way:

> Our model calculations show that the large number of common attributes of globular proteins reflect a deeper underlying unity in their behavior. . . . This landscape is *(pre)sculpted* by general considerations of geometry and symmetry. . . .
>
> . . . Protein folds do not evolve: rather, the menu of possible folds is determined by physical law.[18] (italics in original)

It turns out that the $X{\sim}1$ state for flexible tubes (and polypeptide chains) represents a type of critical state, precisely balancing self-attraction of the tube with the repulsive forces caused by steric hindrance. In this critical phase, polypeptide chains possess a set of just-right properties that make them ideally suited for life. For example, the geometric and thermodynamic constraints on the number of permissible folds (when the folding parameter approximates 1) make it possible for polypeptides to fold *rapidly* into their three-dimensional native state.

This behavior helps resolve Levinthal's paradox, which was articulated in the late 1960s by molecular biologist Cyrus Levinthal, who determined that a single polypeptide chain can fold into a vast number of possible three-dimensional structures. According to Levinthal's calculations, the number of possible folds is so extensive that if a polypeptide sampled all the possible folds it could adopt, it would take longer than the universe's age for it to adopt its native state. Yet when proteins are produced at ribosomes, they quickly and reproducibly adopt their native state within a second or so. Biochemists have long wondered how proteins find their native structures so rapidly. Part of the solution to this puzzle can be found in the physicochemical constraints that restrict the number of possible folds. Polypeptide chains can assume their native structures in seconds because physicochemical constraints limit the sampling process to only a few thousand folds. If not for these constraints, it becomes difficult to envision how life would be possible due to Levinthal's paradox.

The $X{\sim}1$ critical state of folded polypeptide chains also benefits life in two other ways. First, it allows proteins to withstand the potentially deleterious effects of mutations, which cause changes in the amino acid sequences of polypeptide chains. It is true that, in many instances, changing a single amino acid in a protein's primary sequence can have devastating effects on its function (and is the source of many genetic disorders). The situation could be far worse. If not for the constraints on polypeptide chain folding, most mutations would unduly disrupt the protein's three-dimensional structure and, hence, function. The physicochemical restraints that allow protein folds to tolerate amino acid changes in the primary structure also make it possible for the same fold to perform different functions. It is fascinating that the constraints on protein folding yield tertiary structures with such versatility.

The $X{\sim}1$ critical state can also be envisioned as a transitional state between few folds for $X>>1$ and a near-infinite number of folds for $X<<1$. Because $X{\sim}1$ is a transitional state, it means that the permissible protein folds in this state have an inherent instability that imparts conformational flexibility to the polypeptide chain's three-dimensional structure. This flexibility plays a critical role in protein function. For example, conformational flexibility allows proteins that operate as enzymes to proceed through their catalytic cycles as they convert substrate molecules into reaction products. Conformational flexibility also makes it possible for the cell's machinery to regulate protein function through allosteric interactions. These types of interactions are mediated by small molecules called effectors that bind to allosteric sites on the surface of proteins. Oftentimes, this binding triggers a change in the protein's conformation, which either activates or inhibits its function. Likewise, conformational flexibility impacts signaling pathways in the cells. In many instances, these pathways get triggered when a small molecule binds a receptor protein on the cell surface. This binding event triggers conformational changes in the receptor protein that then triggers its interaction with other proteins, initiating the signaling pathway.

Random Coils

Up to this point, we focused our attention on the higher ordered secondary, supersecondary, and domain structures that exist within proteins as we have sought to assess the case for the biochemical anthropic principle. However, our emphasis on the ordered regions of protein structure runs the risk of giving a false impression. Biochemists have discovered that up to 40 percent of a protein's amino acids locate to regions of the polypeptide chain that fail to display any secondary and, consequently, supersecondary structural features. In fact, these amino acids take part in three-dimensional regions of the protein structure which lack any discernible order whatsoever. The amino acid residues that make up these types of regions randomly orient in space. For this reason, biochemists appropriately refer to these regions as random coils.

Does the existence of random coils in proteins undermine the case for the biochemical anthropic principle? I don't think so. Here's why: even though random coils display no discernible order, each time a protein folds into its final three-dimensional shape, the random coil regions within its structure reproducibly adopt the same geometry. This discovery makes sense. Biochemists have learned that the omega and psi bond angles are restricted due to steric hindrance, just as they are for amino acids that take part in alpha-helix and beta-pleated sheet structures. In fact, the omega and psi bond angles for amino acids found within random coils are restricted to the exact same bond angles as the corresponding amino acids that appear in alpha-helices and beta-pleated sheet secondary structures.[19] This prompts some questions: If these amino acids adopt the same bond angles as amino acids in alpha-helices and beta-pleated sheets, then why don't they form these types of secondary structures? Why do random coils exist at all?

The answers to these questions can be found, in part, in the pioneering work of biochemists Peter Chou and Gerald Fasman.[20] In the 1970s, these two researchers discovered that each amino acid has characteristic propensities to take part in both alpha-helix and beta-pleated sheet secondary structures. For example, some amino acids (such as methionine, alanine, leucine, glutamate, and lysine) have strong propensities to form alpha-helices. On the other hand, other amino acids (such as tyrosine, phenylalanine, tryptophan, threonine, valine, and isoleucine) have strong tendencies to form beta-pleated sheet structures. Still other amino acids (such as glycine and proline) disrupt secondary structures.

As it turns out, a region of a polypeptide chain will adopt an alpha-helical geometry only when a sequential run of amino acids (around four to six in length) occurs along the length of the polypeptide, such that each amino acid in the run displays a high propensity to form an alpha-helix. This core sequence is called a nucleation site. The alpha-helix will extend in both directions from the nucleation site, as long as the amino acids added to the sequence have a relatively high propensity to take part in an alpha-helix. The same scenario applies for beta-pleated sheet secondary structures, except that the nucleation site only requires a run of about three to five amino acids. If neither type of nucleation site exists for a particular segment of the polypeptide chain, then it will form a random coil structure.[21]

The existence of random coils in no way undermines the case for the biochemical anthropic principle because even though these regions don't form a regular, higher-order structure that conforms to a pattern, the precise structure of a random coil is dictated by constraints on the omega and psi bond angles of the amino acids in it. Biochemists have come a long way toward understanding these constraints, to the point of being able to predict where random coils will occur in a polypeptide chain. They are slowly gaining the ability to even predict the exact geometry of the individual random coils that occur within proteins. In other words, though random coils don't conform to a pattern, they are specified by the laws of nature and display the just-right set of physicochemical properties that are needed to support protein structure and function.

Constraints on the Structures of Protein Binding Sites

The function of many proteins depends on their interactions with small molecules. These interactions often occur in pockets (or concave regions) at the protein surface. These interactions display a high degree of specificity that arises from the precise interactions between the small molecules and the chemical groups that line the protein pocket. These interactions make it possible for a protein to discriminate among the vast number of small molecules in the cellular environment. This discrimination is central to protein function.

A team of life scientists from Georgia Institute of Technology (Georgia Tech) carried out a series of theoretical studies designed to characterize protein binding sites. In one study, the researchers compared the binding pockets for all the known proteins found in nature (housed in the Protein Data Bank) with the concave surface structures of artificial protein-like materials found in a database of randomly generated structures.[22] They discovered that only around 400 different types of binding pockets exist in naturally occurring proteins. Remarkably, all these binding pockets also occur in the randomly generated database of artificial protein-like materials. In a follow-up study, these investigators surveyed the Protein Data Bank for distinct small-molecule binding sites on protein surfaces and discovered about 1,000 distinct binding pockets.[23] Based on these studies the researchers concluded that highly similar pockets occur in a variety of protein structures and that a protein's ability to bind small molecules results simply from its structural and physicochemical properties. These researchers state:

> Overall, there is a remarkable coincidence of the properties of native protein pockets with those of artificial proteins, whose sequences are selected purely for fold stability and not function. We find that the structural space of pockets is remarkably small and likely complete. . . .
>
> . . . The clear implication is that the fundamental physical–chemical properties of proteins are sufficient to explain many of their structural and molecular functional properties.[24]

Even though physicochemical constraints limit the number of binding pockets, the sites that do exist appear to be fit for life. The Georgia Tech scientists discovered that the small-molecule binding sites on protein surfaces are promiscuous. These binding sites inherently form structures that bind large numbers of small molecules with a wide range of chemical features. In turn, researchers demonstrate that the specificity characteristic of protein binding sites in biochemical systems can be achieved by small alterations of the amino acid residues that line the pocket.

This survey demonstrates that the higher-order, three-dimensional structures that constitute protein supersecondary and tertiary structures appear to be fundamentally dictated by some deep, underlying principles that arise out of the laws of nature. As a corollary, in light of the physicochemical and geometric constraints that shape the folds of polypeptide chains, it seems unlikely that proteins are the outworking of an historically contingent evolutionary process. As Banavar and Maritan conclude:

> Protein folds do not evolve—rather, the menu of possible folds is determined by physical law. In that sense, it is as if evolution acts in the theater of life and shapes sequences and functionalities, but does so within the fixed backdrop of the Platonic folds.[25]

It is provocative to think that the constraints placed on the supersecondary and tertiary structures of proteins by the laws of nature produce three-dimensional structures with the just-right properties that make proteins central to the biochemical operations that sustain cellular life. The higher-order structures of proteins are precisely the types of structures life needs to exist.

At this point, I have identified anthropic coincidences for protein primary, secondary, supersecondary, and tertiary structures, including the small-molecule binding pockets of protein surfaces. That leaves us with one level of protein structure to examine: its quaternary structure.

Protein Quaternary Structure and the Anthropic Principle

I doubt that many biochemists thought that any underlying, organizing principles existed to account for the wide range of protein quaternary structures observed in nature—until a few years ago. Over the years, biochemists have determined the structure of thousands of protein complexes. Because they assumed there was no framework to guide their investigation, they approached the structure of protein complexes on a case-by-case basis. According to biochemist Joe Marsh, "Evolution has given rise to a huge variety of protein complexes, and it can seem a bit chaotic."[26]

The historical contingency of the evolutionary process seemed to provide the best explanation for the bewildering array of protein complexes. Presumably, unguided processes transformed preexisting proteins into complexes that, in turn, have been shaped by natural selection to perform key functions in the cell. Yet, in 2015, life scientists from the United Kingdom discovered a way to create a biochemical periodic table that explains and predicts protein quaternary structure, suggesting that physicochemical constraints even influence the architecture of protein complexes.[27]

Instead of analyzing the architecture of protein complexes to uncover structural principles, the research team focused on the protein complex assembly process. They noted that protein complexes assemble in an orderly fashion that seems to be conserved across the biological realm. They identified three basic steps in the assembly process:

1. Dimerization, in which two identical protein subunits interact to form a complex
2. Cyclization, in which a ring of three or more identical subunits forms
3. Addition, in which two different subunits join to form a dimer

The researchers also recognized that the nature of the interactions between the subunits is important. They identified two types of interactions:

1. Isologous, in which identical corresponding surfaces on the protein subunits interact ("head-to-head")
2. Heterologous, in which different surfaces between the subunits interact ("head-to-tail")

By systematically going through multiple iterations of these steps, they discovered a wide range of structures for protein complexes. The researchers discovered that by organizing the quaternary structures into columns (by number of unique subunits) and rows (by number of times the subunit combinations repeated), a periodic table of protein complexes emerged. To their surprise, this table accounts for 92 percent of all known protein complex structures. (Four percent of the unaccounted-for complexes was due to assignment errors.) Using their periodic table, the researchers explained the frequency of occurrence for protein complex structures. They even predicted the structure of yet-to-be-discovered protein complexes.

The existence of a periodic table for protein quaternary structure strongly implies that physicochemical constraints shape the structure of protein complexes, more so than historically contingent evolutionary forces. Instead of a near-infinite, chaotic array of quaternary structures, only a limited number of protein complexes exist in reality. And those structures that exist aren't determined by natural selection, but instead manifest from an underlying set of principles related to the assembly process. Again, it is a bit eerie to think that

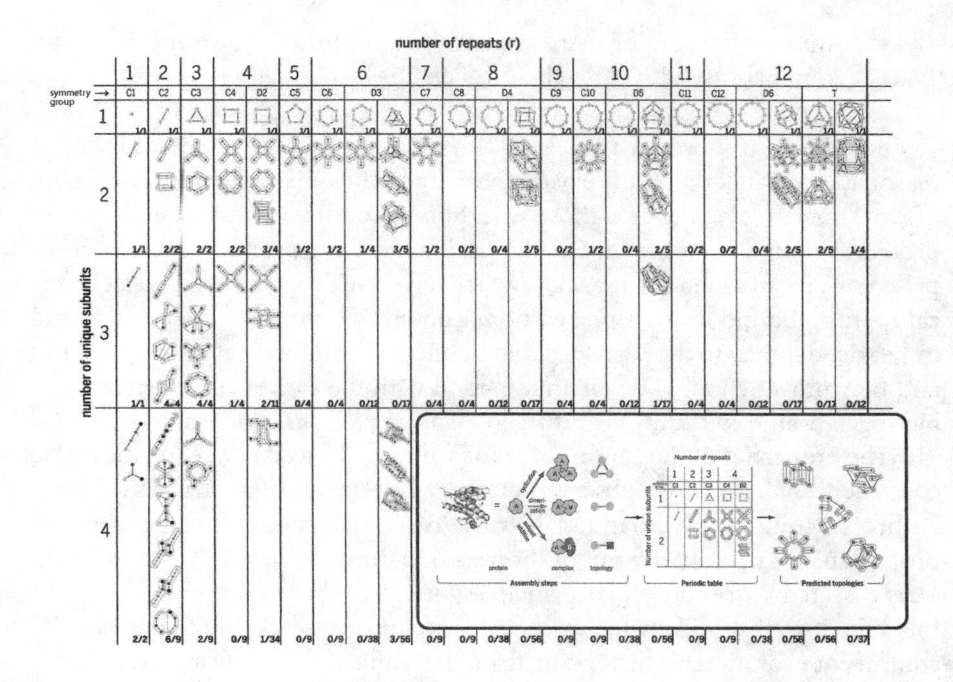

Figure 4.14: The Periodic Table of Protein Quaternary Structure

these physicochemical constraints yield a set of protein structures with the necessary properties to support life, independent of the influence of natural selection.

Does the Biochemical Anthropic Principle Apply to Proteins?

The goal of this chapter is to determine if the anthropic principle applies to proteins. As noted, I expected to find anthropic coincidences based on the prescient and pioneering work of a handful of biochemists. But, instead of just pointing to these initial insights, I wanted to address this question in a more complete and systematic way by determining if these three criteria are satisfied:

- Physicochemical constraints arising out of the laws of nature exist, which dictate the structural features of amino acids and proteins.
- A molecular rationale exists that explains the structure of amino acids and proteins, accounting for their fitness for life.

- Amino acids and proteins appear to be unusually fit for life, based on comparisons with close chemical analogs.

Based on our survey, it appears all three criteria have been met. In fact, it is reasonable to think that if life exists anywhere in the universe, it must make use of workhorse molecules and those workhorse molecules must be just like the proteins that make up life on Earth. Specifically, these proteins must be comprised of a set of alpha amino acids, which have a hydrogen atom and a chemical R group bound to the alpha carbon. The set of R groups must be identical or nearly identical to the R groups that define the canonical set of amino acids, and they must be joined by an amide bond with the carbonyl oxygen and the hydrogen bound to the amide nitrogen assuming a *trans* geometry.

The properties of the canonical set of amino acids lead to a set of anthropic coincidences that can be observed at every level of protein structure. These anthropic coincidences manifest because of physicochemical constraints on protein structure that arise out of the laws of nature and appear to be based on a deep-seated set of molecular principles—a sort of physicochemical logic. In this vein, it seems unlikely that proteins reflect the outworking of a historically contingent evolutionary history. Instead, they appear to largely arise out of the dictates of the laws of nature.

The amide linkage that joins amino acids together and forms the primary structure of polypeptide chains is a highly unusual chemical bond. The planarity and geometry of the amide linkage (and the steric effects they cause) constrain the conformation of the polypeptide backbone, so that, for the most part, alpha-helices and beta-pleated sheet structures are the only permissible secondary structures. Because of topological effects, these two secondary structures are forced to interact to form a limited collection of supersecondary structures. In addition, the domains that result from the interactions of supersecondary structures can only interact in a restricted number of ways to form the domain architecture that comprises protein tertiary structures. The restrictions on polypeptide three-dimensional structures also lead to a limited number of small-molecule binding sites on the protein surfaces. In large measure, the physicochemical constraints that dictate protein tertiary structures stem from the geometric parameters (such as diameter) of the polypeptide chain. Topological constraints that emerge from the assembly process also appear to dictate the quaternary structures of protein complexes.

Remarkably, this cascade of constraints produces structural features in proteins that make these biomolecules well-suited for life. These anthropic coincidences aren't inevitable. The physicochemical constraints could have been different—if they even existed at all. It is just as easy to envision a universe where the physicochemical constraints that dictate protein structure render life untenable. Yet the fitness of the resulting protein structures is openly displayed, making it difficult to deny.

What about the Need for Chaperones?

Key to the case for the biochemical anthropic principle is the idea that physicochemical constraints, to a large degree, dictate the three-dimensional structures of proteins. Yet someone who is skeptical about the existence of the biochemical anthropic principle might object, pointing out that while physicochemical constraints indeed dictate protein architecture, these constraints may not be all that important because proteins often misfold when they are produced at the ribosome. (See chapter 6, pages 195–196.) In fact, only a limited number of polypeptide chains will spontaneously fold into their proper three-dimensional structure. For those that do, the process is slow and inefficient.

In the cell's environment, improperly folded proteins, along with proteins that fold slowly and inefficiently, represent a potential catastrophe. In the crowded cellular interior, misfolded proteins aggregate and form massive clumps that hinder cell operations. To sidestep this potential disaster, practically every cell throughout the biological realm relies on a family of proteins called chaperones to facilitate efficient and accurate protein folding. Two types of chaperones exist: molecular chaperones and chaperonins.[28] Once produced at the ribosome, some polypeptide chains will properly fold only with the assistance of several different chaperones that work collaboratively. These chaperone proteins bind to these polypeptides and help stabilize the partially folded protein, preventing it from aggregating with other proteins in the cell. When these chaperones de-bind from the polypeptide chain, it folds into its intended three-dimensional shape.

Other proteins need more help to fold than the chaperones can provide. These proteins are ferried to chaperonins after the chaperones disassociate from the partially folded polypeptide chain. Chaperonins are large complexes that consist of several protein subunits. Perhaps the best-understood chaperonin is found in the bacterium *Escherichia coli*. Biochemists call this chaperonin GroEL-GroES. The GroEL component of the *E. coli* chaperonin consists of 14 subunits that organize into two ring-like structures that stack on top of one another. The stacked rings form a barrel-like ensemble with a large open cavity. The partially folded polypeptide is ushered into the GroEL cavity. Another protein complex, GroES, serves as a cap that covers the GroEL cavity. The cavity provides the optimal environment for protein folding.[29] Once properly folded, the polypeptide chain is released from the GroEL cavity after the GroES lid disassociates from the barrel.

The necessary role that chaperones play in protein folding doesn't threaten the case for the biochemical anthropic principle in the least, nor does it undermine the claim that physicochemical constraints arising out of the laws of nature largely determine protein structure. In the case for all enzymes, chaperones don't alter the thermodynamics of biochemical processes. They only impact the rate at which they occur. This means, as a rule of thumb, that chaperones don't determine the most stable three-dimensional structure for the protein. Rather, the laws of nature dictate a protein's native state. Chaperones assist the nascent polypeptide chain to find the native state as rapidly as possible by either stabilizing folding intermediates (in the case of chaperones) or by creating a milieu that helps proteins quickly find their native state (in the case of chaperonins), all while inhibiting partially folded proteins from aggregating with other proteins and preventing them from becoming kinetically trapped in a non-native state.

In short, amino acids and proteins appear to be a manifestation of the biochemical anthropic principle. We could end our investigation of the biochemical anthropic principle with proteins because of their central importance to life. As biochemist Michael Denton rightly pointed out: if the anthropic principle extends into the realm of biochemistry, it should be most evident in proteins. Indeed, we see strong evidence for the anthropic principle at all levels of protein structure. But if a biochemical anthropic principle does exist, then I think it should also be evident in other biochemical systems, including DNA—the molecule that harbors life's blueprint.

Chapter 5

The Nucleic Acids

Truth be told, I am not much of a fan of contemporary Christian music (CCM).

Without question, many musicians who write and perform in this musical genre are talented, world-class songwriters and performers. However, much of CCM seems derivative to me and feels a bit bland and uninspired. (No pun intended.) Yet, over the years, a handful of highly innovative CCM artists and musical groups have captured my attention, such as AD, the Resurrection (Rez) Band, the 77s, Three Crosses, Caedmon's Call, and Third Day to name a few. The group that tops my list of all-time favorite CCM artists is White Heart.

Founded in 1982 by guitarist Billy Smiley and keyboard player Mark Gersmehl, White Heart recorded and performed through 1997, when they went on a permanent hiatus, though never formally disbanding. White Heart experienced a never-ending change in lineups, with Smiley, Gersmehl, and lead singer Rick Florian (who joined in 1986) serving as the group's core members.

With this nucleus in place, White Heart hit its stride as a band, blending hard-driving '80s rock with progressive rock elements. Florian's distinct vocal style and impressive range became an integral part of White Heart's signature sound. There was no other band quite like White Heart in either the CCM or secular music scenes.

I lived on the outskirts of Cincinnati, Ohio, in the 1990s when White Heart enjoyed its heyday as a band. I remember dragging my wife and some friends all the way to Richmond, Indiana, on a cold, rainy Saturday night to see White Heart perform live. I'm not sure how much my companions enjoyed the experience, but for me, hands down, it was one of the best rock 'n' roll shows I ever witnessed.

I had an opportunity to hang out with Smiley, Gersmehl, and Florian a couple of years later—thanks to a happy accident. My family and I made a trip to our local Berean Christian Bookstore one Saturday afternoon and, to my absolute amazement, the three of them were there promoting their latest album *Inside*. As part of the promotion, Berean was giving away a signed copy of *Inside* to a lucky raffle winner. Everyone in our family entered the raffle. Why not? And as fate would have it, my four-year-old daughter won. Though she had no idea who White Heart was, she proudly had them sign her CD and took a photo with the band. Then she spent the next several minutes hanging out with White Heart. It was quite a picture in contrast to see a cute little girl talking to the three veteran rockers. They were so kind and gracious to my daughter and our family.

Every time I hear a White Heart song or even pick up a White Heart CD off the shelf, memories of that incredible afternoon come flooding back, putting a smile on my face.

It is hard for me to pick my favorite White Heart song, but one that is near the top of my list is "Letter of Love." This song didn't originally appear on any of their studio albums. Instead, it was released for the first time in 1994 on *Nothing but the Best: Rock Classics*, one of their two official compilation albums.

This song is a quintessential White Heart offering. Florian was at the top of his game when the group recorded this song, a straight-ahead rocker propelled by guitar and synthesizer riffs playing off one another. This song is a euphoric expression of the joy felt by someone who has discovered love—a forever love. But the love described in the song is not the love between a man and a woman. Rather, it is the love that Jesus Christ extends to each of us as our Savior, if we will accept it. The song is a letter of love returned to the Savior. It is also a call to everyone, inviting them to experience Christ's everlasting love.

This song energizes me, in part, because it is a rocker. I become filled with joy and euphoria, knowing that I have received the never-ending love of my Savior. And it reminds me that when I share the gospel, I am sending my own letter of love to those whose lives intersect with my own.

• • • • • • • • • • • •

If there is a class of biomolecules that serves as a letter of love to biochemists and molecular biologists, it is the nucleic acids DNA and RNA. These molecules are rich with the information instantiated within their structural makeup. In fact, the chief function of DNA (deoxyribonucleic acid) is to harbor the instructions the cell needs to directly and indirectly build all its component parts and to then regulate the individual production of cellular components at key stages of the cell cycle. Sometimes life scientists refer to DNA as life's instruction manual or blueprint.

Life scientists have strong motivation to characterize the wealth of information found within DNA. By decoding this information, molecular biologists and biochemists position themselves to gain fundamental understanding of the cell's biology. DNA represents a scientific letter of love for investigators who want nothing more than to understand how life works at its basic, most fundamental level.

The importance of this insight for life scientists was highlighted on June 22, 2000, when President Bill Clinton announced the sequencing of the human genome at a press conference held in the East Room of the White House. President Clinton was flanked by Francis Collins, the head of the National Human Genome Research Institute, and Craig Venter, then-president of Celera Genomics, a private sector competitor to the publicly funded effort. In his remarks, President Clinton marveled at the significance of this scientific achievement when he stated:

> Today, we are learning the language in which God created life. We are gaining ever more awe for the complexity, the beauty, the wonder of God's most divine and sacred gift. With this profound new knowledge, humankind is on the verge of gaining immense, new power to heal. Genome science will have a real impact on all our lives—and even more, on the lives of our children. It will revolutionize the diagnosis, prevention, and treatment of most, if not all, human diseases.[1]

Collins echoed the president's remarks when it was his turn at the microphone: "It is humbling for me, and awe-inspiring, to realize that we have caught the first glimpse of our own instruction book, previously known only to God."[2]

Given the central role that nucleic acids play in living systems as a biochemical instruction manual—particularly DNA—it serves to reason that, if the anthropic principle extends into the biochemical arena, it should also be evident in the structure and function of nucleic acids. I might even go so far as to say that perhaps the signal for the anthropic principle might be even more pronounced in the architecture and operation of DNA and RNA than in proteins.

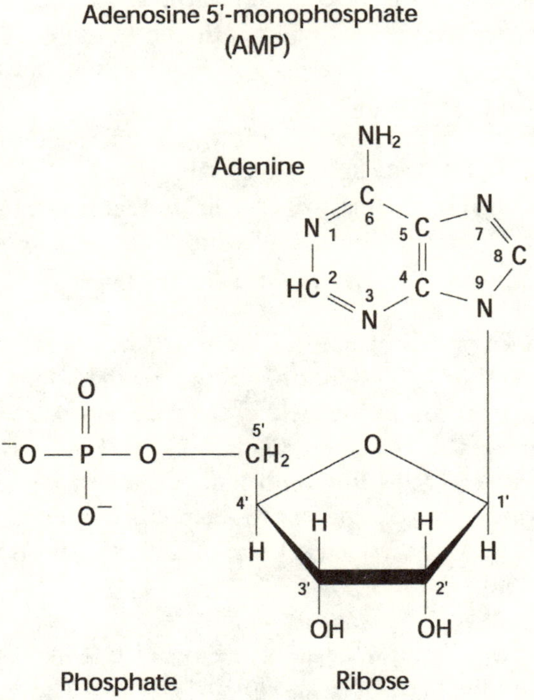
Adenosine 5'-monophosphate
(AMP)
Adenine
NH2
C
N
N
C
C
C
HC
N
N
1
2
3
4
5
6
7
8
9
O
P
O
O
O
CH2
5'
4'
1'
3'
2'
H
H
H
H
OH
OH
Phosphate
Ribose

Figure 5.1: The Structure of a Typical Nucleotide

The central question of this chapter is this: do nucleic acids show evidence of anthropic coincidences? To pursue an answer, we will adopt a similar approach to the one used in previous chapters to demonstrate the application of the anthropic principle to chemical systems and proteins:

- Determine if a molecular rationale undergirds the structure of nucleic acids, accounting for their fitness for life.
- Determine if physicochemical constraints arise from the laws of nature that dictate the structural features of nucleic acids, influencing their fitness for life.
- Compare the structural features of nucleic acids with close chemical analogs to determine if the characteristics possessed by DNA and RNA make them fit for life.

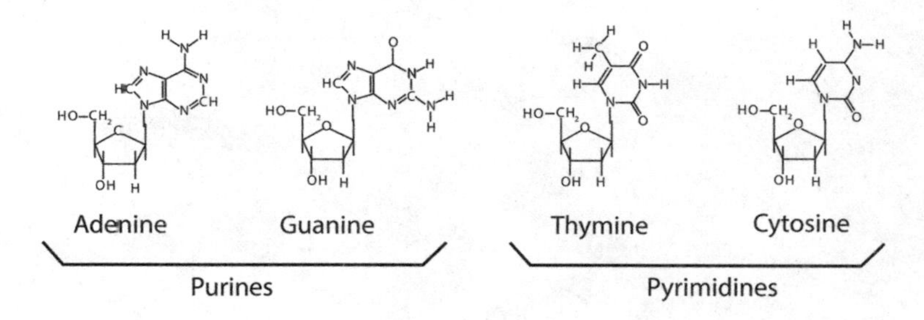

Figure 5.2: The Structure of the Four Nucleobases in DNA

If these three criteria are met, then we can reasonably conclude that the structural and functional features of DNA and RNA aren't the contingent out-working of evolutionary processes, but are instead specified by constraints that arise from the laws of nature—constraints that produce biomolecular systems with a suitable set of properties for life.

Before we begin our quest, a brief review of DNA and RNA structure is in order. (If you are already familiar with the basics of nucleic acid biochemistry, please feel free to skip ahead to "Nucleic Acids and the Anthropic Principle.")

An Overview of Nucleic Acid Biochemistry

The best place to begin a review of nucleic acid structure is with DNA. This biomolecule is a polymer built from four different subunit molecules (monomers) called nucleotides—adenosine (A), guanosine (G), cytidine (C), and thymidine (T).

The Nucleotides

The nucleotides that form DNA's double strands are complex molecules. Each one consists of both a phosphate moiety (part of a molecule) and a nucleobase (either adenine, guanine, cytosine, or thymine) joined to deoxyribose, which is a five-carbon sugar, via a chemical bond called an N-glycosidic linkage.

In the 1950s, biochemist Erwin Chargaff discovered something unusual about DNA's compositional makeup. By characterizing the chemical composition of DNA from a wide range of organisms, Chargaff learned that while

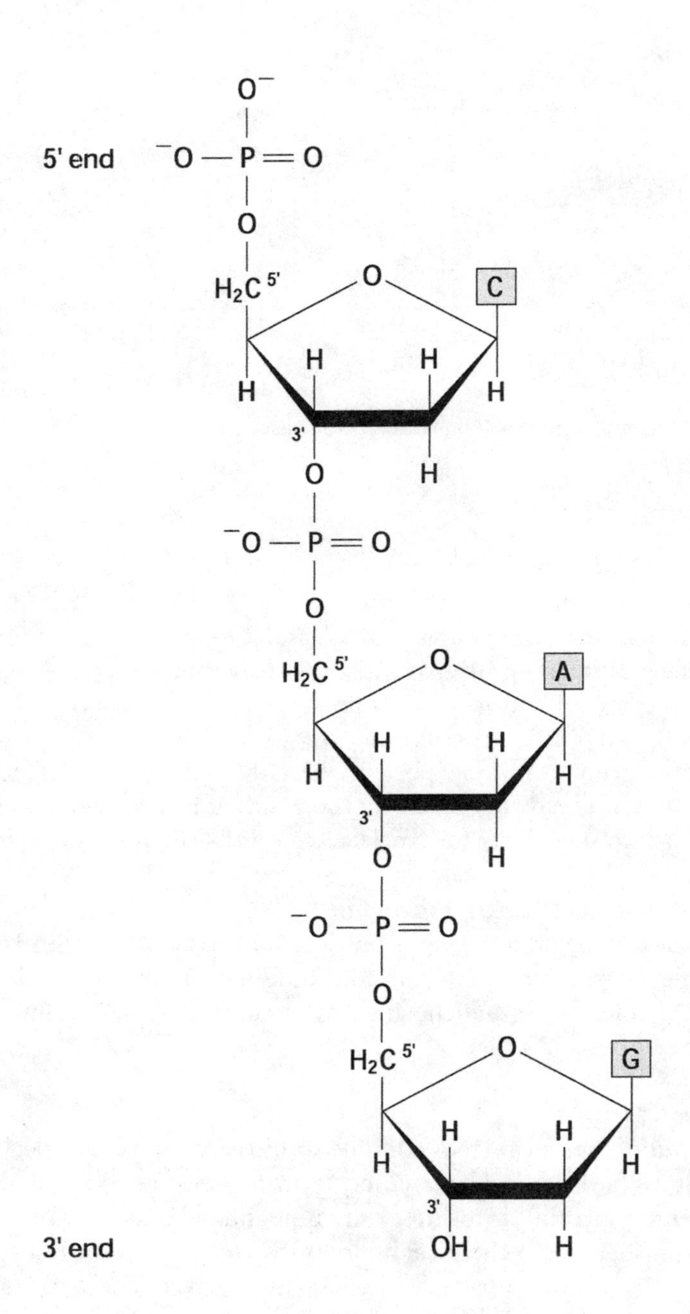

Figure 5.3: The Structure of a Polynucleotide

DNA composition differs from organism to organism, the amount of adenine in an organism's DNA always equals the amount of thymine and the amount of guanine always equals the amount of cytosine. This relationship, known as Chargaff's rule, served as an important clue that helped James Watson and Francis Crick solve the structure of DNA in 1952.

DNA

The primary structure of DNA consists of a polynucleotide strand formed when the cell's machinery repeatedly joins the phosphate group of one nucleotide to the deoxyribose unit of another nucleotide. Biochemists call this linkage a 5′ to 3′ phosphodiester bond. The alternating sugar-phosphate moieties form the backbone of the polynucleotide strand. The nucleobases extend as side chains from the sugar-phosphate backbone.

Each polynucleotide strand displays polarity because of the 5′–3′ phosphodiester linkages. For this reason, the phosphodiester bond serves as the basis for the convention used by biochemists to designate the directionality of a polynucleotide strand as either 5′–3′ or 3′–5′.

The secondary and tertiary structures of DNA result from the interaction of two polynucleotide strands. These strands align in an antiparallel fashion, with the two strands arranged alongside one another with the starting point of one strand (the 5′ end) in the polynucleotide duplex located next to the ending point of the other strand (the 3′ end) and vice versa. The paired polynucleotide chains resemble a ladder, with the sugar-phosphate backbones serving as the ladder's uprights. The nucleobase side groups extend from the polynucleotide backbone into the center of the molecule, with nucleobases from each polynucleotide serving as interaction points. In effect, the nucleobases bridging the two strands form the ladder's rungs.

The sequences of the paired polynucleotide chains have a special relationship to one another, which accounts for Chargaff's rule. When the two DNA strands align, the adenine (A) side chains of one strand always pair with thymine (T) side chains from the other strand. Likewise, the guanine (G) side chains from one DNA strand normatively pair with cytosine (C) side chains from the other strand. On occasion, a mismatch in the base pairing occurs, which is far from ideal. Only low levels of mismatched base pairing can be tolerated without disrupting the DNA structure.

Because of these base-pairing rules, the nucleotide sequences of the two DNA strands are complementary. Looking at it another way, if we know the nucleotide sequence of one DNA strand in the double helix, we can deduce the

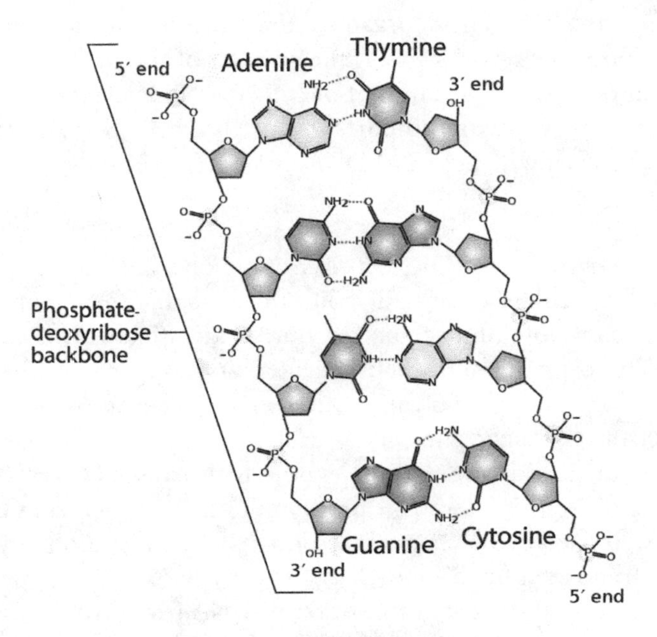

Figure 5.4: The Structure of Paired Polynucleotides

nucleotide sequence of the other strand using the base-pairing relationships.

When the side chains pair, they form cross bridges between the two DNA strands. The lengths of the A-T and G-C cross bridges are nearly identical. Adenine and guanine are both composed of two rings, and thymine and cytosine are composed of one ring. Each cross bridge consists of three rings.

When A pairs with T, two hydrogen bonds mediate the interaction between the two nucleobases. Three hydrogen bonds accommodate the interaction between G and C. The specificity of the hydrogen-bonding interactions accounts for the A-T and G-C base-pairing rules. The specificity of the hydrogen-bonding interactions serves as a positional force, ensuring that the two polynucleotide chains of DNA perfectly align. Many biochemists regard DNA as the gold standard for molecular recognition because of the precise pairing between the two polynucleotide strands.

The double helix of DNA forms when the coupled polynucleotide chains twist in a right-handed direction. The pitch (the height of one complete turn) of the DNA double helix is 34 angstroms (Å), consisting of 10 paired bases, with

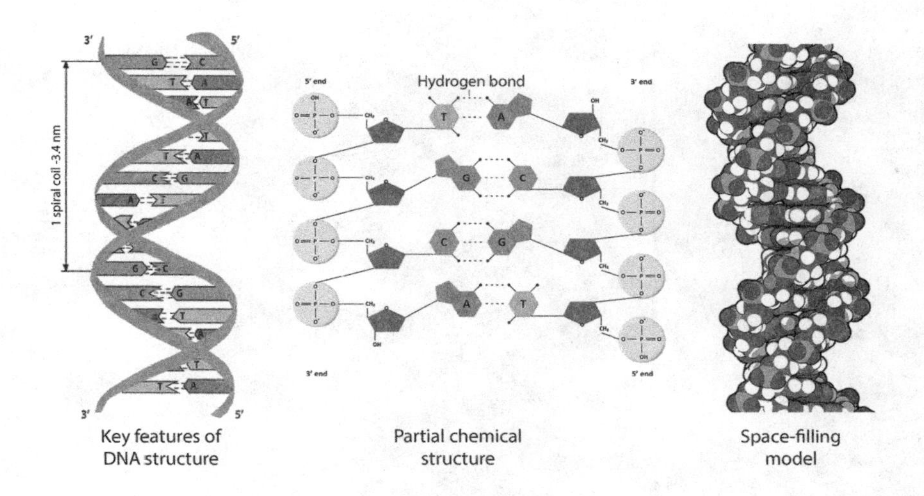

Figure 5.5: The DNA Double Helix

each base pair offset from the base pair beneath it by 3.4 Å. The sugar-phosphate backbone is positioned on the exterior of the DNA double helix. The nucleobases are sequestered into the helix's interior. Locating the backbone of the polynucleotide strand on the exterior of the double helix makes sense because the phosphate moiety bears a negative charge and the deoxyribose residue is a polar compound. Both properties lead to favorable interactions between this portion of the molecule and the water molecules surrounding DNA. On the other hand, the nucleobases are hydrophobic and water insoluble. Orienting the nucleobases toward the interior of the double helix allows them to avoid contact with water. This arrangement imparts stability to the double helix architecture by taking advantage of the hydrophobic effect (See chapter 3, pages 68–70.)

Within the interior of the double helix, the nucleobases adopt a perpendicular orientation with respect to the double helix's axis. Because of their planar geometry, the nucleobases of adjacent nucleotides "stack" on top of one another within the interior of the helix. This stacking leads to a special type of noncovalent interaction among the nucleobases, called pi stacking, in which the electron clouds that form the chemical bonds of one nucleobase overlap with the electron clouds of the adjacent nucleobases above and below it. These interactions result in a molecular orbital that spans the entire length of the

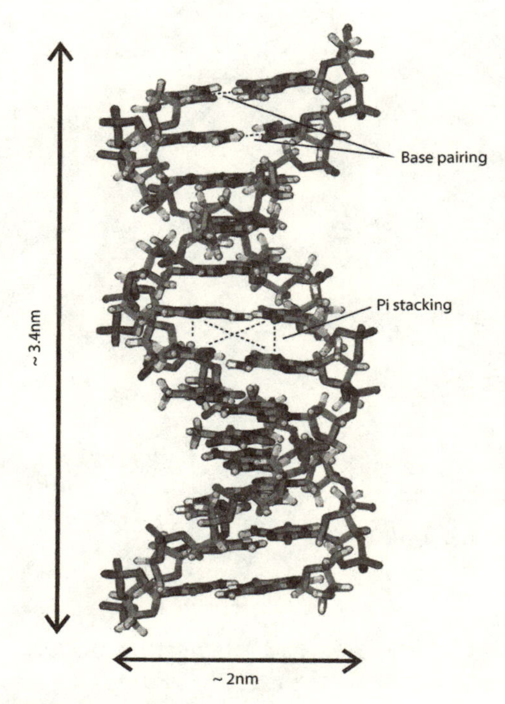

Figure 5.6: Pi Stacking Interactions

double helix. In effect, the molecular orbital functions like a wire running through the DNA interior. These pi stacking interactions also serve to stabilize the double helix architecture.

Another important structural feature of the DNA double helix is the grooves found on the surface of the double helix. Examining the double helix reveals two distinct types of grooves, called the major and minor grooves, that differ in their width and depth. The major groove is 12 Å wide and 8.5 Å deep. The minor groove is 6 Å wide and 7.5 Å deep. The major and minor grooves alternate along the length of the double helix. The orientation of the N-glycosidic linkage that joins each nucleobase to deoxyribose sugar residue in the polynucleotide backbone explains this pattern of alternating grooves.

The major and minor grooves serve as contact points for proteins to bind along the length of the DNA double helix. These interactions are necessary for key metabolic processes involving DNA, such as replication, repair, and transcription. (See chapter 6.) Because of its larger size, the major groove is much

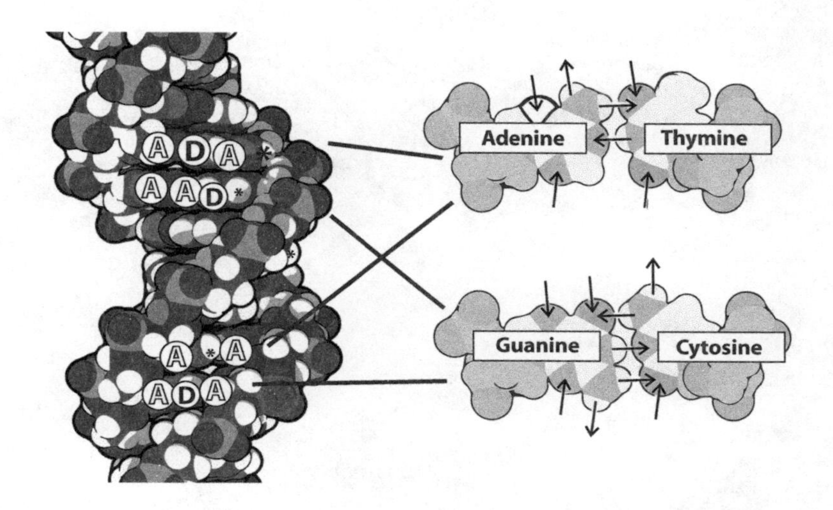

Figure 5.7: The Major and Minor Grooves of DNA

more accessible to proteins than the minor groove. The inner concave surface of both grooves is lined with chemical groups from the nucleobases located in the helix interior. These groups serve as hydrogen bond donors and acceptors. Different chemical groups line the inner surface of the major and minor grooves. The types and positioning of the chemical groups also differ for adenine-thymine base pairs compared to guanine-cytosine base pairs. These differences allow proteins to bind to specific DNA sequences.

DNA's Information Content

As noted, DNA's central function is to harbor the information the cell needs to directly and indirectly build its component parts. Specifically, DNA contains the information necessary to make all the proteins that the cell uses. Contiguous nucleotide sequences along polynucleotide strands in the DNA specify the sequences of amino acids used by the cell's machinery to assemble the polypeptide chains that comprise proteins. These nucleotide sequences are referred to as a gene (or coding sequence). In this sense, the nucleotides can be likened to the letters of an alphabet used to form words—in this case biochemical words (i.e., genes). For this reason, nucleotides in the DNA strands are sometimes called genetic letters.

Biochemists have long recognized that the nucleotide sequences comprising

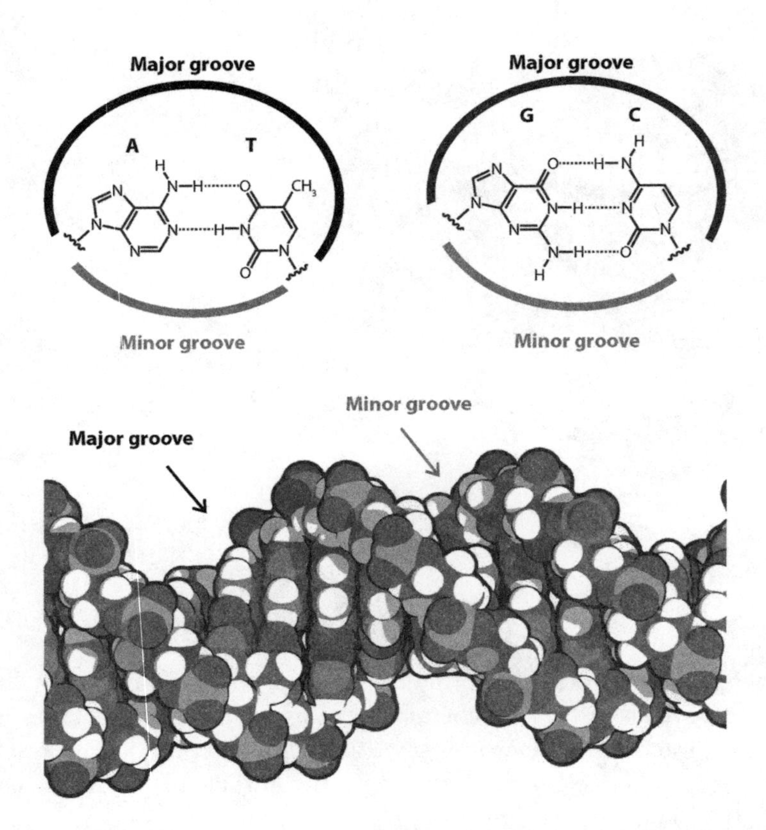

Figure 5.8: The Chemical Groups Located in Major and Minor Grooves of DNA

DNA molecules encode digital information. The nucleotide sequences represent a succession of discrete units (i.e., nucleotides), just like the binary digits 1s and 0s that encode digital information. In this framework, a gene consists of an isolated piece of code specifying the digital information (i.e., amino acid sequence) used by the cell's machinery to build a particular protein.

Recently, two researchers discovered that DNA also houses analog information.[3] In contrast to digital information, which exists in discrete units, analog information consists of a continuously varying quantity. This analog information is housed in the supercoiled double helix and other higher-order DNA structures.

The DNA double helix becomes supercoiled when enzymes called topoisomerases twist the double helix into a superhelix. When not supercoiled, the

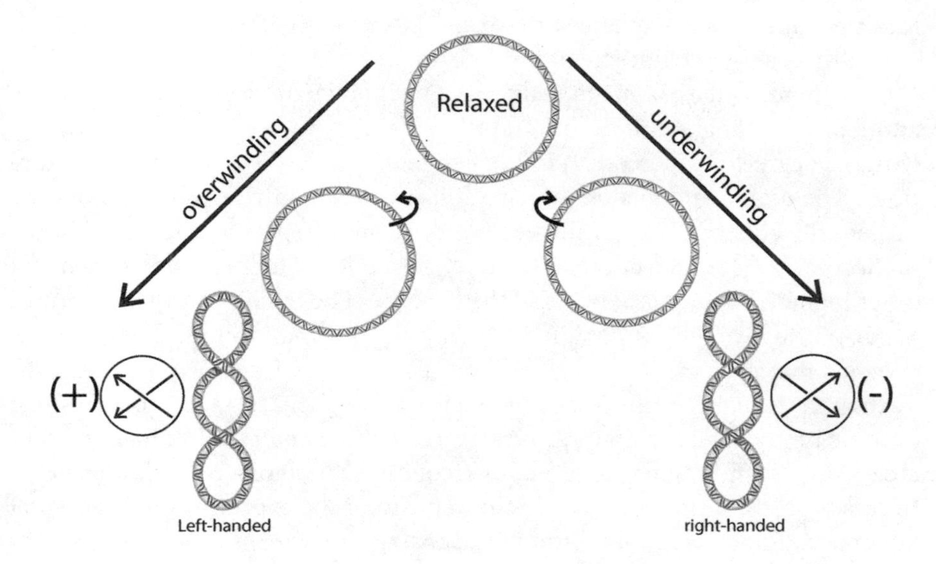

Figure 5.9: DNA Supercoiling

DNA double helix exists in a relaxed state, consisting of 10 base pairs per each turn of the DNA double helix. In this configuration, there is no torsional stress on the DNA molecule. Topoisomerases introduce torsional stress into the DNA double helix, causing it to supercoil by cutting the DNA backbone, then unwinding the double helix so that the helix pitch decreases to less than 10 bases per helix turn (which leads to negative supercoiling) or winding the double helix even more so that the pitch increases to more than 10 bases per turn (which causes positive supercoiling). After altering the pitch of the double helix, the topoisomerases reseal the broken backbone. The net result is that the DNA double helix now experiences stress. To relieve this stress, DNA supercoils, which distributes the torsional stress through the entire double helix.

Several parameters can describe supercoiling: (1) positive or negative, (2) twist, and (3) writhe. These parameters vary continuously, just like the grooves cut into a vinyl record. In other words, supercoiling is an analog property.

As it turns out, the nucleotide sequence dictates the supercoiling parameters. Certain localized nucleotide sequences render some regions of the double helix more prone to supercoiling than others. On the other hand, some localized nucleotide sequences cause the DNA double helix to possess high flexibility, readily unwinding and untwisting.

Biochemists have come to realize that the higher-order structures of DNA, and hence the analog information, play a significant role in protein binding, with the degree of supercoiling influencing the extent of DNA-protein interactions. The digital and analog information is coupled intrinsically through the nucleotide sequence. That is, the nucleotide sequence specifies the digital information of the gene and, at the same time, the tendency to form supercoiled and other higher-order architectures of the genome. These higher-order structures in turn influence the expression of the digital information found in the gene. Proteins that interact with DNA aid in the coupling of digital and analog information. Proteins bind to specific nucleotide sequences. Once bound, some of these biomolecules help to promote supercoiling and stabilize higher-order architectures, and others relax the supercoiling or destabilize the double helix. In other words, proteins play a role in regulating the expression of the digital information in the genome through the analog component.

RNA

Like DNA, RNA is a polymer built using four different nucleotides as subunits. These four nucleotides share many common features with the nucleotides used to assemble DNA. They also display some key differences. Like DNA, the nucleotides that make up RNA consist of a sugar that binds a phosphate moiety in the 5′ position and a nucleobase joined to the sugar via a N-glycosidic linkage. Instead of making use of deoxyribose, ribose is the sugar that helps form the RNA nucleotides. The difference between these two sugars is the absence and presence, respectively, of an -OH (hydroxyl) chemical group in the 2′ position of the sugar. RNA shares three of its four nucleobases with DNA (adenine, guanine, and cytosine), but employs uracil (U) as the fourth nucleobase (in place of thymine). Uracil is a close structural analog to thymine, differing only by the absence of a methyl group bound to the pyrimidine ring system.

Like DNA, RNA is a polymer assembled by the cell's machinery, which joins together the phosphate group of one nucleotide to the ribose unit of another nucleotide through a 5′–3′ phosphodiester linkage. Alternating sugar-phosphate moieties form the backbone of the RNA polynucleotide strand with nucleobases extending away from the backbone as side chains.

Unlike DNA, RNA generally exists in the cell as a single-stranded molecule. Still, RNA molecules can form elaborately complex secondary and tertiary structures. These higher-order structures form when the RNA chain interacts

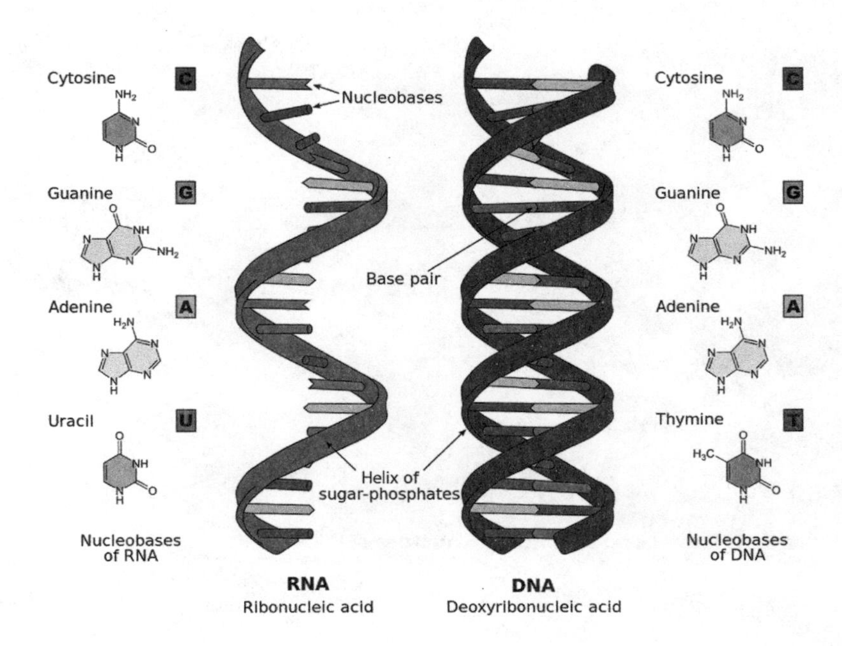

Figure 5.10: Comparison of RNA and DNA Nucleotides
Credit: https://commons.wikimedia.org/wiki/File:Difference_DNA_RNA-EN.svg

with itself through hydrogen-bonding interactions mediated by the nucleobases. The same base-pairing rules found in DNA apply to RNA, with uracil taking thymine's place and associating with adenine.

One of the most common secondary structures in RNA molecules is the stem-loop motif. This structure forms when short segments of the same RNA interact to form a small localized double helix. This interaction is possible because these segments have complementary nucleotide sequences that allow for base-pairing interactions. In turn, the RNA secondary structures can interact with one another through hydrogen-bonding interactions to form complex three-dimensional tertiary structures. Oftentimes, positively charged magnesium ions help mediate the intrachain interactions in RNA, stabilizing its tertiary structure.

RNA, like DNA, also harbors digital (and quite possibly analog) information in its nucleotide sequences. However, RNA does not serve as an information

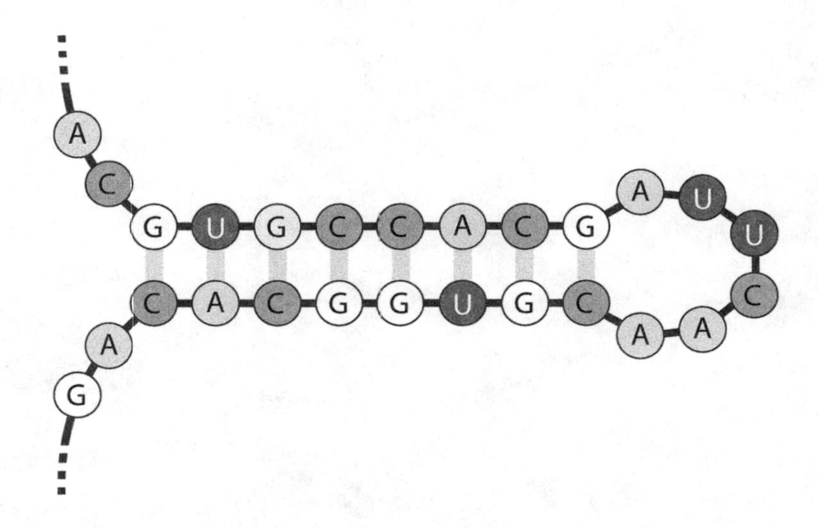

Figure 5.11: Stem-Loop Secondary Structure of RNA

storage system. Instead, it functions as a critical intermediary, making it possible for the cell's machinery to access and read the information in DNA and then use this information to synthesize proteins at ribosomes.

Now we're ready to tackle the central question of this chapter: Do we see evidence for anthropic coincidences in the structural features of the nucleic acids?

Nucleic Acids and the Anthropic Principle

Because of the central importance of nucleic acids, biochemists and molecular biologists have spent significant effort trying to understand the relationship between the structural features of DNA and its biochemical role as an information storage system. Over the years, they have made good progress toward this end. At least two key properties are necessary for DNA to store information: molecular stability and protein binding. If DNA lacked stability, then the information it houses, vital for the cell's existence, would be lost. Protein binding is equally important. It is through protein binding that (1) the information in DNA is accessed (through processes called transcription and gene regulation), so that the cell's components can be built and the cell's life-sustaining operations can be executed; (2) damaged information within DNA is repaired, preserving its integrity amid ongoing chemical and physical insults; and (3) DNA replication proceeds, allowing the cell's information to be passed on to the next generation of cells.

By comparing the structure of DNA and RNA with close chemical analogs, biochemists and molecular biologists have come to recognize that the design of these two nucleic acids appears to be optimal for their role inside the cell, undergirded by an exquisite molecular rationale displayed by nearly every aspect of their molecular architecture. The structural optimality of these two nucleic acids also points to their fitness for life and raises the possibility that DNA and RNA are uniquely suited for their roles. If so, then no other molecular systems could feasibly replace DNA and RNA.

Life scientists have long wondered why the nucleotide subunits of DNA and RNA consist of phosphates, adenine, guanine, cytosine, thymine, uracil, deoxyribose, and ribose. Myriad sugars and numerous other nucleobases could have conceivably been used to construct the cell's information storage (DNA) and information processing systems (RNA).

Why Phosphate?

Thanks to the pioneering work of biochemist Frank H. Westheimer in 1987, biochemists have a good understanding as to why phosphates appear in the backbone structures of DNA and RNA.[4] Phosphate is perfectly suited to bridge sugar residues together into a chain-like sequence because it can form diesters. In theory, phosphite, sulfate, and arsenate—all close chemical analogs—could have been used in place of phosphate.

Like phosphates, both phosphites and sulfates can form two ester linkages with sugars. This capability allows these compounds, in principle, to serve a bridge role in the backbone of a polynucleotide analog. But the phosphate bridge is unique. When phosphate forms a bond with two sugars to bridge two nucleotides, the phosphate moiety retains a negative charge. On the other hand, phosphites and sulfate bridges both lack a negative charge. This distinction makes a significant difference. The negative charge on the phosphate group imparts stability to the DNA and RNA backbones by warding off hydroxide (HO-) ions (and other negatively charged compounds) through electrostatic repulsion. If not for these repulsive interactions, these materials would attack the diester linkages and cleave the polynucleotide backbone. The negative charge on the phosphate also facilitates protein binding through electrostatic attraction with positively charged amino acids in proteins. These types of interactions wouldn't be possible for phosphite or sulfate bridges. As a consequence,

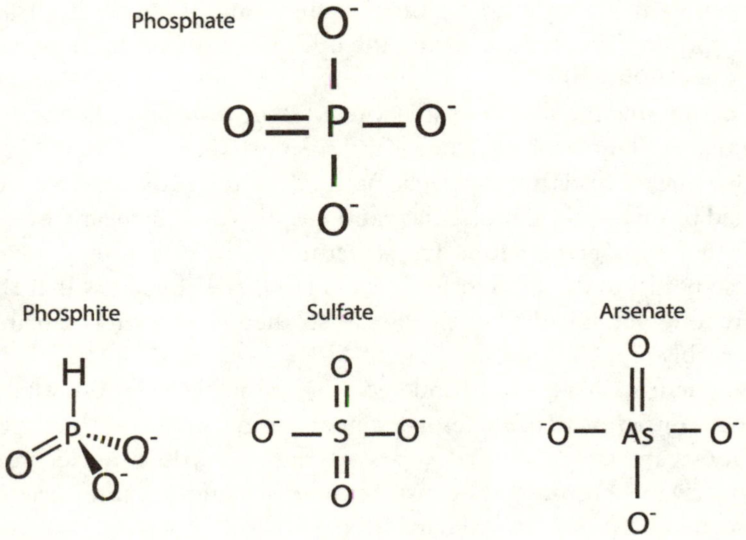

Figure 5.12: Comparison of the Structures of Phosphate with Phosphite, Sulfate, and Arsenate

these two moieties would limit the DNA-protein interactions critical for DNA replication, DNA repair, transcription, and gene regulation. Finally, the negative charge on the phosphate forces the backbone to the exterior surface of the DNA double helix in the aqueous cellular environment. This, in turn, drives the nucleobases into the helix's interior, contributing to the hydrophobic effect, which stabilizes the double helix.

Arsenate is the closest chemical analog to phosphate. In principle, arsenate bridges would also bear a negative charge. However, as a rule of thumb, the bonds arsenate forms with organic materials are unstable, including arsenate esters and diesters. Additionally, the phosphorus in phosphate and the arsenic in arsenate are both in the +5 oxidation states, which is a necessary feature for these two compounds to form esters and diesters. However, arsenic in the +5 oxidation state readily undergoes chemical reduction to arsenic in the +3 oxidation state. In contrast, phosphorus in the +5 oxidation state is far more resistant to the corresponding chemical reduction. In other words, hypothetical nucleic acids formed with arsenate diesters would be much less stable than nucleic acids that incorporate phosphate into their structures.

Despite these differences in stability, in 2010 researchers claimed to have recovered a novel salt-loving microbe (called *Halomonadaceae* GFAJ-1) from Mono Lake in California that could grow on arsenate instead of phosphate.[5] It should be noted that this claim has met with widespread skepticism within the scientific community because of the instability of arsenate esters and diesters. Still, it does open the possibility that arsenate could be used instead of phosphate in the structure of nucleic acids.[6]

In 2012, a research team from Israel carried out a computational analysis to compare nucleic acids with phosphate and arsenate esters.[7] They discovered that nucleic acids made with arsenate esters interacted to a much lesser degree with proteins than nucleic acids made with phosphate esters. These weaker interactions have a wide-ranging impact, frustrating processes such as DNA replication and repair, as well as transcription and the regulation of gene expression. These weaker interactions also compromised the stability of ribosomes, undermining protein synthesis.

Through this comparison, we learn that the phosphate moiety in DNA appears to be uniquely suited for its role in the backbones of DNA and RNA. It also unveils the exquisite molecular rationale for the inclusion of phosphate in the DNA and RNA backbones, though under special conditions, in which phosphate levels in the environment are low and arsenate levels are high, some microbes may be able to substitute arsenate for phosphate.

Why a 5′–3′ Phosphodiester Linkage?

Is there something special about the 5′–3′ linkage in DNA and RNA? To answer this question, we need to focus our attention on RNA. Because deoxyribose lacks an -OH (hydroxyl) group in the 2′ position of deoxyribose, the 5′–3′ linkage is the only possible one for DNA. On the other hand, the phosphodiester linkage that bridges RNA's ribose sugars could involve the 5′ -OH of one ribose molecule with either the 2′ -OH or the 3′ -OH of the adjacent ribose moiety. In nature, RNA makes use of 5′–3′ bonding exclusively, because 5′–3′ linkages impart much greater stability to the RNA molecule than 5′–2′ bonds.[8] This means that the 5′–3′ phosphodiester linkage in both DNA and RNA is a maximally stable bond, providing a rationale for this structural feature. It is fortuitous that deoxyribose has the just-right structural features that permit only the formation of the more stable 5′–3′ linkage.

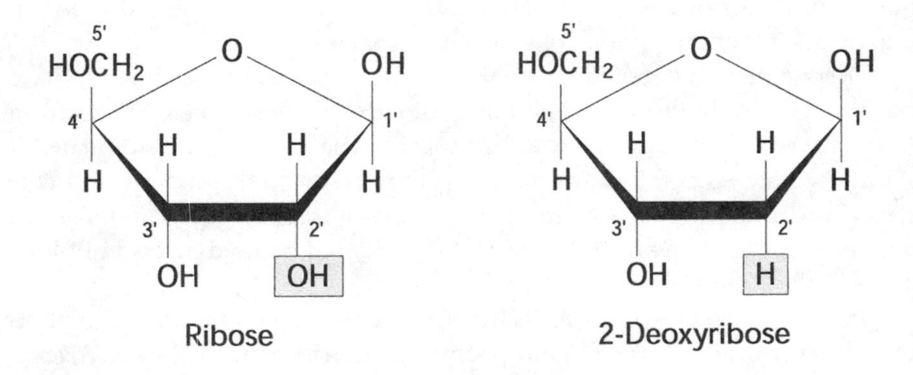

Figure 5.13: Comparison of the Structures of Ribose and Deoxyribose

Why Deoxyribose and Ribose?

Biochemists and molecular biologists have wondered why the backbones of DNA and RNA make use of two different sugars, deoxyribose and ribose, respectively. As noted, the only difference between these two sugars is the presence of a 2′ -OH group in ribose and an absence of this group in deoxyribose. Life scientists who have compared the role these two sugars play now have some explanation for the differences in the DNA and RNA backbones.

Biochemists have good insight into why deoxyribose was selected for DNA and ribose for RNA. Information storage requires molecular stability. But incorporation of ribose into DNA would make the polynucleotide chain inherently unstable. Unlike deoxyribose, ribose possesses a 2′ -OH group, which can catalyze the self-cleavage of DNA's sugar-phosphate backbone.

On the other hand, the self-cleavage of the RNA backbone is not a problem. In fact, some measure of instability is preferable in RNA, making ribose ideally suited as a backbone component in this nucleic acid. To understand why, we need to consider one of the roles RNA plays in the cell. RNA mediates the transfer of information from the nucleotide sequences of DNA to the amino acid sequences of proteins. (The details of the processes associated with RNA's intermediary role in information transfer in the cell are unpacked in the next chapter.) When the cell needs the protein encoded by a particular gene, the cell's machinery copies the information encoded in DNA by making a messenger RNA (mRNA) molecule. Once produced, mRNAs continue to direct

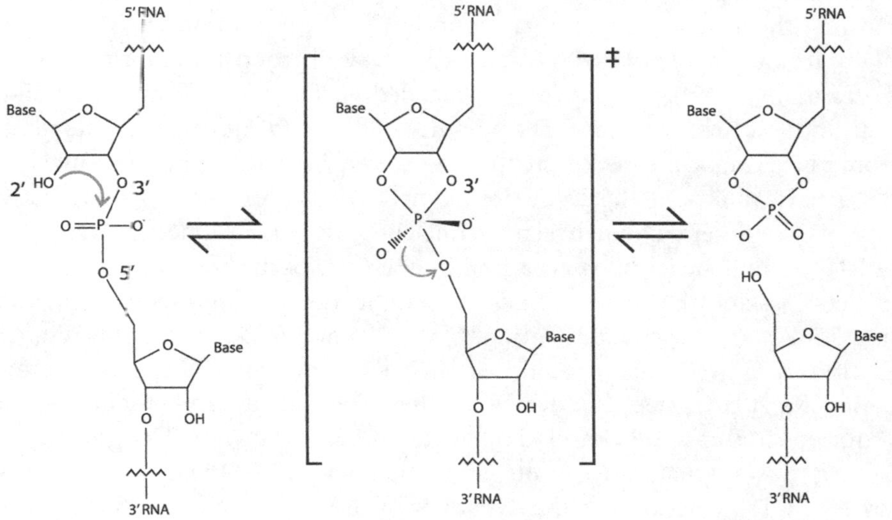

Figure 5.14: 2′ OH Catalyzed Cleavage of the Sugar-Phosphate Backbone

the production of proteins at the ribosome until the cell's machinery breaks down the mRNA molecules. Fortunately, mRNA molecules can exist intact for only a brief period, in part because of the breakdown of the sugar-phosphate backbone mediated by the 2′ OH group. The short lifetime of mRNAs serves the cell well. If mRNAs persisted unduly, then these molecules would continue to direct the production of proteins at the ribosome, producing more copies of the protein than the cell needs and wasting cellular resources. Plus, the excess proteins would clutter the cell's interior.

Biochemists have discovered another reason for the incorporation of deoxyribose in the DNA backbone, instead of ribose. Because deoxyribose lacks an -OH group in the 2′ position, the necessary space within the interior of the DNA double helix becomes available to accommodate the large nucleobases. No other sugar, including ribose, fulfills this requirement.[9]

In 2016, a team of collaborators from several US institutions discovered that deoxyribose provides the DNA backbone with unexpected structural flexibility that appears critical for its role as an information storage system. These researchers also showed that if ribose were to replace deoxyribose in the DNA backbone, the resulting structure would become rigid and inflexible. It turns out that the absence of the -OH group in the 2′ position of deoxyribose explains the flexibility of DNA. Without a 2′ hydroxyl group, the deoxyribose ring freely adopts alternate conformations (called puckering), leading to flexibility to the double helix structure. The presence of the 2′ -OH group prevents ribose from puckering, which would render the double helix rigid and inflexible.[10]

The flexibility of the DNA double helix allows for a type of alternative base-pairing interaction between the nucleobases. As mentioned, in DNA (and in RNA double helices), base-pairing interactions occur between the A and T nucleobases and the G and C nucleobases. Biochemists refer to these interactions as Watson-Crick base pairing. However, in 1959, six years after Francis Crick and James Watson published their structure for DNA, a biochemist named Karst Hoogsteen discovered another way—albeit, rare—that the A and T nucleobases and the G and C nucleobases pair. Known as Hoogsteen base pairing, this alternate way results when the nucleobase attached to the sugar rotates by 180°. Because of the dynamics of the DNA molecule, this nucleobase rotation occurs occasionally, converting a Watson-Crick base pair into a Hoogsteen base pair. However, the same dynamics will eventually revert the Hoogsteen base pair to a Watson-Crick pairing. Hoogsteen base pairs aren't preferred because they distort the DNA double helix. For a "naked" piece of DNA in a test tube, at any point in time, about 1 percent of the base pairs are of the Hoogsteen variety.

Biochemists have recently discovered that the Hoogsteen configuration, while rare in naked DNA, occurs frequently when (1) proteins bind to DNA, (2) DNA is methylated, and (3) DNA is damaged. Biochemists now think that Hoogsteen base pairing is important for maintaining the stability of the DNA double helix, thus ensuring the integrity of the information stored in the DNA molecule.

It looks like the capacity to form Hoogsteen base pairs is a unique property of DNA. Researchers are unable to detect any evidence for Hoogsteen base pairs in double helices made up of two strands of RNA. When they chemically attached a methyl group to the nucleobases of RNA to block the formation of Watson-Crick base pairs and force Hoogsteen base pairing, they discovered that the RNA double helix fell apart. Unlike the DNA double helix—which is flexible—the RNA double helix is rigid and cannot tolerate a distortion to its structure. Instead, the RNA strands can only dissociate.

Several other studies provide additional insight as to why deoxyribose and ribose were selected as the sugar molecules that make up the backbones of

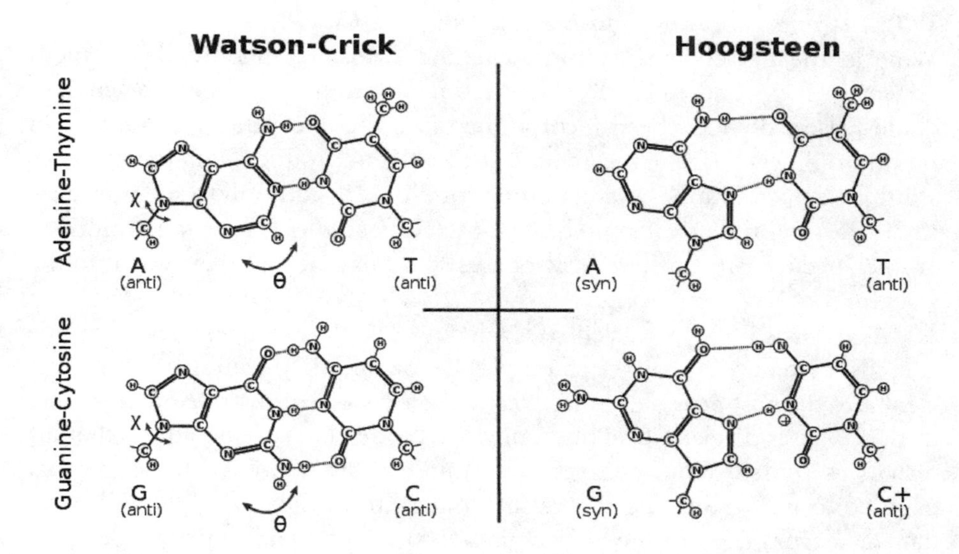

Figure 5.15: Comparison of Watson-Crick and Hoogsteen Base Pairing
Credit: https://commons.wikimedia.org/File:Hoogsteen_Watson_Crick_pairing-en.svg

DNA and RNA. Deoxyribose and ribose are 5-carbon sugars that form 5-membered rings. Researchers have demonstrated that it is possible to make DNA analogs using a wide range of different sugars that contain 4, 5, or 6 carbons that can adopt 5- and 6-membered rings. But they have also shown that these DNA variants have undesirable properties compared to DNA and RNA built with deoxyribose and ribose. For example, some of these DNA analogs don't form double helices. Others do, but the nucleotide strands interact either too strongly or too weakly or display inappropriate selectivity in their associations.[11] Additionally, other studies have shown that DNA analogs made from sugars that form 6-membered rings adopt too many structural conformations.[12] This is untenable. If DNA assumes multiple conformations, it makes it extremely difficult for the cell's machinery to properly execute DNA replication, repair, and transcription.

As in the case for phosphate, impeccable molecular reasoning undergirds the use of deoxyribose and ribose, respectively, in the DNA and RNA backbones.

Why Adenine, Guanine, Cytosine, and Thymine/Uracil?

Why do the nucleobases adenine, guanine, cytosine, and thymine (uracil) appear in DNA and RNA? By some estimates around 16 other nucleobases could reasonably have been incorporated into the nucleic acids. As a case in point, origin-of-life researchers have succeeded in producing diaminopurine, xanthine, hypoxanthine, and diaminopyrimidine, in addition to adenine, guanine, cytosine, and uracil in prebiotic simulation studies.[13] From an evolutionary perspective, any of these nucleobases could have found their way into the structures of DNA and RNA.

As it turns out, the nucleobases adenine, guanine, cytosine, and thymine (uracil) have an optimal set of properties, making this combination of nucleobases the best possible choice. Researchers have demonstrated that this set of nucleobases displays ideal photophysical properties. Compared to other alternatives, these five nucleobases suffer minimal photophysical damage when exposed to ultraviolet (UV) radiation coming from the Sun.[14] Photophysical damage occurs because the nucleobases absorb electromagnetic wavelengths. Yet the destructive effects aren't as bad as they could be. Fortuitously, the nucleobases found in DNA and RNA have a UV radiation absorption maximum that corresponds to the wavelengths most effectively shielded by ozone. Plus, the nucleobases possess structural features that cause the UV radiation to efficiently radiate away as heat after absorption, thus limiting the extent of the damage.

Work by a chemist from Trinity College Dublin in Ireland provides an additional rationale for the incorporation of adenine, guanine, cytosine, and thymine (uracil) into DNA and RNA.[15] This set of nucleobases appears to impart the double helix with a unique structural property that allows the DNA repair machinery of the cell to detect errors when the polynucleotides of the two strands fail to align properly and pair with one another. No other conceivable combination of nucleobases gives the DNA double helix this special quality.

As discussed, when the two DNA strands align, the adenine pairs with thymine and guanine pairs with cytosine. Occasionally, mispairing of the bases can occur. This undesirable event can alter the information stored in DNA, leading to degradation and loss of information. Fortunately, when mispairing occurs, the DNA repair machinery can recognize it because the mispairing distorts the DNA double helix. If not for the deformation of the DNA double helix caused by mispairing, it would be impossible for the cell to reliably harbor information in DNA and successfully transmit that information to the next generation of cells. If any of the other conceivable nucleobases would have been incorporated into DNA, then this error detection property wouldn't exist, because mispairing involving other nucleobases wouldn't cause the double helix to become deformed.

Why Thymine and Not Uracil?

Biochemists also understand why thymine is used in place of uracil in DNA. Uracil is chemically unstable; it undergoes a chemical change (called a deamination) to produce cytosine. Thymine is resistant to that chemical change. Deamination would be problematic for DNA because of its role in information storage. Using thymine instead of uracil staves off this problem. Deamination is not a problem for RNA, which exists transiently inside the cell. As with DNA's other structural features, the selection of nucleobases for DNA makes chemical sense.

Why Uracil and Not Thymine?

So, if thymine is chemically more stable than uracil, then why isn't thymine used in RNA as well? As it turns out, it costs the cell more energy to make thymine (due to the additional methyl group) than it does to make uracil. Because the chemical instability of uracil isn't a problem for RNA, which is short-lived, it makes sense that uracil would be used in place of thymine in RNA molecules.

Why Antiparallel Strands?

Biochemists have come to recognize that even the antiparallel arrangement of the nucleotide strands is an optimal feature of DNA. Researchers have been able to prepare DNA analogs with a parallel orientation of nucleotide strands in the laboratory. By comparing these novel DNA systems with native DNA, it becomes clear that antiparallel alignment of the nucleotide strands leads to greater stability of the DNA double helix than a parallel alignment.[16]

Why Supercoiling?

Even the supercoiling of DNA has an undergirding molecular rationale. When torsional stress is introduced into the DNA backbone and the double helix supercoils, it forms a much more compact structure than a relaxed double helix. These more compact structures allow DNA to store information more efficiently within the limited space of the cell.

A good reason also exists for why DNA, in nature, is most often *negatively* supercoiled. Negative supercoiling arises from an underwinding of the DNA double helix. In this state, the DNA double helix is poised to further unwind, so that its two strands can locally separate from one another during processes such as DNA replication and transcription. An overwound or positively supercoiled DNA double helix would be equally compact as when the DNA double helix is negatively supercoiled, but the two polynucleotide strands would be exceedingly difficult to separate, frustrating DNA replication and transcription.

From this analysis, it becomes evident that an exceptional molecular justification accounts for virtually every structural feature of DNA and RNA. It is also clear that these structural features impart the nucleic acids with a fitness for life. The structural properties contribute to the stability profiles of DNA and RNA and their capacity to bind proteins. Both properties are critical for the role these biomolecules play as an information storage system and an intermediary of biochemical information during transcription and translation, respectively.

This recognition leads to the next question. Are DNA and RNA uniquely suited for their biochemical roles? It seems to me that the answer to this question is yes, based on comparisons of the properties of DNA and RNA, with both hypothetical analogs and biochemical analogs prepared and characterized by researchers in the lab.

But someone who is skeptical about the existence of the biochemical anthropic principle and the uniqueness of DNA and RNA may protest, pointing to work by synthetic biologists and origin-of-life researchers who have prepared artificial, nonnatural versions of DNA.

Artificial DNA and the Biochemical Anthropic Principle

In recent years, it has become vogue for life scientists to seriously entertain the prospect that life as "we don't know it" might exist somewhere else in the solar system or beyond. Part of this speculation includes the possibility that these hypothetical organisms don't use DNA as their genetic material. Instead, as the speculation goes, these alien life-forms utilize a completely different biochemical system to store the information needed to direct life operations and pass that information to offspring.

The interest in alien life has inspired synthetic biologists and origin-of-life investigators to try to create alternative genetic materials, distinct from DNA and RNA. In fact, the quest for nonnatural analogs to DNA and RNA has become a robust and active arena of research. Investigators working in this area hold out the hope that their efforts will lead to novel biochemical systems that can store and transmit information just like RNA and DNA, and yet possess little or no structural relationship to these two nucleic acids. This work has bearing on the applicability of the biochemical anthropic principle to nucleic acids, but it is simply not practical to summarize all the work that has been accomplished. Instead, I am going to focus on work done by Steven Brenner's group. Arguably, no other researcher has done more important work in this area than Brenner, so much so that his name is closely associated with the synthesis of artificial DNAs.

Brenner and his teams have designed variants of DNA with nonnatural nucleobases incorporated into their structures, along with the four naturally occurring nucleobases (A, G, C, and T).[17] These nonnatural nucleobases have base-pair partners just like adenine/thymine and guanine/cytosine. Brenner and his team have shown that when nucleotides containing these nonnatural nucleobases are incorporated into DNA, they don't distort the double helix. In fact, the resulting double helix still retains major and minor grooves.

Even more interestingly, they have demonstrated that laboratory-modified DNA polymerases (see chapter 6, pages 204–208) can use a DNA strand containing nonnatural nucleobases as a template to generate a complementary strand that also includes nonnatural bases. In other words, once they have synthetically made DNA containing the nonnatural nucleobases, these molecules can be replicated using modified DNA polymerases, just like the natural form of DNA. They have also shown that artificial RNAs can be transcribed from these DNAs by a laboratory-altered RNA polymerase.

Oftentimes, DNA polymerases make mistakes when assembling a DNA strand. And Brenner's modified DNA polymerase is no exception. These mistakes alter the DNA sequence and in effect constitute mutations. Because of these types of mutations, the artificial DNAs can evolve. (Many life scientists consider self-replication and evolvability to be two key properties of life.)

Studies such as these indicate to life scientists that DNA and RNA aren't unique, and that it is quite possible that life could exist with different types of nucleic acids—and, possibly, even with biochemical information systems that are nothing like nucleic acids at all. Still, up to this point, the chemical space explored by synthetic biologists and origin-of-life researchers has been limited to regions around the naturally occurring nucleic acids found in biological systems. Most artificial DNAs and RNAs are structural variants of the naturally occurring nucleic acids. Yet recent theoretical studies (executed in 2015 and 2019) indicate that the chemical space for artificial nucleic acids available for synthetic biologists and origin-of-life researchers to explore may well be vaster than anyone imagined.[18] Using a structure-generating algorithm, a large team of collaborators—which included scientists from the Earth-Life Science Institute in Tokyo and the German Aerospace Center—generated a library of nucleic acid analogs. They searched the compounds in the library, looking for ones that had structural features suitable for information storage and transmission. According to one collaborator, Markus Meringer:

> We were surprised by the outcome of this computation. It would be very difficult to estimate a priori that there are more than a million nucleic acid-like scaffolds. Now we know, and we can start looking into testing some of these in the lab.[19]

Because this is a theoretical study that generated a library of hypothetical molecules, it is not clear, at this point, how many of these compounds can actually be made and, of those successfully synthesized, how many will actually be suitable as information storage and transmission systems.

Clearly, the work by synthetic biologists and origin-of-life researchers raises questions about the uniqueness of DNA and RNA, with respect to their capacity to store and transmit biochemical information. But, for several reasons, it would be premature to conclude that DNA and RNA are among millions of compounds that can serve as the genetic material for living systems. First, it should be noted that many of the artificial DNAs created by Brenner and his collaborators (as well as other research teams) over the years are, in effect, modified DNA and RNA molecules. In this sense, it isn't surprising that they share structural and functional attributes with the naturally occurring nucleic acids. It should also be noted that many of the alternative DNAs and RNAs conceived and prepared in the laboratory turn out to be abysmal failures as genetic systems, even when they are close chemical analogs to the naturally occurring nucleic acids. In fact, the synthesis and characterization of these analogs has, in many instances, helped biochemists and molecular biologists to identify the molecular rationale undergirding the structural features of DNA and RNA.

Second, just because some of the artificial DNAs created by Brenner's teams share functional attributes with naturally occurring nucleic acids—as impressive as these compounds might be—doesn't mean that these alternatives share *all* the necessary structural and functional attributes. While we currently don't have the data to conclude if this concern is legitimate, we do have good reason to suspect this might well be the case. For example, we have already seen this to be the case for water. (See chapter 3, pages 74–77.) Scientists have proposed several alternatives to water as the solvent system for life, such as ammonia, because these compounds share *some* of the same physicochemical properties of water. In the end, none of these compounds can replace water because they don't possess *all* the attributes possessed by water that are indispensable if a substance is to serve as the matrix for living systems. Given the structural complexity of DNA and RNA—plus the elaborate nature of the biochemical processes these molecules mediate—it seems reasonable to me that nearly every alternative genetic system will come up short in one or more of the attributes needed for life, when all aspects of nucleic acid biochemistry are considered.

Futhermore, even if these artificial DNA systems do display all the attributes of naturally occurring DNA, they still lack the relative simplicity of the nucleic acids found in living systems. If the added complexity of biomolecules offers no functional advantage to the cell, it stands as an unnecessary cost, requiring an excess of cellular resources and energy to synthesize and maintain.

Finally, prebiotic availability is also a factor. From an evolutionary standpoint, there should be prebiotic routes operating on early Earth to generate the building-block molecules that, in turn, would assemble into information-harboring molecules. Origin-of-life researchers have, indeed, identified plausible prebiotic reactions that could, in principle, yield the components of RNA molecules. However, it is questionable if the necessary prebiotic routes even exist for the building-block components of most of the artificial DNA and RNA molecules. Just because a highly skilled organic chemist can produce these building blocks in a laboratory and assemble them into artificial nucleic acids, doesn't mean that these materials could ever be generated through abiotic chemical processes. In fact, it is a bit eerie to think that origin-of-life investigators have, in fact, discovered plausible prebiotic routes that produce the just-right building-block materials that can assemble into the just-right biomolecules with all the necessary properties for life. In other words, anthropic coincidences seem to extend even to prebiotic chemistry.

DNA: One of the Strangest Molecules in Nature

Whether DNA and RNA are unique remains an open question. What is clear is that a remarkable molecular logic accounts for nearly all their structural features. These features render the nucleic acids highly optimal for their role as biochemical information storage and transmission systems. There is something else about DNA and RNA that often gets overlooked: the nucleic acids are unusual molecules. In fact, evolutionary biologist Simon Conway Morris referred to DNA as the strangest of all molecules.[20]

In some respects, DNA is like water. Because we are so familiar with water, we fail to appreciate how odd and unusual this everyday substance actually is. No other liquid behaves like water. And yet, as discussed in chapter 3, water's odd and unusual properties are precisely what make it uniquely suited for its role as life's solvent system. By the same token, life scientists have become so accustomed to the nucleic acids DNA and RNA that we fail to appreciate how odd and unusual these biomolecules actually are. The very properties which make DNA so strange are the ones that make this molecule so ideally suited for its role as an information storage system.

Simon Conway Morris describes DNA's strangeness this way:

> Consider again the nucleotides. As individual building blocks these molecules are rather unremarkable, yet when they combine to form DNA, they suddenly show some very strange properties indeed. . . .
>
> . . . The action of DNA is almost uncannily effective.[21]

In other words, given the structure and physicochemical properties of the nucleotide building blocks of DNA and RNA, very few biochemists would anticipate that polymers formed from these subunits would display the emergent properties shown by the nucleic acids. One of the most remarkable properties of DNA is its capacity for molecular recognition, arising from the capacity of the nucleobases to pair. This base pairing serves as the positional force that enables DNA's polynucleotide chains to align precisely. The capacity for molecular recognition in DNA and RNA serves as the foundation for processes such as transcription, protein synthesis, and DNA replication. Remarkably, the basis for molecular recognition involves hydrogen-bonding interactions. I doubt that few, if any, chemists would base molecular recognition on hydrogen bonding for chemical systems dissolved in an aqueous environment. Yet hydrogen-bonding interactions between the base pairs can serve in this capacity because of the hydrophobic nature of the nucleobases and the hydrophilicity from alternating negative charges (associated with the phosphate moieties) of the DNA backbone. These two properties force the nucleobases into the DNA's interior, creating a milieu in which hydrogen-bonding interactions between the bases can't be disrupted by water.

The flexible backbone of the polynucleotide chains results in a highly twisted complex when the two polynucleotide chains align, locking the DNA strands in a rigid, inflexible conformation. Yet this conformation provides the precise set of topological properties that allow for molecular recognition between the DNA strands, as well as for protein-DNA interactions needed for transcription and regulation of gene expression, along with DNA replication and repair.

As unusual as these two properties may be, perhaps the most bizarre property displayed by the DNA double helix is its capacity to conduct electrical current through its interior. In other words, DNA functions as a molecular-scale wire.

DNA Wires

Caltech chemist Jacqueline Barton and collaborators discovered the wire property of DNA in the mid-1990s.[22] Barton and her collaborators attached two different chemical groups to each of the ends of the DNA double helix. Both compounds possessed redox centers (metal atoms that can give off and take up electrons). When they blasted one of the redox centers with a pulse of light, it ejected an electron that was taken up by the redox center attached to the opposite end of the DNA molecule, causing the compound to emit a flash of light. The researchers concluded that the ejected electron must have traveled through the interior of the double helix from one redox center to the other. Shortly after this discovery, Barton and her team learned that electrical charges move through DNA only when the double helix is intact. Electrical current won't flow through single-stranded DNA, nor will it flow if the DNA double helix is distorted due to damage or misincorporation of DNA subunits during replication.

These and other observations indicate that the conductance of electrical charge through the DNA molecule stems from pi-pi stacking interactions of the nucleobases in the double helix interior. These interactions produce a molecular orbital that spans the length of the double helix. In effect, the molecular orbital functions like a wire running through DNA's interior.

Does DNA Function as a Wire in the Cell?

While the charge conductance through the DNA double helix is an interesting and potentially useful property, biochemists long wondered if DNA functions as a nanowire in the cell. In 2009, Barton and her team discovered the answer. DNA's capacity to transmit electrical charges along the length of the double helix plays a key role in the DNA repair process. In 2017, Barton's collaborators demonstrated that DNA's wire property also plays an important role in the initiation of DNA replication. Both processes are important for DNA to function as an information storage system. Repairing damage to DNA ensures the integrity of the information it houses and replication makes it possible to pass this information on to the next generation.

Detecting Damage to DNA

As mentioned, damage to DNA distorts the double helix. In a process called base excision repair, the cell's machinery recognizes and removes the damaged portion of the DNA molecule, replacing it with the correct DNA subunits.

For some time, biochemists puzzled over how the DNA repair enzymes located the damaged regions. In the bacterium *E. coli*—a model organism often used by biochemists to study cellular processes—two repair enzymes, dubbed EndoIII (Endonuclease III) and MutY (adenine DNA glycosylase), occur at low levels. Biochemists estimate that fewer than 500 copies of EndoIII and around 30 copies of MutY exist in the cell. These are low numbers considering the task at hand. These repair enzymes bear the responsibility of surveying the *E. coli* genome for damage—a genome that consists of over 4.6 million base pairs (genetic letters).

Barton and her team discovered that the two repair enzymes possess a redox center consisting of an iron-sulfur cluster ($4Fe_4S$) that has no enzymatic activity.[23] They speculated that the $4Fe_4S$ cluster functions just like the compounds they attached to the DNA double helix in their original experiment in the 1990s. Barton and her team were right. These repair proteins bind to DNA. Once bound, they send an electron from the $4Fe_4S$ redox center through the interior of the double helix, establishing a current through the DNA molecule. Once the repair protein loses an electron, it cannot dissociate from the double helix. Other repair proteins bound to the DNA pick up the electrons from the molecule interior at their iron-sulfur redox centers. When they do, they dissociate from the DNA and resume their migration along the double helix. Eventually, the migrating repair protein will bind to the DNA again, sending an electron through the DNA's interior.

This process is repeated. However, if the DNA becomes damaged and the double helix distorted, then the DNA wire breaks, interrupting the flow of electrons. When this happens, the repair proteins remain attached to the DNA close to the location of the damage—thus, initiating the repair process.

Initiating DNA Replication

In 2017, Barton and her team discovered that charge conductance through DNA also plays a critical role in the early stages of DNA replication. DNA replication—the process of generating two "daughter molecules" identical to the "parent" molecule—serves an essential life function.

DNA replication begins at specific sites along the double helix, called replication origins. Typically, prokaryotic cells, such as *E. coli*, have only a single origin of replication. The replication machinery unwinds the DNA double helix at the origin of replication, producing a replication bubble. (See chapter 6, pages 204–208.) Once the individual strands of the DNA double helix become unwound and exposed within the replication bubble, they are available to direct the production of the daughter strand.

Before the newly formed daughter strands can be produced, a small RNA primer must be produced. DNA polymerase—the protein that synthesizes new DNA by reading the parent template strand—can't start production from scratch. It must be primed. The primosome—a massive protein complex that consists of over 15 different proteins (including the enzyme primase)—produces the RNA primer. From there, DNA polymerase takes over and begins synthesizing the daughter DNA strand.

Barton and her team discovered that the handoff between primase and DNA polymerase relies on DNA's wire property.[24] Both primase and DNA polymerase possess $4Fe_4S$ redox clusters. When the primase's $4Fe_4S$ redox center loses an electron, this protein binds to DNA to produce the RNA primer. When the primase's $4Fe_4S$ redox center picks up an electron, the protein detaches from the DNA to end the production of the RNA primer.

When DNA polymerase binds to the DNA to begin the process of daughter strand synthesis, it sends an electron from its $4Fe_4S$ redox center along the double helix formed by the parent DNA-RNA primer. When the electron reaches the $4Fe_4S$ redox center of primase, it brings production of the RNA primer to a halt.

Once again, we discover a purpose to all aspects of DNA's structural properties—a method to the madness. This strange property of DNA is precisely needed for this molecule to serve its role in the cell as an information storage system.

Are the Structures of Nucleic Acids Constrained by the Laws of Physics?

Indeed, DNA is a strange molecule. And it appears as if those properties that make DNA so peculiar are the very properties that make this molecule ideally fit for life. But are these strange properties of DNA (and RNA) constrained and dictated by the laws of nature? Or are they the outworking of a contingent evolutionary history?

At this juncture, there is no clear-cut answer to these questions. Yet some clues hint at the possibility that the structural features of nucleic acids are stipulated by the laws of nature. For example, in 2015, a research team from the University of Oxford published a report that indicates that the laws of nature constrain and dictate the secondary structure of RNA molecules.[25] These investigators sought to identify all the possible secondary structures for RNA molecules with sequence lengths between 20 and 126 subunits. To do this, they first delineated the available sequence space, which is unimaginably vast. If each RNA molecule represented by sequence space were synthesized, the total mass of all the RNA molecules would exceed the known mass of the visible universe. The researchers then mapped RNA secondary structures to the RNA nucleotide sequences. They discovered that a relatively limited number of secondary structures exist. They also discovered that RNA molecules with different sequences adopt the same secondary structure. These two observations indicate that physicochemical constraints have influenced RNA secondary structures in nature far more than evolutionary processes. In other words, the RNA secondary structures found in nature don't appear to reflect a contingent evolutionary history shaped by natural selection. Instead, it appears that constraints imposed by laws of nature shaped RNA secondary structures. We employed the same type of reasoning to conclude that protein folds reflect constraints imposed by the laws of nature rather than the outworking of natural selection.

This interesting result with RNA molecules suggests that similar constraints also may have shaped DNA structure, though, to my knowledge, no study has explored this possibility directly. In the absence of any focused work on this question, efforts to design and synthesize artificial DNAs offer us some insight. In their attempts to create artificial DNAs, synthetic biologists and origin-of-life researchers have discovered key structural features required by biochemical systems if they are to serve as information storage molecules. These structural features serve as a sort of blueprint that guides researchers' efforts to rationally design alternative genetic systems. For example, researchers have discovered that to function as an information storage system, the molecule must be a linear polymer made up of subunits. These subunits must have specific properties that, when combined, yield a polymer with a hydrophilic backbone with an identical repeating charge along its length. These properties ensure formation of a water-soluble polymer with chains that don't clump together, but remain in an extended geometry. The artificial DNAs must consist of a duplex formed by two polymers with complementary sequences. As we will see in the next chapter, this property makes it possible for the cell's machinery to detect any degradation in the information stored in the molecule and makes it possible to recover the lost information. This property also ensures the fidelity of information transmission during replication. The complementarity of the sequences of the two polymer chains in the duplex requires positional forces that are mediated by hydrogen bonding. To achieve these types of interactions in an aqueous system, the nucleobases (or their analogs) must be hydrophobic so that they are driven to the interior of the duplex, sequestered away from water. The chemical groups that mediate the interactions between the two chains must display a structural uniformity to avoid distorting and destabilizing the duplex.

In effect, these structural requirements reflect constraints that, arguably, arise out of the laws of nature. If so, it is remarkable that the constraints imposed on RNA and DNA structure result in molecular systems with the just-right properties required for life.

Does the Biochemical Anthropic Principle Apply to Nucleic Acids?

This chapter sought to determine if the anthropic principle applies to nucleic acids. Through analysis of nucleic acid structure, we discovered that:

- A molecular rationale undergirds the structures of DNA and RNA, which makes the naturally occurring nucleic acids well-suited for life.
- The properties of DNA (and RNA) are odd and unusual, yet those features that make these materials so unusual are precisely the properties that make the nucleic acids well-suited for life.
- The structures of RNA, and possibly DNA, appear to be dictated and constrained by the laws of nature. Again, these constraints yield molecules with the precise properties that make them well-suited to serve in their roles as information storage and transmission systems.
- Based on comparisons to their close chemical analogs, the naturally occurring nucleic acids appear to be unique in their suitability for life. However, the strength of this conclusion is weakened by other work, such as the creation of artificial DNAs in the laboratory and a theoretical survey of the chemical space for nucleic acid-like molecules.

Based on this survey of DNA and RNA structures and comparisons to chemical analogs of naturally occurring nucleic acids, it is evident that the first of our three criteria for a biochemical anthropic principle is easily met. Clearly, an exquisite molecular rationale underlies the structural features of DNA and RNA. As for the second criterion, it appears that the laws of nature constrain RNA secondary and tertiary structures and may well limit the structure of DNA (and other DNA-like information storage systems). Along these lines, it is intriguing that the plausible prebiotic chemical processes identified by origin-of-life researchers suggests that if life emerged through chemical evolution, this process fortuitously led to the production of the ideally suited (if not unique) components of DNA and RNA (third criterion), and rendered irrelevant many possible alternatives to DNA and RNA created in the lab by organic chemists or identified as possible nucleic acid-like analogs through surveys of chemical space. As for the final criterion, it may well be that DNA and RNA uniquely possess the just-right set of properties that make them well-suited for life. Though, to be fair, one could rightly question whether this criterion has actually been met, because of the creation of artificial DNAs in the laboratory.

All in all, it appears that the biochemical anthropic principle applies to nucleic acids, though admittedly, the case is not as compelling as it is for proteins. That is, we can detect the signal for the biochemical anthropic principle for the nucleic acids—it is just not as strong as the signal for proteins. And it is noisier.

In the next chapter, we will continue our query by moving beyond the structural features of proteins and nucleic acids, asking the question: Do we see anthropic coincidences in the processes used by the cell to produce the two classes of biomolecules used in DNA replication and transcription (which produces RNA), and translation (which produces proteins)? As we will see, the metabolic processes involving nucleic acids and the synthesis of proteins are intertwined with each other. For this reason, we expect that the anthropic principle should be evident here as well, given the importance of these two biomolecular classes.

Chapter 6

The Synthesis of Proteins and Nucleic Acids

One of my all-time favorite recording artists is singer-songwriter Johnny Cash. Over the span of his career, which traversed nearly five decades, Cash sold more than 90 million records. His songs bridged several musical genres including country, rock 'n' roll, rockabilly, blues, folk, and gospel. Thanks to his versatility as a performer and recording artist, he won widespread appeal, leading to his induction into the Country Music, Rock and Roll, and Gospel Music Halls of Fame.

Living in West Virginia, I couldn't recall a day when I didn't hear one of Cash's songs on the radio. One of the first Johnny Cash songs that struck a chord with me was his novelty track "A Boy Named Sue." Though Cash made this song popular, it was written by Shel Silverstein for his own 1969 album *Boy Named Sue (And His Other Country Songs)*. Cash recorded the song live at the San Quentin State Prison in California for his 1969 record *At San Quentin*. This song turned out to be one of Cash's most popular recordings, spending three weeks at No. 2 on the Billboard Hot 100 chart.

As the title denotes, the song is about a boy named Sue. He was given that name by a mean and abusive father who abandoned him at three years old. Because of his name, Sue experiences a lifetime of ridicule and mistreatment. The never-ending harassment causes Sue to grow up tough and smart. Sue vows to find his father one day and make him pay for the lifetime of misery he caused by naming him Sue. Sue finally meets his father and comes to learn that he gave Sue his name as an act of love, with the hope it would cause him to grow up to be strong and capable.

As someone with an unusual name, I identified with the teasing and ridicule Sue experienced. Though my father was stern and no-nonsense, he was far from mean and abusive. In fact, he was a loving, devoted father who did everything he could to help me succeed in life. He named me Fazale, a common name in the Islamic world loosely meaning a gift from God. From my father's vantage point, my name was a deep expression of his love for me and the pride he felt for his Muslim heritage. But, in West Virginia, my name singled me out for both good-natured teasing and mean-spirited ridicule.

I dreaded being introduced to new people, too, especially my friends' parents. I was met with the same reaction that usually involved a snide remark and an unkind reference to my family's ethnicity and origin. My name was a constant reminder to everyone that I was the kid whose father was a foreigner, and that our family were outsiders to the community.

Like Sue, my experiences shaped me and made me strong in many ways. But unlike Sue, I took a different tack than fighting with those who made fun of my name (and heritage). I laughed at the hurtful comments, rolling with the punches. I refused to let others' ridicule get to me. Like Sue, I vowed that if I ever had a son, I would name him Bill or George. Anything but Fazale!

As a young kid, I didn't fully grasp Cash's genius as an entertainer. That appreciation came later in life when I first listened to his albums *At Folsom Prison* and *At San Quentin* in their entirety. It is a marvel to behold the way Cash immediately identified with and connected to the audiences at these two prisons. Cash performed at prisons for over a decade before he convinced the decision makers at Columbia House to allow him to record *At Folsom Prison* in 1968.

Cash's motivation to perform at prisons stemmed from his unwavering commitment to live out his Christian faith and honor the teachings of Christ, who instructed his disciples to serve and care for those neglected and forgotten by society, including those in prison.

Many of Cash's songs identify with those who are downtrodden, celebrating their heroism and defiance in the face of unwinnable odds. Another Cash favorite of mine exemplifies this ideal: his 1976 novelty song, "One Piece at a Time" from the album of the same name. This recording was Cash's last song to reach the top of Billboard's country charts. He sang about a poor man who left his home in the foothills of Appalachia (near the place I was raised) to get a job at a Detroit automotive manufacturing plant. As one of the working poor, the song's protagonist can't afford to buy the Cadillacs that he spends his days building on the assembly lines, piece by piece—until he hatches a scheme. Even though he knows it's wrong, he plans on sneaking out the parts of a Cadillac, one piece at a time, with the hope of one day assembling his own car in his garage from the parts he pilfered. But because the parts he took from the factory spanned twenty years, the car he put together was a hodgepodge. It was a "'49, '50, '51, '52, '53, '54, '55, '56, '57, '58, '59, '60, '61, '62, '63, '64, '65, '66, '67, '68, '69, '70 automobile." So, instead of riding around town in style, as he dreamed, he draws the laughter of people on the street as he cruises by in his psychobilly Cadillac.

This song frequently came to mind when I worked in R&D for a Fortune 500 company in the 1990s. Though I was involved mostly in upstream projects, from time to time I would get called to one of our plants to troubleshoot a problem. It was awe-inspiring to watch products being made on the assembly lines—an engineering marvel. And no matter how I tried to shake it from my thoughts, the soundtrack that played in my mind whenever I watched our manufacturing operations was Cash's "One Piece at a Time."

• • • • • • • • • • • •

I also distinctly remember humming and singing "One Piece at a Time" when I studied protein synthesis as both an undergraduate and a graduate student. This song still plays in my mind today when I review the details of this biochemical process—and for good reason. The production of proteins at ribosomes bears a peculiar resemblance to the best, most sophisticated assembly line operations devised by human engineers. So does the replication of DNA and the production of RNA molecules, through the process of transcription. But, far from being a hodgepodge of mismatched car parts, the products of these assembly line operations are elegant, highly optimized biomolecules.

As we saw in the previous two chapters, the structural features of proteins and nucleic acids evince the existence of the biochemical anthropic principle. So, does the synthesis of nucleic acids and proteins display anthropic coincidences as well? This question seems reasonable to ask. Following the reasoning of Michael Denton, I predicted that the anthropic principle would be on full display for proteins and nucleic acids because of the central role that these two biomolecular classes play in living systems. In like manner, it seems reasonable to expect that we would also be able to identify anthropic coincidences in the processes used by cells to synthesize these two classes of molecules: namely, transcription and translation (which constitute the process of protein synthesis) and DNA replication (which is the biochemical process that synthesizes DNA).

This expectation leads us to the central question of this chapter. Do we see anthropic coincidences when we examine the biochemical operations that produce proteins and nucleic acids?

To pursue an answer to this question, we will adopt a similar approach to the one used in previous chapters. Specifically, we will:

- Determine if a molecular rationale undergirds transcription, translation, and DNA replication, accounting for the fitness of these processes for life.
- Determine if there are physicochemical constraints that arise from the laws of nature that dictate the synthesis of proteins and DNA, influencing their fitness for life.
- Compare the features of protein synthesis and DNA replication to alternative processes to determine if these biochemical systems have the appearance of being well-suited for life.

If these three criteria are met, then we have justification to conclude that protein synthesis and DNA replication aren't solely the outworking of contingent evolutionary processes. Instead, their design must be largely influenced by constraints that arise from the laws of nature—constraints that produce biomolecular systems with a suitable set of properties for life.

The processes of transcription, translation, and DNA replication are complex. It is beyond the scope of this book to examine these processes in detail. Instead, we will sample some of the most important aspects of these biochemical operations, with an eye toward identifying features of these processes that evince the existence of a biochemical anthropic principle. Perhaps the best place to begin our survey and evaluation of the three criteria is with the central dogma of molecular biology—the framework that describes the intimate interrelationships between the structure and synthesis of the nucleic acids (DNA and RNA) and the proteins. In many respects, the central dogma of molecular biology could well be considered *the* chief organizing principle in biochemistry. And the central importance of this idea gives us more reason to expect to see the signatures of the biochemical anthropic principle in its design and operation, if such a principle indeed exists.

The Central Dogma of Molecular Biology

Francis Crick (codiscoverer of the DNA double helix) conceived this principle in 1958. As the story goes, he soon came to regret the term. In his autobiographical account, *What Mad Pursuit*, Crick writes:

> I called this idea the central dogma, for two reasons, I suspect. I

> had already used the obvious word hypothesis in the sequence hypothesis, and in addition I wanted to suggest that this new assumption was more central and more powerful. . . .
>
> As it turned out, the use of the word dogma caused almost more trouble than it was worth. Many years later Jacques Monod pointed out to me that I did not appear to understand the correct use of the word dogma, which is a belief *that cannot be doubted*. . . . I used the word in the way I myself thought about it, not as most of the world does, and simply applied it to a grand hypothesis that, however plausible, had little direct experimental support.[1]

Even though Crick came to rue labeling his idea as *dogma*, the term seems to fit, all connotations aside, because of its singular importance to molecular biology.

The central dogma of molecular biology describes the directional flow of information in the cell. As discussed in chapter 5, DNA harbors the information the cell's machinery needs to make proteins. That information is contained within the nucleotide sequences that constitute DNA's primary structure. The information needed to specify a single polypeptide chain is called a gene and it occurs as a contiguous nucleotide sequence along a defined region of the DNA molecule.

According to the central dogma of molecular biology, biochemical information moves from DNA to RNA to proteins. Information can flow from DNA to DNA during DNA replication, from DNA to RNA during transcription, and from RNA back to DNA during reverse transcription. However, biochemical information can't flow from proteins to either RNA or DNA. In other words, informational flow only proceeds in a single direction from nucleic acids to proteins. It can never flow in the reverse direction (as I will soon discuss). The strict directional flow of information arises from the structure of the genetic code (pages 178–181).

Another way to think about the central dogma is that the information stored in DNA is expressed functionally through the activities of proteins. When it is time for the cell's machinery to produce a particular protein, it copies the appropriate information from the DNA molecule through a process called transcription and produces a molecule called messenger RNA (mRNA). Once assembled and processed, mRNA migrates to the ribosome, where it

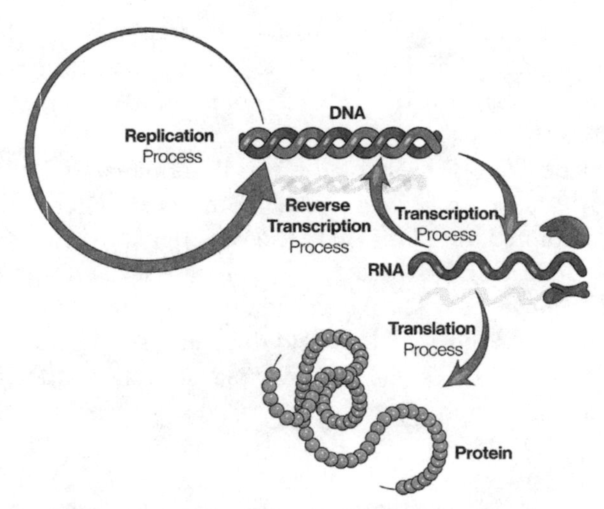

Figure 6.1: The Central Dogma of Molecular Biology

directs the assembly of a polypeptide chain through translation. The process of translation is mediated by two other types of RNA molecules: transfer RNAs (tRNA) and ribosomal RNAs (rRNA).

Chemical Evolution and the Central Dogma of Molecular Biology

In the time since Crick first proposed this principle, many life scientists have regarded the central dogma of molecular biology as the accidental outworking of the evolutionary process that led to the genesis of Earth's first life. In fact, many life scientists regard the central dogma as prima facie evidence for the "RNA world hypothesis" for life's origin. This idea posits that the first biochemistry organized exclusively around RNA, and that only later did evolutionary processes transform the RNA world into the familiar DNA-protein world of contemporary organisms. According to this view, the DNA-protein world is merely the consequence of a historically contingent evolutionary outcome—an accident.

The RNA world hypothesis gained important support in the 1980s with the discovery of a class of RNA molecules called ribozymes. These biomolecules possess functional capabilities, just like enzymes (made up of proteins). For origin-of-life researchers, this insight meant that RNA could not only harbor information (like DNA), it could also carry out biochemical functions like proteins. For this reason, these scientific investigators believe that an entire biochemical system could be built around RNA molecules—that is, an RNA world. Presumably, later in the origin-of-life process, the dual capabilities of RNA biochemistry were apportioned between DNA (information storage) and proteins (function).

Though many people consider the RNA world hypothesis to be the leading model for the origin of life, several significant scientific problems confront this idea. (Hugh Ross and I detail these problems in *Origins of Life*, and I further elaborate on the shortcomings of the RNA world hypothesis in *Creating Life in the Lab.*) These problems have prompted origin-of-life researchers to investigate other approaches to the origin of life, such as metabolism-first and membrane-first models. Still, most origin-of-life investigators think that chemical evolution must have proceeded through an RNA world, for at least two reasons. First, it resolves the so-called chicken-and-egg paradox of which came first, DNA or proteins. As we will see later in this chapter, DNA and proteins are interdependent. While DNA specifies the information used to build proteins, that information can't be expressed without proteins. And DNA cannot replicate, if not for proteins. However, according to RNA world proponents, if life proceeds through an RNA world first, then this paradox finds resolution, because RNA would have assumed the roles of both DNA and proteins. Additionally, they claim that the RNA world hypothesis is the only model that explains the intermediary role RNA plays in protein synthesis. For this reason, many origin-of-life researchers see the intermediary role of RNA in the central dogma as perhaps the most compelling evidence for the RNA world, regardless of the scientific problems confronting this idea. Accordingly, they view RNA's reduced role as a vestige of evolutionary history. In this sense, RNA is viewed as a sort of molecular fossil. Because of the RNA world model, the central dogma of molecular biology came to be considered the happenstance, unintended outcome of chemical evolution that proceeded through an RNA world.

The Biochemical Anthropic Principle and the Central Dogma of Molecular Biology

Considering this perspective on the central dogma of molecular biology, it is hard to envision how protein synthesis could contribute to the case for the biochemical anthropic principle. Yet work published in 2013 by Ian S. Dunn suggests otherwise.[2] Dunn wasn't directly interested in questions related to the origin of the central dogma of molecular biology. He was interested in determining if the use of "molecular alphabets" (to use Dunn's terminology) by terrestrial organisms is a universal requirement for life. In Dunn's scheme, nucleic acids and proteins are biomolecules comprised of molecular alphabets, with nucleotide and amino acid subunits used to build the nucleic acids (DNA and RNA) and proteins.

Chemical Complexity and Life

One of life's defining characteristics is its chemical complexity. In fact, this complexity is required to support the cellular operations fundamental to biology. According to Dunn, this complexity can only be achieved through a large ensemble of macromolecules, each one carrying out a specific task in the cell. To achieve the wide range of required functions, however, the macromolecules must be assembled from molecular alphabets. Because of the combinatorial possibilities, macromolecules built from molecular alphabets possess the range of structural variability necessary for the functional diversity required for life.

Proteins help illustrate Dunn's point regarding combinatorial potential. As discussed in chapter 4, these biomolecules are built from an alphabet that consists of 20 different amino acids. Each protein carries out a specific role in the cell. A typical protein might consist of 300 amino acids. For a protein that size, the number of possible amino acid sequences is 20^{300}, with each one potentially forming a distinct structure and, consequently, performing a distinct function. Dunn argues that it is difficult to imagine how this level of complexity could be achieved with autocatalytic cycles involving small molecules or, alternatively, uniquely specified macromolecules. If Dunn is correct, then it follows that the use of molecular alphabets appears to be a universal property of living systems.

Two Types of Molecular Alphabets

Another defining feature of life is its ability to replicate. For a cell to reproduce, the information that specifies the alphabet sequences of the functional macromolecules must be duplicated so that it can be passed on to daughter cells. Based on this requirement, Dunn identifies a need for both a primary and secondary molecular alphabet.

Macromolecules comprised of a primary molecular alphabet must be able to replicate themselves. This requirement, however, places constraints on the macromolecules and prevents them from carrying out the full range of functional activities needed to support the chemical complexity life requires. To overcome this restriction, a secondary alphabet, specified by the primary alphabet, is needed. The secondary alphabet is not constrained by the need to replicate, so it possesses the full range of functional possibilities.

In other words, if life is to exist, it seems reasonable to think that it must be built from biochemical systems that are organized around molecular alphabets and that those molecular alphabets must be partitioned into distinct classes of biomolecules that harbor a primary and a secondary molecular alphabet.

Molecular Alphabets and the Central Dogma of Molecular Biology

As noted, DNA harbors the information the cell's machinery needs to produce proteins and it possesses the ability to replicate. The nucleotides that make up DNA, therefore, serve as a primary molecular alphabet. The amino acid sequences of proteins comprise a secondary molecular alphabet that enables proteins to serve as the cell's workhorse molecules. In effect, the roles assumed by the primary and secondary molecular alphabets are embodied by the central dogma of molecular biology. For the information from the primary molecular alphabet to be translated to the information in the secondary alphabet, a decoding apparatus is required. In the central dogma, the decoding machinery is comprised of mRNA, along with tRNAs and rRNAs.

The key point is, if Dunn is correct that molecular alphabets are a universal requirement for life, it follows that the central dogma most likely isn't the accidental outcome of chemical evolution. Instead, it reflects an underlying molecular logic dictated by life's requisite chemical complexity. As we saw in the previous chapter, an exquisite molecular rationale undergirds the structure of the nucleic acids. We also learned that DNA appears to be well-suited for its role as an information storage molecule—a macromolecule that consists of a primary molecular alphabet, if you will. Likewise, RNA appears to be fit for its role as an intermediary between the information stored in DNA and the information expressed operationally through proteins, the secondary molecular alphabet. In chapter 4, we also saw that the laws of nature constrain and dictate every level of protein structure, imparting these biomolecules with a unique fitness for their role as workhorse molecules. In other words, for life to exist, it must be built from proteins and, most reasonably, DNA and RNA. And if life is built from these biomolecular classes, according to Dunn's insight, they must interrelate through a process much like the one described by the central dogma of molecular biology. In other words, the central dogma not only displays an exquisite molecular logic, it seems to be a fundamental dictate of the laws of nature. It meets at least two of the three criteria for the biochemical anthropic principle.

As pointed out, the RNA molecule's intermediary role is an essential feature of the central dogma of molecular biology. RNA molecules serve to transmit and translate the information stored in DNA to the information functionally

Molecular Alphabets Undermine the RNA World Hypothesis

Dunn's insight into the universal character of molecular alphabets unwittingly undercuts the RNA world scenario. In the RNA world scenario, the molecular alphabet that comprises RNA is both the primary alphabet and the secondary alphabet. But based on Dunn's work, RNA's need to replicate would have constrained its range of functional capabilities—meaning the ribozymes of the RNA world cannot provide the chemical complexity necessary to sustain life. The central dogma of molecular biology would have to be in place at the point that life originated. Perhaps this restriction explains why origin-of-life researchers have been unable to design or evolve a true self-replicating ribozyme.

expressed by proteins. But for this translation to occur, there must be a well-defined relationship between the nucleotide sequences in the nucleic acids and the amino acid sequences in proteins. To state it differently, a set of rules must exist that specifies the secondary alphabet based on the primary alphabet. And there is. In fact, biochemists have deciphered these rules, dubbing them the genetic code.

The Genetic Code

The importance of the genetic code to the central dogma of molecular biology becomes evident when comparing the molecular alphabet that makes up DNA (which consists of 4 nucleotides) with the molecular alphabet that makes up proteins (which consist of 20 amino acids). A one-to-one relationship cannot exist between the 4 nucleotides of DNA and the 20 amino acids that comprise proteins. Hypothetically, a two-to-one relationship cannot exist either. Combinations of 2 nucleotides can yield only 16 possible coding units ($4^2 = 16$). However, if the coding units consist of combinations of 3 nucleotides, then 64 coding units result ($4^3 = 64$), which is ample to specify 20 amino acids. This is precisely how the genetic code is constructed. The genetic code utilizes groupings of 3 nucleotides to specify the 20 different amino acids used to build proteins. These nucleotide triplets are called codons and represent the fundamental coding units of the genetic code.

In effect, a molecular rationale undergirds the number of nucleotides utilized to construct the basic coding units of the genetic code. Coding triplets stand as the simplest codon structure that can specify 20 different amino acids. (As discussed in chapter 4, there is a molecular rationale undergirding the use of the canonical amino acids. These 20 amino acids form an optimal set, with the just-right range of properties.) A coding quartet of nucleotides could also be used to build a genetic code. This would yield 256 (4^4) codons. Yet using coding quartets of nucleotides introduces unnecessary complexity. It would also require more cellular resources to maintain the information stored in DNA and transmitted via RNA molecules. Assuming a typical protein consists of 300 amino acids, if coding triplets are used to construct the genetic code, the corresponding gene would require 900 nucleotides. On the other hand, 1200 nucleotides would be required to specify the same gene, if coding quartets were used to build the genetic code.

In actuality, of the 64 codons that make up the genetic code, only 61 signify amino acids. Three of the codons serve as stop codons. Because the genetic code needs to encode only 20 amino acids, some of the codons are redundant. That is, different codons can code for the same amino acid. In some instances, up to six different codons specify the same amino acids. There are some amino acids that are specified by only a single codon. As we will soon discuss, this redundancy is one of the most important features of the genetic code.

Table 6.1 depicts the universal genetic code. Based on convention, life scientists present the genetic code based on how the information appears in transcribed mRNA molecules. Accordingly, the first nucleotide of the coding triplet begins at the 5′ end of the sequence. Table 6.1 is arranged so that the nucleotide in each codon's first position (5′ end) can be read from the left-most column and the nucleotide in the second position can be read from the row across the top of the table. The nucleotide in each codon's third position (termed the 3′ end) can be read within each box. For example, the two codons, 5′ UUU and 5′ UUC, which specify phenylalanine (abbreviated Phe), are listed in the box located at the top left corner of table 6.1.

As noted, biochemists designate some codons as stop codons (or nonsense codons). These coding units don't specify any amino acids. (For example, the codon UGA is a stop codon.) These codons always occur at the end of the mRNA molecule to delineate the end of the gene and, hence, mRNA transcript. The stop codon informs the ribosome's protein-manufacturing machinery

Table 6.1: The Genetic Code

5′ End	U	C	A	G
U	UUU Phe UUC Phe UUA Leu UUG Leu	UCU Ser UCC Ser UCA Ser UCG Ser	UAU Tyr UAC Tyr UAA End UAG End	UGU Cys UGC Cys UGA End UGG Trp
C	CUU Leu CUC Leu CUA Leu CUG Leu	CCU Pro CCC Pro CCA Pro CCG Pro	CAU His CAC His CAA Gln CAG Gln	CGU Arg CGC Arg CGA Arg CGG Arg
A	AUU Ile AUC Ile AUA Ile AUG Met(Start)	ACU Thr ACC Thr ACA Thr ACG Thr	AAU Asn AAC Asn AAA Lys AAG Lys	AGU Ser AGC Ser AGA Arg AGG Arg
G	GUU Val GUC Val GUA Val GUG Val(Start)	GCU Ala GCC Ala GCA Ala GCG Ala	GAU Asp GAC Asp GAA Glu GAG Glu	GGU Gly GGC Gly GGA Gly GGG Gly

where the polypeptide chain ends. Some coding triplets play a dual role in the genetic code. These codons not only encode amino acids, but also "tell" the protein-manufacturing machinery where a polypeptide begins. Biochemists refer to them as start codons. For example, the codon GUG not only encodes the amino acid valine, it also specifies the starting point of the polypeptide chain.

To a first approximation, all life on Earth possesses the same genetic code. To say it another way, the genetic code is, in effect, universal. However, there are organisms that possess a genetic code that deviates from the universal code. These codes are referred to as nonuniversal codes. In reality, these codes are variants of the universal code, typically deviating in only 1 or 2 coding assignments. Presumably, these variant codes originated when the universal genetic code evolved, altering coding assignments. The reassignments frequently involved redesignating stop codons to specify an amino acid. Mitochondrial genomes frequently employ variant codes. One of the most common differences between the universal genetic code and the one found in mitochondrial genomes is the reassignment of one of the codons that specifies isoleucine (in the universal code) so that it specifies methionine. (I will discuss the rationale for this reassignment in pages 191–192.) So even though there are "nonuniversal" genetic codes, in my view, it is reasonable to refer to the genetic code in nature as universal.

The Genetic Code and the Biochemical Anthropic Principle

The genetic code resides at the heart of the central dogma of molecular biology, serving as the conduit for information to flow from DNA through RNA to proteins. The genetic code defines the information harbored in DNA. Given the genetic code's centrality, it is reasonable to think that if there is a biochemical anthropic principle, it should also manifest in the design of the genetic code. This expectation leads us to ask: is there any rhyme or reason that explains the genetic code's coding assignments?

Shortly after the rules of the genetic code were deciphered, biochemists began to speculate on the code's origin. One idea that achieved widespread acceptance early on was advanced by Francis Crick in 1968.[3] According to this view, the codon assignments were dictated primarily by chance as the code expanded from a primitive state, in which only a few amino acids were specified, to the contemporary genetic code that specifies 20 amino acids. Once established in the earliest life-forms on Earth, the coding assignments were frozen in place because evolutionary changes to the code would be lethal, explaining the genetic code's universal stature. This is how Crick describes the frozen nature of the genetic code:

> The code is universal because at the present time any change would be lethal, or at least very strongly selected against. This is because in all organisms (with the possible exception of certain viruses) the code determines (by reading the mRNA) the amino acid sequences of so many highly evolved protein molecules that any change to these would be highly disadvantageous unless accompanied by many simultaneous mutations to correct the "mistakes" produced by altering the code.[4]

What about the discovery of nonuniversal codes? Doesn't their existence undermine Crick's point? Not at all. It is true that organisms can tolerate limited evolutionary changes to the genetic code, but only under a highly specific set of circumstances, such as reassignments that involve (1) a stop codon or (2) a codon that occurs at a low frequency within the organism's genome.

But if the reassignment involves most other codons, catastrophe is sure to follow, as Crick rightly argued. It means that every amino acid encoded by the reassigned codon at every position in every protein encoded by the organism's genome will be altered. And while proteins can tolerate some changes in their amino acid sequences, many changes are deleterious. The net effect becomes global disruption of the biochemical processes taking place in the cell. So, Crick appears to be correct. Once the coding assignments in the genetic code became instantiated, wide-scale evolution of the genetic code appears to be unlikely, if not impossible.

This restriction on code evolution has important implications. For if, indeed, the coding assignments reflect a frozen accident, then it is difficult to maintain that the structure of the genetic code makes any contribution to the case for the biochemical anthropic principle. However, simple inspection of the genetic code reveals something unusual about the codon assignments. It doesn't appear that the coding assignments are the haphazard result of a contingent history. Instead, a visual inspection of the genetic code leaves us with the impression that a rationale based on molecular properties undergirds the genetic code's design. Qualitative investigation of the genetic code's assignments indicates that the genetic code has the capacity to minimize the deleterious effects of substitution mutations.

Mutations

A mutation refers to any change that takes place in the DNA nucleotide sequence. DNA can experience several types of mutations. Substitution mutations are one common type. In a substitution mutation, one or more of the nucleotides in the DNA strand is replaced by another nucleotide. For example, an A may be replaced by a G, or a C may be replaced by a T. If the substitution mutation takes place within a gene it will alter the corresponding codon at that point. As a result, the amino acid specified by that codon may or may not be changed (depending on the redundancy of the genetic code). If a different amino acid is specified by the new codon, it holds the potential to alter the structure and function of the protein, particularly if the substituted amino acid possesses physicochemical properties dramatically different from the native amino acid.

Another type of mutation is a frameshift mutation. These mutations are much more devastating than substitution mutations. Frameshift mutations result when nucleotides are inserted into or deleted from the DNA sequence of the gene. If the number of inserted or deleted nucleotides is not divisible by three, then the added or deleted nucleotides cause a shift in the gene's reading frame and alter the codon groupings. Frameshift mutations change all the original codons to new codons at the site of the insertion or deletion and onward toward the end of the gene.

For example, six codons encode the amino acid leucine (Leu). If, at a particular amino acid position in a protein chain, Leu is encoded by 5′ CUU, then substitution mutations in the 3′ position from U to C, A, or G would produce three new codons—5′ CUC, 5′ CUA, and 5′ CUG, respectively—all of which would code for Leu. (See table 6.1.) The net effect leaves the amino acid sequence of the polypeptide unchanged. For this scenario, the cell avoids the negative effects of a substitution mutation. Likewise, a change of C in the 5′ position to a U would generate a new codon, 5′ UUU, which specifies phenylalanine, an amino acid that has physical and chemical properties similar to Leu's.

Table 6.2: The Different Types of Mutations

normal	AUG met	GCC ala	TGC cys	AAA lys	CGC arg	TGG trp
silent	AUG met	GC**T** ↓ ala	TGC cys	AAA lys	CGC arg	TGG trp
nonsense	AUG met	GCC ala	TG**A** ↓ ---	AAA ---	CGC ---	TGG
missense	AUG met	GCC ala	**G**GC ↓ arg	AAA lys	CGC arg	TGG trp
frameshift (deletion -1)	AUG met	GC- ↓	TGC ala	AAA glu	CGC asn	TGG ala
frameshift (insertion +1)	AUG met	GCC **C** ↓ ala	TGC leu	AAA gln	CGC thr	TGG leu
insertion +1 deletion -1	AUG met	GCC **C** ↓ ala	TGC leu	AAA gln	-GC thr	TGG trp

Changing C to an A or to a G would produce codons that code for isoleucine and valine, respectively. As these two amino acids also possess chemical and physical properties similar to leucine, again, the harmful effects of a substitution mutation are sidestepped. As biochemist Stephen J. Freeland describes it:

> The standard genetic code is decidedly non-random in this respect: although redundancy is distributed between most amino acids, synonymous codons (i.e. those that code for the same amino acid) are typically clustered together, sharing two out of the three nucleotides. . . . [There is] a more subtle pattern: that amino acids sharing similar biochemical properties are also clustered within the code, such that even where single point mutations *do* lead to a change in amino acid meaning, they tend to substitute an amino acid similar to that specified by the unmutated codon.[5]

The nonrandom nature of the coding assignments raises suspicions and invites the prospect that the universal genetic code may be unduly fit for its role in mediating the flow of biochemical information. Evidence toward this end comes from several studies in which biochemists have quantitatively compared the error-minimization capacity of the universal genetic code with alternative genetic codes made up of random codon assignments. This work is far from trivial. Information theorist Hubert Yockey has estimated that 10^{70} possible coding assignments exist, and 10^{18} possible coding assignments exist for genetic codes with the same type of redundancy as the universal genetic code.[6] More recent work places the estimate for the number of possible genetic codes at around 10^{84}.[7]

One of the first studies toward this end reported, in 1991, that results indicated the universal genetic code could withstand the potentially harmful effects of substitution mutations better than all but 0.02% of 10,000 randomly generated genetic codes.[8] To arrive at this conclusion, the researchers calculated the effect of changing a codon by a single base for all possible single-base changes in the genetic code separately for the first, second, and third codon positions, using values for an amino acid's polarity, hydropathy (a measure of the side chain's hydrophilic or hydrophobic properties), molecular volume, and isoelectric point (the pH in which the side chain is neutral for acidic and basic R groups). Then, they did the same calculations for randomly generated codes that retained the same level of redundancy as the natural code.

This initial work did not account for the fact that, in nature, some types of substitution mutations occur more frequently than others. For example, an A-to-G substitution occurs more frequently than an A-to-C or an A-to-T mutation. Also, during translation, errors occur more frequently in the first and third codon positions. When researchers incorporated these corrections into their analysis, they discovered that the naturally occurring genetic code performed better than 1,000,000 randomly generated genetic codes and that the genetic code in nature resides near the global optimum for all possible genetic codes with respect to its error-minimization capacity.[9]

According to these results, the genetic code's error-minimization properties appear to be far more dramatic than initially indicated. When the researchers calculated the error-minimization capacity of the one million randomly generated genetic codes, they discovered that the error-minimization values formed a distribution with the naturally occurring genetic code lying outside the distribution. (See table 6.2.) As noted, Yockey estimated the existence of 10^{18} possible genetic codes possessing the same type and degree of redundancy as the universal genetic code. All these codes appear to fall within the error-minimization distribution. This means, of 10^{18} possible genetic codes, few, if any, have an error-minimization capacity that approaches the code found universally throughout nature.

In 2001, a team of biochemists from Belgium extended these two earlier studies by examining the effect of a different parameter on error minimization: the frequency in which the different amino acids occur in proteins. And, instead of using just four amino acid properties, these researchers developed a different measure of error minimization. They examined changes in protein folding behavior (using a standard set of proteins) for all possible point mutations. They discovered that when they added amino acid frequency to the mix, the genetic code that occurs in nature outperforms random genetic codes with the same type and degree of redundancy as the universal genetic code, at a frequency of 2 out of 1 billion.[10]

These studies support the idea that the genetic code appears to be well-suited for its role in decoding information stored within DNA, displaying a remarkable degree of error minimization toward the effects of substitution mutations. But to what extent is the universal genetic code unique in this capacity?

Most biochemists agree that the genetic code appears to be highly nonrandom regarding its error-minimization capacity. But some would argue that the universal genetic code is far from unusual. As a case in point, some modeling studies designed to explore hypothetical evolutionary pathways of the universal genetic code starting from a random code indicate that alternative genetic codes exist that are more robust against the harmful effects of substitution mutations than the code found in nature. To put it another way: it appears as if the genetic code is only partially optimized.[11]

Yet other studies indicate that the genetic code resides near a global

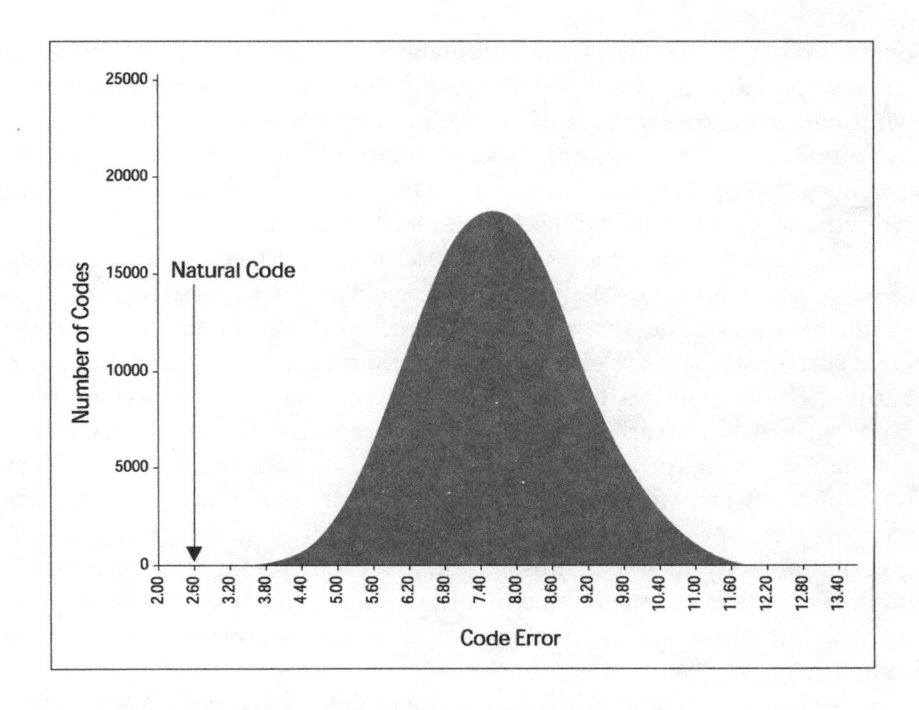

Figure 6.2: Error-Minimization Capacity of the Genetic Code

optimum in an evolutionary sense. For example, a team of collaborators from the University of Bath (in the UK) and Princeton University demonstrated that the genetic code found in nature resides near a global optimum when considering biological constraints that arise out of the relationship between the metabolic routes that make amino acids and codon assignments.[12] Amino acids group into metabolic families. Those that belong to the same family share the same metabolic routes. As it turns out, biosynthetically related amino acids are often assigned to similar codons. These related codons have the same nucleotide in the first position. When this restriction is factored into the comparisons, few alternative genetic codes match the error-minimization capacity of the naturally occurring genetic code.

All these early studies focused exclusively on the error-minimization capacity of the genetic code with respect to substitution mutations. More recently, researchers have discovered that the universal genetic code appears to be optimized to tolerate other types of mutations. For example, researchers from Germany demonstrated that the universal genetic code is highly optimized to withstand errors that result from frameshift mutations. They determined this by comparing the error-minimization capacity of the code found in nature with 1 million random codes, discovering that only a few codes in the random set outperformed the universal genetic code.[13]

With these previous studies serving as a backdrop, a German research team more deeply probed the genetic code's optimality.[14] These researchers focused simultaneously on the optimization of three properties of the genetic code: (1) resistance to harmful effects of substitution mutations, (2) resistance to harmful effects of frameshift mutations, and (3) capacity to support overlapping genes. As with earlier studies, the team assessed the optimality of the naturally occurring genetic code by comparing its performance with sets of random codes that would be conceivable alternatives. For all three property comparisons, they discovered that the natural genetic code displays a high degree of optimality. In gaining this insight, the research team discovered an additional dimension to the optimality of the genetic code that extends beyond its error-minimization capacities: its facility to support overlapping genes. The researchers write, "We find that the SGC's [standard genetic code] optimality is very robust, as no code set with no optimised properties is found. We therefore conclude that the optimality of the SGC is a robust feature across all evolutionary hypotheses."[15]

In addition to possessing the information the cell's machinery requires to synthesize proteins, DNA harbors information for other processes. This information dictates protein binding and is critical for the interaction of DNA with histones (which leads to the formation of chromosomes in eukaryotic cells) and transcription factors (which regulate gene expression). DNA also holds the information that dictates how RNA molecules fold into higher-order structures, once they are transcribed. And DNA holds the information needed for RNA splicing reactions, which also influences gene expression—information that is critical to RNA function. Each set of information requires its own specific code. In other words, DNA harbors multiple layers of overlapping information within its nucleotide sequence that has to be separately decoded for the cell's machinery to access it. As it turns out, when compared to a set of randomly generated codes, the genetic code also appears to be optimized to accommodate these additional codes.[16] This optimization arises out of the capacity of the naturally occurring genetic code to withstand the deleterious effects of frameshift mutations and, specifically, appears to be dictated by the identity of the stop codons. These codons are unusual. When a frameshift occurs, the original coding triplet is often changed into a new coding triplet, which corresponds to a codon that specifies an amino acid, which occurs with high frequency in proteins.

In 2020, investigators from Israel also learned that the genetic code is optimized for resource conservation.[17] This optimization is vital, contributing to life's ability to persist during times when carbon- and nitrogen-containing resources in the environment become depleted.

Carbon and nitrogen hold immense value for living organisms, serving as two of life's key raw materials. In fact, ecologists regard the carbon and nitrogen levels in an ecosystem as limiting factors. In other words, the amount of available carbon and nitrogen in the environment controls an ecosystem's biomass. When the levels of nutrients containing these two elements are low, the ecosystem can't sustain the number of organisms that it can when the levels of carbon- and nitrogen-containing nutrients are high.

Life-forms use carbon and nitrogen atoms to assemble the building-block materials that, in turn, are used to construct important biomolecules such as proteins, the nucleic acids (DNA and RNA), and cell membrane components (such as phospholipids). Additionally, organisms use carbon-containing compounds as a source of energy to power their operations.

Not all biomolecules are created equal. Some require more carbon and nitrogen atoms to assemble than others. Consider proteins as a case in point. These biomolecules are built from amino acids that have been linked together in a chain-like manner by the cell's biosynthetic machinery. To first approximation, the cell's machinery uses twenty chemically and physically distinct amino acids to build proteins. The assembly of each of these amino acids requires differing amounts of carbon and nitrogen atoms, with some necessitating more carbon atoms or, conversely, more nitrogen atoms than others.

Biochemists have learned that the composition of biomolecules is influenced by nutrient availability. Organisms that live in low carbon environments tend to have proteins that are predominantly made up of amino acids relatively rich in nitrogen. On the other hand, biochemists have discovered that organisms that reside in a low nitrogen environment possess proteins that are relatively low in nitrogen-rich amino acids. They've also observed that composition of an organism's genome also reflects the availability of nutrients in the environment. The DNA that makes up an organism's genome consists of linear chains of nucleotides. The cell's biosynthetic apparatus assembles DNA from four nucleotides: adenosine, guanosine, cytidine, and thymidine (abbreviated

A, G, C and T, respectively). Organisms that live in low carbon environments have genomes made up of a higher guanosine and cytidine (G+C) content, both of which have a higher ratio of nitrogen-to-carbon atoms in their molecular makeup. Conversely, organisms that reside in a low nitrogen environment have genomes made up of a higher fraction of adenosine and thymidine (A+T), both of which have a higher ratio of carbon-to-nitrogen in their molecular compositions.

In light of this understanding, the researchers from Israel speculated that the genetic code may be optimized to minimize the excess resource cost that results when one amino acid is replaced with another as a result of a substitution mutation. Toward this end, the research team developed parameters that measured the cost of substituting one amino acid for another based on the number of additional carbon and nitrogen atoms the new amino acid contains compared to the original one. For example, the change of the codon 5′ CCA (which specifies proline) to 5′ CGA (which specifies arginine) results in an amino acid with 1 additional carbon and 3 additional nitrogen atoms for an N cost of 3 and a C cost of 1.

Using this set of parameters, the research team calculated that the expected random mutation cost for the genetic code found in nature is 0.44 for carbon, 0.16 for nitrogen, and 0.16 for oxygen. Comparing these parameters for the genetic code found in nature with 1 million randomly generated codes indicated that the naturally occurring code has a lower expected random mutation cost than all but 128 random codes. They also discovered that this optimization is independent from the optimization that resists the harmful effects of substitution and framework mutations.

The bottom line is that the universal genetic code stands out when compared to randomly generated alternative genetic codes. It displays multidimensional optimality. The genetic code found in nature appears to be highly optimized to tolerate the harmful effects of both substitution and frameshift mutations. It is also optimized to allow DNA to encode overlapping genes and encode multiple sets of information that direct protein binding, as well as RNA processing and folding, once transcribed. So, while there may well exist alternative genetic codes that are better suited than the universal genetic code for a single parameter, when all the parameters are considered in unison, the universal genetic code just might be unique. As biochemists Stefan Wichmann and Zachary Ardern write:

> The idea of trade-offs could be very important in understanding

> the nature of SGC optimality. Every property has a cost versus some other property and assuming that the genetic code had some freedom in its evolution, some trade-offs were plausibly experienced as constraints.[18]

What about the Nonuniversal Genetic Codes?

In addition to making comparisons between the universal genetic code and randomly generated genetic codes, we can also compare the code found in nature with the nonuniversal genetic codes.

As previously noted, nonuniversal codes are best understood as variants of the universal code. Typically, the rules of these deviants are almost identical to the universal genetic code, with only a few minor exceptions. Code deviations almost always occur in relatively small genomes, such as mitochondrial genomes. They involve either (1) reassignments that have a minimal, low frequency impact in that particular genome; or (2) reassignments that involve stop codons.

A comparison of the universal genetic code with deviant codes found in nature indicates that the nonuniversal codes display an error-minimization capacity (with respect to substitution mutations) that was either equivalent to or less effective than that possessed by the universal genetic code.[19] This result affirms the fitness of the universal genetic code.

There may well be a rationale based on molecular principles that undergirds the alternative coding assignments of the nonuniversal codes. For example, in the genetic code of mitochondria, the codon AUA is often reassigned. Instead of specifying isoleucine, AUA denotes methionine. This fortuitous reassignment protects proteins in the inner membrane of mitochondria from oxidative damage.[20] (See chapter 8.) Metabolic reactions that take place in mitochondria during the energy-harvesting process generate high levels of reactive oxygen species (ROS). These highly corrosive compounds damage the lipids and proteins of the mitochondrial inner membranes. The amino acid methionine is also readily oxidized by ROS to form methionine sulfoxide. Once this happens, the enzyme methionine sulfoxide reductase (MSR) reverses the oxidation reaction by reconverting the oxidized amino acid to methionine. This conversion and reconversion protect the inner membrane from oxidative damage.

Let me elaborate. As a consequence of reassigning the isoleucine codon, methionine replaces isoleucine in the proteins encoded by the mitochondrial genome. Many of these proteins reside in the mitochondrial inner membrane. Interestingly, many of the isoleucine residues of the inner mitochondrial membrane proteins are located on the surfaces of the biomolecules. The replacement of isoleucine by methionine has minimal effect on the structure and function of these proteins because these two amino acids possess a similar size, shape, and hydrophobicity. But because methionine can react with ROS to form methionine sulfoxide and then be converted back to methionine by MSR, the mitochondrial inner membrane proteins and lipids are protected from oxidative damage. To put it another way, the codon reassignment results in a highly efficient antioxidant system for mitochondrial inner membranes in organisms at higher risk for oxidative damage.

The discovery of this antioxidant mechanism leads to another question. Why is the codon reassignment not universally found in the mitochondria of all organisms? As it turns out, this codon reassignment occurs in active animals that place a high metabolic demand on mitochondria (and with it, concomitantly elevated production of ROS). This codon reassignment does not occur in Platyhelminthes (flatworms, which live without requiring oxygen) and inactive animals, such as sponges and echinoderms. So, even though the nonuniversal genetic codes are not as optimal as the universal genetic code, an exquisite rationale undergirds the codon reassignment.

In short, a survey of the scientific literature leads to the conclusion that an impeccable logic characterizes the universal genetic code. When compared to alternative codes, its multifaceted optimization makes it ideally suited for its role in biochemical systems. That is, the universal genetic code displays a fitness for purpose, adding to the evidence for the biochemical anthropic principle. Ultimately, the genetic code's optimization arises out of its redundancy. What, then, is the source of the code's redundancy?

Transfer RNAs and the Genetic Code

To understand the source of the genetic code's redundancy, we first need to understand how it is physically instantiated in biochemical systems. Transfer RNA (tRNA) molecules play an important role in this regard, serving as molecular adapters physically linking amino acids to the codons residing in mRNA.

Transfer RNA

Transfer RNA (tRNA), like mRNA, consists of a single RNA strand. Once transcribed, tRNA adopts a precise three-dimensional structure critical for its role as an adapter molecule. As the single tRNA strand folds to form its three-dimensional shape, four segments of the tRNA strand pair. This gives tRNA a clover-leaf shape in two dimensions. The paired regions of the clover leaf, in

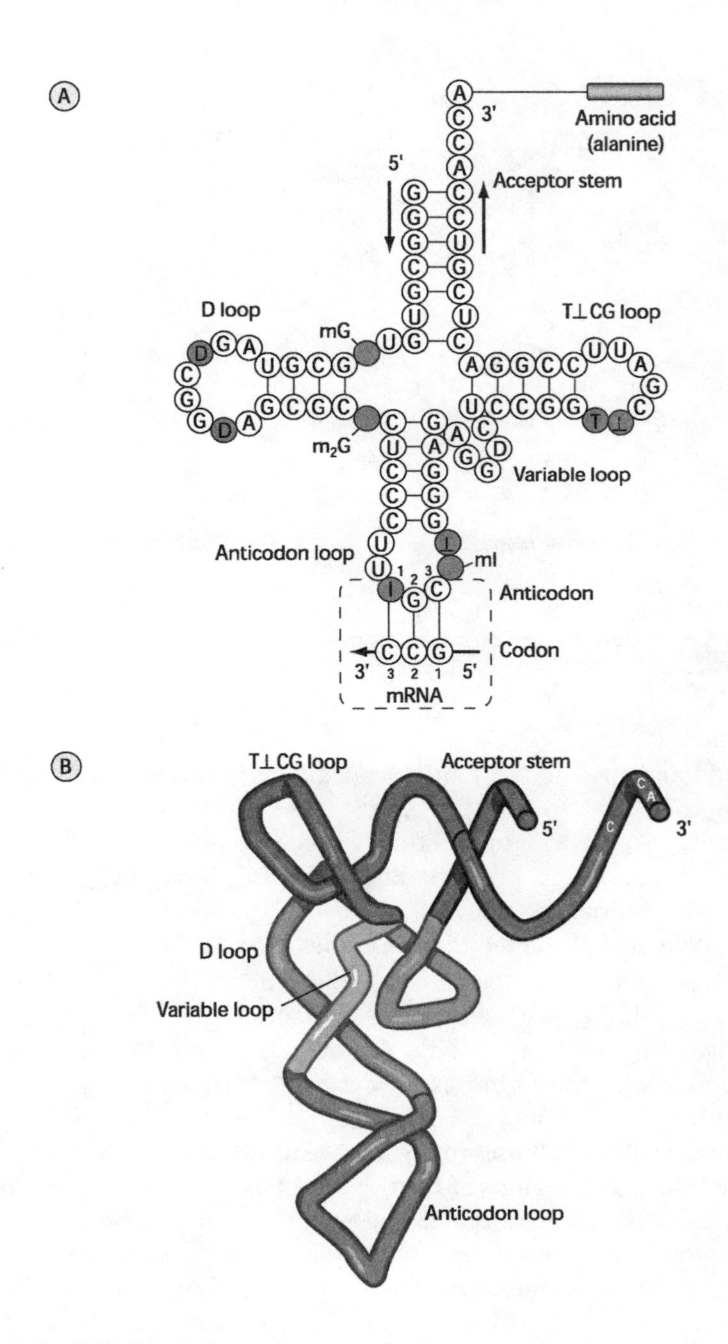

Figure 6.3: tRNA Structure

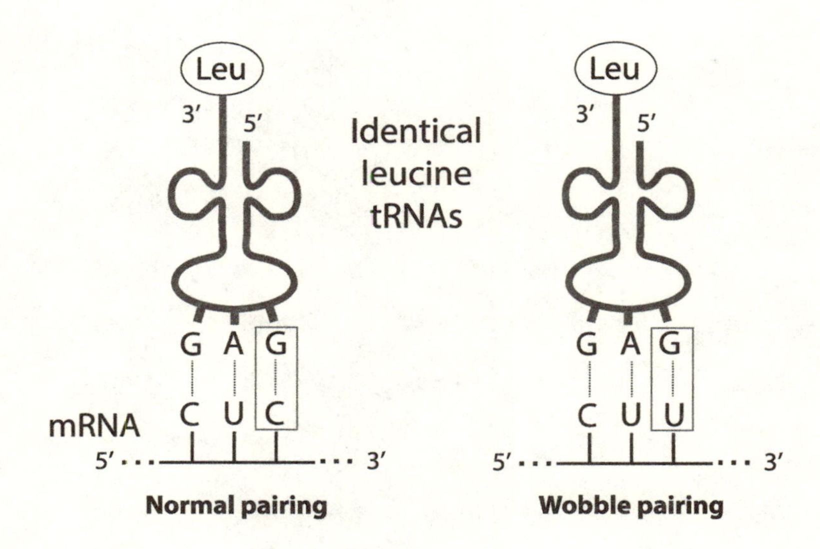

Figure 6.4: Codon-Anticodon Interactions

turn, bend and twist to yield an L-shaped architecture in three-dimensional space.

Like mRNA, tRNA is formed from ribonucleotides that contain either adenine, guanine, cytosine, or uracil. But tRNA also contains some unusual nucleobases in its structure. The cell's machinery introduces these nucleobases into the tRNA structure after it is transcribed. That is, the unusual nucleobases are produced by post-transcriptional modifications. These modifications help to form and stabilize the tRNA tertiary structure.

Transfer RNA molecules bind amino acids and ferry them to the ribosome. This process makes the amino acids available to the protein synthetic machinery. In this way, tRNA molecules serve as adapter molecules. Each of the 20 amino acids that the cell uses to form proteins has at least one corresponding tRNA molecule. Two regions of the tRNA tertiary structure are of exceptional importance to tRNA's adapter role: the acceptor stem and the anticodon loop. The acceptor stem occurs at the 3′ end of the tRNA molecule and serves as the attachment site for amino acids. An enzyme called aminoacyl-tRNA synthetase (sometimes called activating enzymes) links each amino acid to its specific tRNA carrier. Each tRNA and amino acid partnership has a corresponding activating enzyme specific to that pair. The activating enzymes recognize and bind the appropriate amino acid to the appropriate tRNA molecule, in part, through their interactions with the anticodon loop of the tRNA.

As the name implies, the anticodon loop functions as the site on the tRNA molecule that interacts with the codons in mRNA molecules at the ribosome. (See figure 6.4.) The anticodon consists of a three-nucleotide sequence that pairs with the complementary three-nucleotide sequence of the corresponding codon residing in mRNA. Some tRNA molecules harbor an unusual nucleobase in the anticodon loop called inosine, which plays an important role in the wobble hypothesis. (See pages 197–199.) This nucleobase is formed from adenine through a deamination reaction after the primary transcript is produced.

Ribosomes

Ribosomes also play a central role in protein synthesis. These subcellular entities bind and manage the interactions between tRNA and mRNA. They also catalyze the chemical reactions that form the peptide bonds that join amino acids together in polypeptide chains. The ribosome is a massive complex formed from proteins and another type of RNA molecule called ribosomal RNA (rRNA). Ribosomes consist of two subunits dubbed the large (LSU) and small (SSU) subunits. These organelles are dynamic structures that assemble in the presence of mRNA, aided by proteins called initiation factors. Ribosomes disassemble upon the completion of protein synthesis, again, aided by proteins called release factors.

Protein Synthesis

The ribosome, mRNA, and tRNA molecules collaborate to synthesize polypeptide chains (that constitute proteins), one amino acid at a time, making use of an assembly line process. (Cue up Johnny Cash.) This synthetic apparatus can add 3–5 amino acids to the growing polypeptide chain per second. Ribosomes can assemble the cell's smallest proteins, about 100–200 amino acids in length, in under one minute.

At the onset of protein synthesis, the ribosome complex assembles around mRNA. The rRNAs located within the ribosome bind to mRNA, exactly positioning it in the ribosome. The tRNA-amino acid complex that corresponds to the first amino acid position in the polypeptide chain (and the first codon of mRNA) becomes situated in a site in the ribosome called the P site. The tRNA-amino acid complex corresponding to the second amino acid in the polypeptide chain (and the second codon of mRNA) takes its place in an adjacent site,

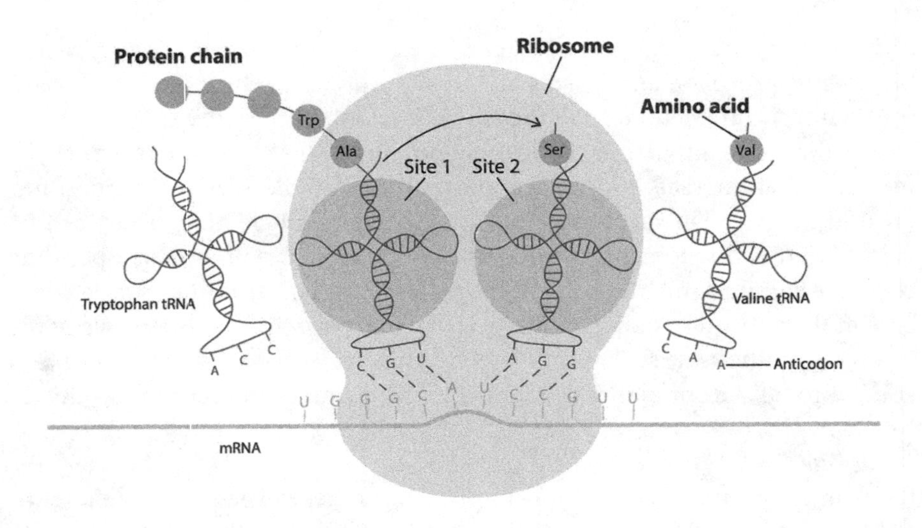

Figure 6.5: Protein Synthesis at the Ribosome

called the A site, located downstream from the P site. The protein synthetic machinery uses the mRNA codon–tRNA anticodon pairing interactions to properly position the tRNA-amino acid complexes.

Once positioned in the A and P sites, a region of rRNA in the LSU (peptidyl transferase) joins together the first and second amino acids in the polypeptide chain through the formation of an amide linkage (chapter 4). After this reaction takes place, the amino acid in the P site dissociates from its tRNA. The tRNA in the P site also exits the ribosome, making itself available to bind another amino acid. The tRNA in the A site, which has the growing polypeptide chain attached to it, moves to the P site. The tRNA-amino acid complex that corresponds to the third position in the polypeptide chain enters the A site and, once again, amide bond formation occurs, followed by tRNA dissociation and the transfer of the tRNA with the nascent polypeptide chain from A to P site. The entire process repeats until all the information in the mRNA is read and the entire polypeptide is synthesized. For each step in this assembly line process, the ribosome complex advances along the mRNA length one codon at a time in a 3′ to 5′ direction.

Transfer RNAs and the Biochemical Anthropic Principle

Because codon-anticodon interactions convert the information harbored in the nucleotide sequences of mRNA (and hence, the nucleotide sequences of the genes located along the length of DNA) into the amino acid sequences of the polypeptide chains synthesized at the ribosome, the tRNA-amino acid adducts represent the physical embodiment of the genetic code. This being the case, it reasonably follows that if the genetic code evinces the biochemical anthropic principle, then it seems reasonable to think that it should manifest in the codon-anticodon interactions that take place at the ribosome. And it appears that this is, indeed, the case. To understand why, we need to examine the nature of the codon-anticodon interactions described by the wobble hypothesis.

The Wobble Hypothesis

To understand the significance of the wobble hypothesis, it is necessary to understand a key observation first made by biochemists in the early 1960s. At that time, these life scientists learned that some tRNA molecules can recognize and bind to more than one codon. Biochemists estimate that only 31 tRNA molecules are needed to recognize the 61 codons that specify amino acids in the genetic code. In 1966, Francis Crick proposed the wobble hypothesis as an explanation for this phenomenon.

Considering the base pair interactions that take place in DNA and RNA, we would expect the codon-anticodon interactions to follow the same base-pairing rules, with each of the three positions in the codon (read in the 5′ to 3′ direction) interacting by forming Watson-Crick base pairs with the complementary positions in the anticodon (read in the 3′ to 5′ direction). According to the wobble hypothesis, however, only the first two codon positions form Watson-Crick base pair interactions with the second and third positions of the anticodon, respectively. The third codon position and the first anticodon position (called the wobble base) interact through a less stringent set of hydrogen-bonding rules that reflect the capability of the wobble base to adopt a range of spatial orientations. Crick referred to these varying spatial positions as "wobble." Contributing to the wobble interactions is the occurrence of inosine in the first position of some anticodon sequences. Inosine can readily form a hydrogen bond with adenine, cytosine, and uracil. In that respect, inosine dramatically expands the range of wobble interactions. Inosine's role as a wobble base provides a rationale for the occurrence of this unusual base in tRNA.

Table 6.3 shows some of the most important wobble rules.

Notably, the capacity of some tRNA molecules to recognize more than one codon, through wobble interactions, is basically responsible for the genetic code's redundancy and helps explain why codons that differ in the third

Table 6.3: Wobble Rules

First Anticodon Base	Third Codon Base
C	G
A	U
U	A or G
G	U or C
I	U, C, or A

position often encode the same amino acid—as in the case for the three codons, 5′ UUU, 5′ UUC, and 5′ UUA, which each specify the amino acid leucine. In other words, wobble interactions offer a physicochemical explanation for the genetic code's remarkable capacity to tolerate the potentially harmful consequences of substitution mutations.

Wobble interactions also restrict the number of possible genetic codes that could conceivably exist in nature. Even though information theorist Hubert Yockey estimated that hypothetically 10^{70} possible alternative genetic codes exist, in reality only 10^{18} are possible—because of the wobble interactions. These interactions dictate the block structure of the coding assignments, which serve in part to limit the number of possible genetic codes. The only way to escape the block structure of the genetic code would be to come up with a different set of wobble rules. But this would require changing the laws and constants of nature. To put it another way, the wobble interactions are predetermined and constrained by the laws and constants of nature, including those that influence the strength of hydrogen-bonding interactions. In this way, the redundancy of the genetic code is specified, in part, by the fundamental design of the universe.

The wobble interactions are also possible because of the incorporation of the unusual base inosine into the anticodon loop. While this structural feature of tRNA molecules could be explained through contingent evolutionary events, there appears to be more to the story. It is quite fortuitous—and a bit surprising—to think that inosine can be so easily derived by simply deaminating adenosine, given inosine's just-right physicochemical properties that make it so versatile as a wobble base. Intuitively, it would be much more reasonable to think that the deamination product of adenosine would be disruptive to codon-anticodon interactions. Instead, it makes redundancy possible in the genetic code, which imparts an optimality to the universal genetic code.

In the end, the wobble interactions, which dictate the nature and the degree of the genetic code's redundancy, which in turn contributes to the multidimensional optimality of the genetic code, ultimately arise out of the laws of nature. It is also true that the optimality of the genetic code arises from the relationship between the tRNAs and their amino acid partners (called cognates). And, as is the case for inosine's occurrence in tRNAs, this relationship could be explained as an outworking of evolutionary contingencies. But, once again, it is suspicious to think that the relationships between tRNA molecules and their amino acid cognates would collectively yield a genetic code that, fortunately, makes the universal genetic code so fit for its role in mediating the translation of biochemical information.

Ribosomes and the Biochemical Anthropic Principle

One of the most important stages of protein synthesis happens at the ribosome. This is where (1) the interactions between the codons of mRNA and the anticodons of charged tRNA molecules dictate the amino acid sequences of proteins, and (2) the peptide bond is formed between amino acid subunits. As mentioned, peptidyl transferase catalyzes the formation of the peptide bonds. This biocatalyst is not a protein enzyme but an rRNA molecule, making peptidyl transferase a ribozyme.

It is hard to imagine a biomolecular machine more important than the ribosome, because of the central roles proteins play in biochemical systems. Because of the ribosome's importance, it is tempting to think that the biochemical anthropic principle would also manifest in the structure and function of this organelle. However, many life scientists would reject this possibility out of hand because of the widespread view that the ribosome stands as the historically contingent product of an evolutionary origin-of-life process that, at some point, proceeded through an RNA world. Accordingly, we wouldn't necessarily expect an underlying rationale for ribosome structure, other than its emergence as the happenstance of a contingent evolutionary history.

Ribosomes and the RNA World Hypothesis

As noted, origin-of-life researchers are quick to point to RNA's intermediary role in protein synthesis as evidence for the RNA world hypothesis. Again, for these investigators, RNA's reduced role in contemporary biochemical systems stands as a vestige of evolutionary history. They view RNA as a molecular fossil and a prime illustration of the RNA world's legacy in contemporary biochemistry. To review, two subunits of different sizes (comprised of proteins and RNA molecules) combine to form a functional ribosome. In organisms such as bacteria, the large subunit (LSU) contains 2 ribosomal RNA (rRNA) molecules and about 30 different protein molecules. The small subunit (SSU) consists of a single rRNA molecule and about 20 proteins. In more complex organisms, the LSU is formed by 3 rRNA molecules that combine with around 50 distinct proteins, and the SSU consists of a single rRNA molecule and over 30 different proteins. Ribosomal proteins are distinct from most other proteins in the cell, being unusually small in size and short but relatively uniform in length.

On the other hand, rRNA molecules are massive in size, functioning as scaffolding and organizing the myriad ribosomal proteins. As noted, the rRNA peptidyl transferase also catalyzes the formation of peptide bonds between amino acids. Again, the ribosome can be understood to be a ribozyme, in essence.

At the ISSOL (International Society for the Study of the Origin of Life) 2002 meeting, I heard origin-of-life researcher Leslie Orgel insist that the RNA world hypothesis must be valid because rRNA catalyzes protein bond formation. Orgel's perspective (and that of other origin-of-life investigators) gains support considering the inefficiency of ribozymes as catalysts. Protein enzymes are better suited for catalyzing reactions than ribozymes. In other words, it would seem to be better and more efficient to design ribosomes so that proteins, and not rRNA, catalyze bond formation between amino acids. This reason convinces origin-of-life researchers that the role rRNAs play in protein synthesis is little more than a relic of life's contingent evolutionary history.

Is There a Rationale for Ribosome Structure?

So, are ribosomes primarily the product of a historically contingent evolutionary process? Or has a logic undergirded by molecular principles largely dictated ribosome structure? Work reported in 2017 by researchers from Harvard University and Uppsala University in Sweden, provides key insight into the compositional makeup of ribosomes and, in doing so, helps answer these questions.[21] The Harvard and Uppsala investigators tried to answer several questions related to the composition of ribosomes, including:

1. Why are ribosomes made up of so many proteins?

2. Why are all ribosomal proteins nearly the same size?
3. Why are ribosomal proteins smaller than typical proteins?
4. Why are ribosomes made up of so few rRNA molecules?
5. Why are rRNA molecules so large?
6. Why do ribosomes employ rRNA, instead of much more efficient proteins, as the catalyst to form bonds between amino acids?

Is There a Rationale for Ribosome Composition?

To appreciate the work of the scientists from Harvard and Uppsala, it is important to keep in mind that before a cell can replicate, ribosomes must manufacture the proteins needed to form more ribosomes. In fact, the cell's machinery needs to manufacture enough ribosomes to form a full complement of these subcellular complexes. This ensures that both daughter cells have the sufficient number of protein-manufacturing machines to thrive once the cell division process is completed. Because of this constraint, cell replication cannot proceed until a duplicate population of ribosomes is produced.

Using mathematical modeling, the Harvard and Uppsala investigators discovered that if ribosomal proteins were larger, or if these biomolecules were variable in size, ribosome production would be slow and inefficient. Building ribosomes with smaller, uniform-size proteins makes it possible to more rapidly duplicate the ribosome population, permitting cell replication to proceed in a timely manner. These researchers also learned that if the ribosomal proteins were any shorter, then inefficient ribosome production would result. This inefficiency stems from the biochemical events needed to initiate protein production. If proteins are too short, then the initiation events take longer than the elongation processes that build the protein chains.

The bottom line: the mathematical modeling work by the Harvard and Uppsala research team indicates that the sizes of ribosomal proteins are optimal to ensure the most rapid and efficient production of ribosomes. The mathematical modeling also determined that the optimal number of ribosomal proteins runs between 50 to 80—matching the number of ribosomal proteins biochemists find in nature.

As part of their mathematical modeling study, these researchers also provided an explanation for why ribosomes are made up of such large RNA molecules. Because the number of steps involved in rRNA production is fewer than the steps required for protein manufacture, rRNA molecules can be made more rapidly than proteins. This being the case, ribosome production is more efficient when these organelles are built using fewer and larger rRNA molecules as opposed to numerous smaller ones. In effect, ribosomes containing more rRNA can be built faster than ribosomes made up of more proteins. This fact also helps explain why rRNA operates as the catalytic portion of ribosomes (linking amino acids together to construct proteins), though it's less efficient as a catalyst than proteins.

These insights also explain the compositional differences among ribosomes found in bacteria, eukaryotic cells, and mitochondria. Bacteria, which typically replicate faster than eukaryotic cells, possess ribosomes that contain proportionally more rRNA and fewer proteins than ribosomes found in eukaryotic cells. Mitochondria—organelles found in eukaryotic cells—possess ribosomes with a much greater ratio of proteins to rRNA than the eukaryotic cells that house these organelles. This observation makes sense because ribosomes in mitochondria don't produce themselves. Instead, they are produced by the cell's machinery found in the cytoplasm.

Ribosome Composition Is Optimal for the Energetics of the Cellular Economy

In 2018, a team from the University of Michigan offered more insight into the molecular rationale for ribosome composition.[22] These researchers point out that around 30 percent of the overall number of proteins produced by the cell are those that make up the ribosome. To put it another way, the cell spends a significant fraction of its energy budget on making ribosomal proteins and, hence, on ribosomes. These investigators discovered that utilizing shorter, smaller proteins of approximately uniform size for ribosomes minimized the synthetic costs for the cell by reducing the number of ribosomal proteins that would be mistranslated and would have to be degraded. The longer a protein, the more likely that an error can occur during its production at the ribosome. Avoiding mistranslation also avoids the production of defective proteins that could potentially escape degradation and, in turn, cause dysfunction when incorporated into ribosomes. Biochemists have learned that large complexes comprised of many proteins, such as the ribosome, don't tolerate mistranslations because of the large number of protein-protein interactions necessary to support their structure and function. These "economic" benefits become increasingly pronounced for proteins produced at high levels, which is the case for ribosomal proteins. In other words, the composition of ribosomes reflects an optimal design that contributes to the efficiency of cellular energetics.

These researchers also provided another explanation for why rRNAs are so large. They determined—just as the investigators from Harvard and Uppsala did—that rRNAs are produced more rapidly and with a much greater degree of fidelity than ribosomal proteins. They estimate that the production of rRNA molecules displays an error rate 10,000 times lower than the error rate for protein synthesis. They also learned that producing fewer and larger rRNA molecules makes it easier for the cell to establish and maintain the appropriate compositional balance among ribosomal proteins and rRNAs, which ensures orderly and accurate ribosome assembly.

Ribosome Composition Is Optimal to Produce a Varied Population

A third study offers another reason to think that ribosome composition is optimal. Biochemists long thought that the ribosome population is homogeneous. In 2017, they learned that a heterogeneous—not a homogeneous—population of ribosomes exists within cells.[23] Instead of every ribosome in the cell being identical, capable of producing each and every protein the cell needs, a diverse ensemble of distinct ribosomes exists in the cell. Each type of ribosome manufactures characteristically distinct types of proteins. Typically, ribosomes produce proteins that work in conjunction with one another to carry out related cellular functions. The heterogeneous makeup of ribosomes contributes to the overall efficiency of protein production and provides an important means to regulate protein synthesis. It wouldn't make sense to use an assembly line to make disparate consumer products, such as antiperspirant sticks and Cadillacs. In the same manner, it doesn't make sense to use the same ribosomes to make the myriad proteins performing different functions for the cell.

Because ribosomes consist of many small proteins, the cell can efficiently produce heterogeneous populations of ribosomes by assembling a ribosomal core and then including and excluding specific ribosomal proteins to generate a diverse population of ribosomes.[24] In other words, the protein composition of ribosomes is optimized to efficiently replicate a diverse population of these subcellular particles.

Ribosomes have an unexpected and unusual composition that, for many biochemists, can largely be explained as the random outcome of the transition from the RNA world to the DNA-protein world. But several studies indicate that the compositional features of ribosomes display a remarkable molecular logic that accounts for the unusual size and distribution of ribosomal proteins and the large size of rRNA molecules. These studies also provide a rationale for why RNA (ribozyme), instead of a protein (enzyme), catalyzes peptide bond formation during protein synthesis, though this is counterintuitive at first blush. Even though ribosomes are unusual, it is their unusual properties that make these organelles nicely suited for their role in protein synthesis. Their optimal composition also raises the possibility that these subcellular particles are uniquely fit for their role in the cell.

So far, our survey of the key stages of protein synthesis lends support to the biochemical anthropic principle. A careful examination of the central dogma of molecular biology—perhaps the organizing principle for all of biochemistry—and a sampling of some of the most important stages in protein synthesis reveal several anthropic coincidences and fitness for purpose. Far from reflecting the historically contingent consequences of an evolutionary history, a deep-seated molecular rationale undergirds: (1) the central dogma of molecular biology, (2) the coding assignments of the genetic code, (3) the codon-anticodon interactions between mRNA and tRNA at the ribosome, and (4) the compositional makeup of ribosomes. The highly optimal nature of the genetic code and ribosomal composition indicates that these two biochemical systems display an unexpected and extraordinary—if not unique—fitness for life. The wobble interactions between codons and anticodons—the physical instantiation of the genetic code—prescribe the character and degree of the genetic code's redundancy, which in turn produces the multidimensional optimality of the genetic code. Ultimately, the wobble interactions arise out of the laws of nature, at least in part. The three criteria laid out for the biochemical anthropic principle are satisfied.

But what about the process that synthesizes DNA? Does it evince the biochemical anthropic principle? To answer this question, we first need to summarize the process of DNA replication.

DNA Replication

Biochemists refer to DNA replication as a "template-directed, semiconservative process." *Template-directed* means that the parent DNA molecule's nucleotide sequences function as a template for the assembly of the two daughter molecules' DNA strands, using Watson-Crick base-pairing rules. *Semiconservative* means that after replication, each daughter DNA molecule contains one newly formed DNA strand and one strand from the parent molecule.

Conceptually, template-directed, semiconservative DNA replication entails the separation of the parent DNA double helix into two single strands. By using the base-pairing rules, each strand serves as a template for the cell's machinery to use when it forms a new DNA strand with a nucleotide sequence complementary to the parent strand. Because each strand of the parent DNA molecule directs the production of a new DNA strand, two daughter molecules result. Each one possesses an original strand from the parent molecule and a

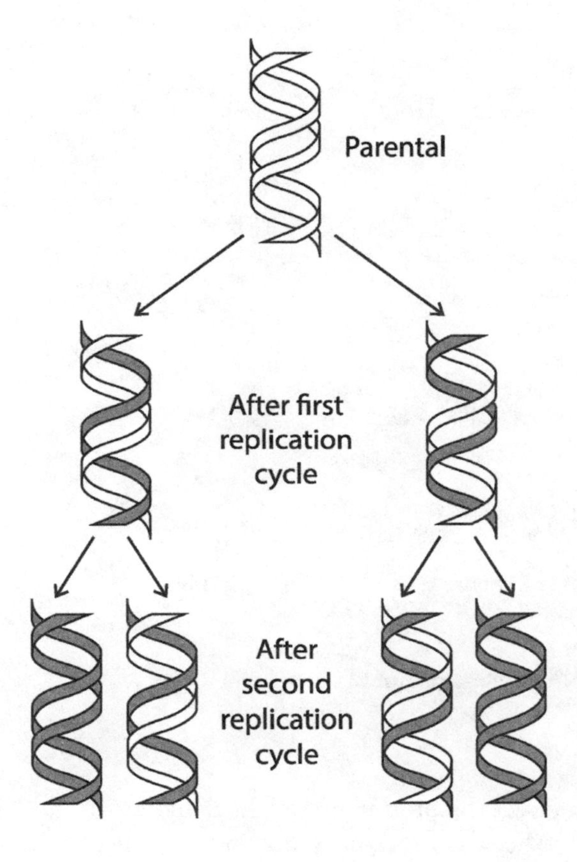

Figure 6.6: Semiconservative DNA Replication

newly formed DNA strand produced by a template-directed synthetic process.

DNA replication begins at specific sites along the DNA double helix, called replication origins. Typically, prokaryotic cells have only a single origin of replication. More complex eukaryotic cells have multiple origins of replication. The DNA double helix unwinds locally at the origin of replication to produce what biochemists call a replication bubble. During replication, the bubble expands in both directions from the origin. Once the individual strands of the DNA double helix unwind and are exposed within the replication bubble, they

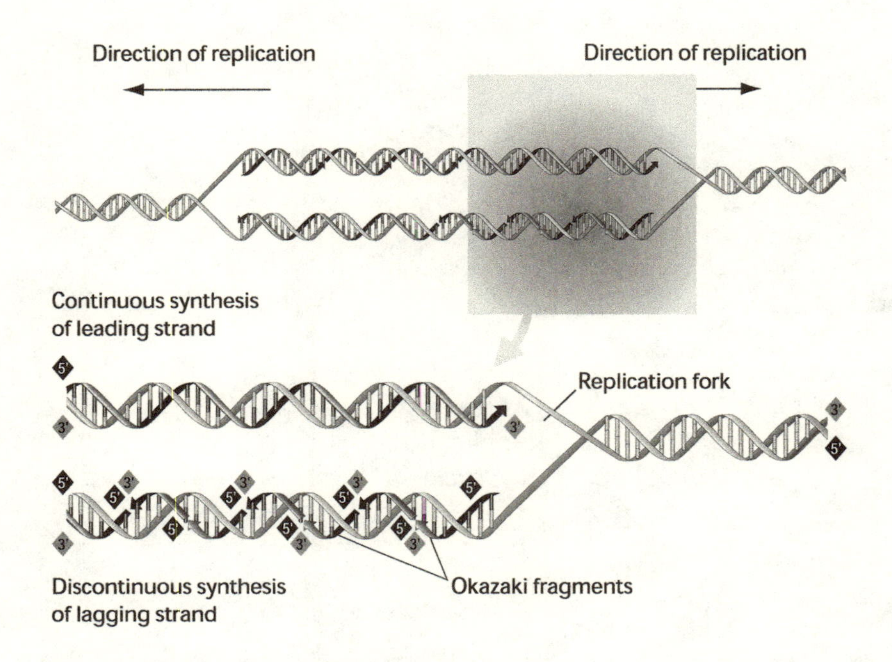

Figure 6.7: DNA Replication Bubble

are available to direct the production of the daughter strands. The site where the DNA double helix continuously unwinds is called the replication fork. Because DNA replication proceeds in both directions away from the origin, there are two replication forks within each bubble.

DNA replication proceeds in a single direction, from the 5′ to 3′ direction. (The template strand is oriented in the 3′ to 5′ direction.) Because the strands that form the DNA double helix align in an antiparallel fashion, only one strand at each replication fork has the proper orientation to direct the assembly of a new strand in 5′ to 3′ direction. For this strand—referred to as the leading strand—DNA replication proceeds rapidly and continuously in the direction of the advancing replication fork.

DNA replication does not proceed along the strand with the 5′ to 3′ orientation until the replication bubble has expanded enough to expose a sizable stretch of DNA. When this happens, replication moves away from the advancing replication fork. Replication can proceed only a short distance along the 5′ to 3′ parent strand before the process has to stop and wait for more of the

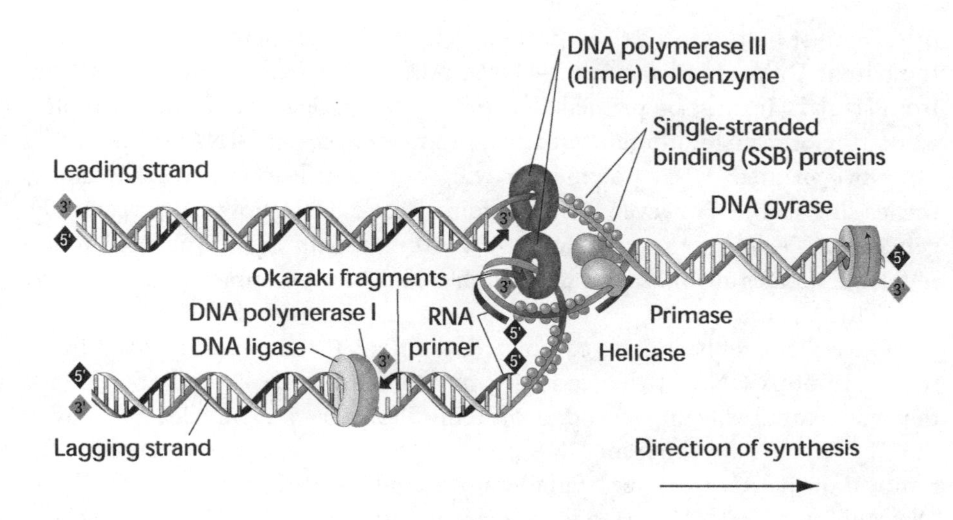

Figure 6.8: DNA Replication Proteins

parent DNA strand to be exposed. When a sufficient length of the parent DNA template is exposed, DNA replication can proceed again, but only briefly before it has to stop again and wait for more DNA to be exposed. The process of discontinuous DNA replication takes place repeatedly, piece by piece, until the entire strand is replicated. Each time DNA replication starts and stops, a small fragment of DNA is produced.

Biochemists refer to these pieces of DNA (that will eventually compose the daughter strand) as Okazaki fragments, named after the biochemist who discovered them. Biochemists call the strand produced discontinuously the lagging strand because DNA replication for this strand lags behind the more rapidly produced leading strand. One additional point: the leading strand at one replication fork is the lagging strand at the other replication fork, since the replication forks at the two ends of the replication bubble advance in opposite directions.

An ensemble of proteins is needed to carry out DNA replication. Once the origin recognition complex (which consists of several different proteins) identifies the replication origin, a protein called helicase unwinds the DNA double helix to form the replication fork.

Once the replication fork is established and stabilized, DNA replication can begin. Before the newly formed daughter strands can be produced, a small RNA primer must be produced. The protein that synthesizes new DNA by reading the parent DNA template strand—DNA polymerase—can't start production from scratch. It must be primed. The primosome, a massive protein complex consisting of over 15 different proteins, produces necessary RNA primer.

Once primed, DNA polymerase will continuously produce DNA along the leading strand. However, for the lagging strand, DNA polymerase can only generate DNA in spurts to produce Okazaki fragments. Each time DNA polymerase generates an Okazaki fragment, the primosome complex must produce a new RNA primer.

Once DNA replication is completed, the RNA primers are removed from the continuous DNA of the leading strand and from the Okazaki fragments that make up the lagging strand. A protein called a 3′–5′ exonuclease removes the RNA primers. A different DNA polymerase fills in the gaps created by the removal of the RNA primers. Finally, a protein called a ligase connects all the Okazaki fragments together to form a continuous piece of DNA out of the lagging strand.

DNA Replication and the Biochemical Anthropic Principle

This cursory description of DNA replication illustrates the complexity of this biochemical operation. (Many details of the process were left out of the discussion.) This description also exposes the reason that many life scientists are wont to view this DNA replication as arising as a chance outcome of a historically contingent process, cobbled together from the biochemical leftovers of the RNA world. DNA replication appears to be cumbersome and unwieldy. There seems to be no obvious reason why replication proceeds as a semiconservative, RNA primer-dependent, unidirectional process involving leading and lagging strands to produce daughter molecules.

Yet, as we will see, careful consideration of the process of DNA replication exposes an impressive molecular logic that undergirds the process. There are good reasons why DNA replication is a semiconservative process. There are good reasons why DNA replication proceeds in a single direction, despite the fact that it adds to the complexity of the process by creating the need for leading and lagging strands. There are also sound reasons why an RNA primer is utilized to kick-start replication. And tantalizing clues that suggest that the process of DNA replication may be fundamentally dictated by the laws of nature.

Why Semiconservative, Template-Directed Replication?

For the information in DNA to be duplicated, the strands of DNA must serve

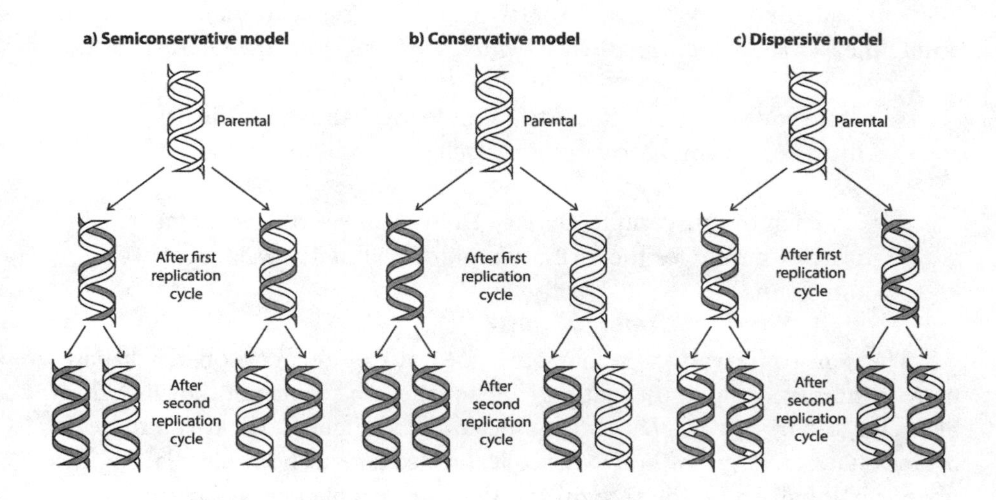

Figure 6.9: Modes of DNA Replication

as a template, by necessity. But precisely, how do they do so? When life scientists were trying to decipher the mechanism for DNA replication in the 1950s, they identified three possibilities:

- **Conservative replication:** In this mode of replication, one of the daughter DNA molecules is made up of two newly synthesized strands of DNA and the other daughter molecule is made up of the two original DNA strands.
- **Semiconservative replication:** In this mode of replication, each daughter DNA molecule consists of one original strand and one newly synthesized strand.
- **Dispersive Replication:** This type of replication involves fragmentation of the parent double helix, with the pieces of the parent strand intermixed with newly synthesized pieces, resulting in two DNA molecules.

Despite these three possibilities, inspection of DNA's structure immediately suggests that the only reasonable mode for DNA replication is the semiconservative mechanism, with each of the DNA strands functioning as a template to direct the synthesis of daughter strands. The structure of DNA doesn't allow for either conservative or dispersive modes of replication. James Watson and Francis Crick recognized this fact when they described the DNA structure:

> Our model for DNA suggests a simple mechanism for the first process [the ability to duplicate itself]. . . .
>
> . . . Each of our complementary DNA chains serves as a template or mould for the formation onto itself of a new companion chain.[25]

The semiconservative mechanism arises out of the Watson-Crick base-pairing interactions that make it possible for the two strands of the DNA double helix to harbor complementary nucleotide sequences. Clearly, a molecular rationale undergirds the DNA replication scheme. The molecular logic is further highlighted by the recognition that the complementary nature of the individual DNA strands not only makes possible template-directed replication, but also allows for error detection in the nucleotide sequences and a means to effect repair by comparing the nucleotide sequence of the newly made strand to the nucleotide sequence of the template strand.

It is a bit peculiar that the same unusual features of DNA that make it ideally suited to store information are also the same features that dictate its capacity to self-replicate. In fact, these structural features of DNA are so elegant and clever that they serve as the gold standard and inspiration for researchers trying to design self-replicating systems.

Why DNA Synthesis Proceeds in a Single Direction

If there is one feature of DNA replication that is largely responsible for the complexity of the process, it's directionality—from 5′ to 3′. At first glance, it would seem as if the process would be simpler and more elegant if replication could proceed in both directions (from 5′ to 3′ and from 3′ to 5′).

Yet work by a team from Sapporo, Japan, indicates that there is an exquisite molecular rationale for the directionality of DNA replication.[26] When these researchers discovered a class of enzymes that adds nucleotides to the ends of tRNA molecules, they recognized an important opportunity to ask why DNA replication proceeds in a single direction. Damaged tRNA molecules cannot carry out their role in protein production. Fortunately, there are repair enzymes that can fix damaged tRNA molecules. One of them is called Thg1-like protein (TLP).

TLP adds nucleotides to damaged ends of tRNA molecules. But instead of adding the nucleotides in the 5′ to 3′ direction, the enzyme adds these subunit molecules in the opposite direction of DNA replication (3′ to 5′). By determining the mechanism employed by TLP during 3′ to 5′ nucleotide addition, the researchers gained important insight into the constraints of DNA replication. It turns out 3′ to 5′ addition is a much more complex process than the normal 5′ to 3′ nucleotide addition. The 3′ to 5′ addition reaction is a cumbersome two-step process that requires an enzyme with two active sites linked together in a precise way. In contrast, 5′ to 3′ addition is a simple, one-step reaction that proceeds with a single active site. In other words, DNA replication proceeds in a single direction (5′ to 3′) because it is mechanistically simpler and more efficient.

One could argue that the complexity that arises by the 5′ to 3′ DNA replication process is a trade-off for a mechanistically simpler nucleotide addition reaction (a single reaction step as opposed to two steps). Still, if DNA synthesis proceeded in both directions the process would be complex and unwieldy. For example, the cell would require two distinct types of primosomes and DNA polymerases, one set for each direction of DNA replication. Employing two sets of primosomes and DNA polymerases is clearly less efficient than employing a single set of enzymes.

Ironically, if DNA synthesis could proceed in two directions, there still would be a leading and a lagging strand. Why? Because 3′ to 5′ synthesis is a two-step process and would proceed more slowly than the single-step, 5′ to 3′ synthesis. In other words, the assembly of the DNA strand in a 3′ to 5′ direction would lag behind the assembly of the DNA strand that traveled in a 5′ to 3′ direction.

Bidirectional DNA synthesis would also cause another complication due to a crowding effect. Once the replication bubble opens, both sets of replication enzymes would have to fit into the bubble's constrained space. This molecular overcrowding would further compromise the efficiency of the replication process. Overcrowding is not an issue for unidirectional DNA synthesis that proceeds in a 5′ to 3′ direction, carrying out its operation on both the leading and lagging strands.

Why Is an RNA Primer Needed?

Another feature that makes DNA replication appear cumbersome and inefficient—qualities that bespeak a contingent evolutionary origin—is the reliance on an RNA primer to initiate the process. Use of an RNA primer to initiate the replication of the leading strand is of minor consequence, but this is not the case when it comes to the lagging strand. Every time an Okazaki fragment is laid down, the primosome has to first generate an RNA primer. And this process occurs repeatedly as the lagging strand is synthesized. Then, once replication is completed on the lagging strand, each RNA primer needs to be excised and replaced with the appropriate DNA sequence before the Okazaki fragments can be ligated together into a continuous DNA strand. Utilizing an RNA primer seems to add unnecessary complexity.

Yet sound reasons exist for the use of an RNA primer to kick off DNA replication. Because DNA polymerase catalyzes the addition of nucleotides to the growing DNA chain in the 5′ to 3′ direction, it requires an available 3′-OH group to initiate the set of reactions that generate the daughter DNA strand. DNA polymerases are optimal for rapidly adding nucleotides to the DNA chain and for performing a proofreading function to ensure that the correct nucleotide is added to the growing chain. But these enzymes can't initiate replication because they don't have the wherewithal to bind a free nucleotide in the first position. On the other hand, the primosome can. This enzyme is a massive protein complex composed of several subunits. This complex is ideally designed to initiate DNA replication. It binds to the replication origin and recruits helicase enzymes to unwind DNA. It then has a separate binding pocket for the nucleotide that corresponds to the first position in the daughter strand. Then it appends additional ribonucleotides to the first ribonucleotide, using DNA as a template to form an RNA strand of about 10 ribonucleotides in length. This short RNA sequence can base pair with the DNA template strand to form a DNA-RNA hybrid duplex. It is the 3′-OH group of the short RNA chain that DNA polymerase utilizes. If the primosome activities were incorporated into DNA polymerase, it would undermine this enzyme's efficiency in adding nucleotides to the growing daughter DNA strand and in performing its proofreading function, both of which are necessary for the rapid, accurate replication of DNA.

The initiation process performed by the primosome is not only slow and cumbersome, it is error-prone, particularly with respect to the addition of the first few ribonucleotides to the RNA primer. This explains the rationale for using an RNA primer generated as a separate and distinct step to inaugurate the replication process. Because the primer has to be removed and replaced, the errors introduced in the first few positions of the DNA sequence can be corrected after the primer is removed. If a DNA primer was used, instead of RNA, this error correction would not take place. Using DNA primers would make DNA replication much more error-prone compared to the use of RNA primers.

Use of an RNA primer to initiate DNA replication serves other purposes. It helps to regulate DNA replication and serves to coordinate the leading and lagging strands. The rate of lagging strand production depends on primer production. It turns out that the rate of primer production is controlled by the primosome concentration in the cell, with primer production increasing as the number of primosome copies increases. The primosome concentration appears to be fine-tuned to ensure that the replication of the leading and lagging strands remains coordinated.

In 2017, a team of biochemists used single-molecule techniques to monitor the behavior of individual DNA polymerase enzymes. They discovered that DNA replication along the leading strand occurs in fits and starts.[27] The speed of the DNA polymerase is highly stochastic, varying as much as a factor of 10. At times, this enzyme even stalls for a time before it restarts the replication process. To compensate for this erratic behavior, helicase will speed up or slow down to match the speed of the DNA polymerase, with the tau protein mediating the interactions between helicase and DNA polymerase. Fortunately, the primosome concentration in the cell is at the just-right level to ensure that when it is time for DNA synthesis to proceed on the lagging strand, the RNA primer can be made without delay. If the concentration of the primosome complex were too high, then it would gum up the replication process; if too low, then the disparity between leading and lagging strand replication would become too great, leaving the single-stranded DNA exposed and vulnerable to cleavage. These researchers even discovered that the rate of primer production exceeds the average rate of DNA replication on the leading strand. This fortuitous coincidence ensures that as soon as enough of the bubble opens for lagging strand replication to continue, the primase can immediately lay down the RNA primer and restart the process. Again, this is all thanks to the fine-tuning of primosome concentration.

The upshot shows an exquisite molecular logic on full display when it comes to DNA replication, accounting for the key features of this critical biochemical operation. And given its structure, the simplest, most clear-cut way DNA can duplicate itself is through template-directed, semiconservative replication. It is fortuitous that the very structural features that make DNA unusually fit as an information storage molecule also make DNA replication possible. From this analysis, it appears reasonable to conclude that two of the three criteria for the biochemical anthropic principle are satisfied, indicating that the anthropic principle applies to DNA replication. But is the process of

DNA replication dictated and constrained by the laws of nature? An interesting study published in 1999 by a team from NIH (National Institutes of Health), offers important insight into this question.

On the Origin of DNA Replication

As noted, many life scientists think that DNA replication is the chance outcome of a historically contingent process, arising out of an RNA world. Likewise, many of these same scientists take the view that DNA replication emerged prior to the appearance of the last universal common ancestor (LUCA), the organism that anchors the evolutionary tree of life. The reason for this perspective stems, in large measure, from the extreme complexity of DNA replication. According to paleontologist J. William Schopf, one of the world's leading authorities on early life on Earth:

> Because biochemical systems comprise many intricately interlinked pieces, any particular full-blown system can arise only once. . . .
>
> . . . Since any complete biochemical system is far too elaborate to have evolved more than once in the history of life, it is safe to assume that microbes of the primal LCA cell line had the same traits that characterize all its present-day descendants.[28]

Another reason for this view relates to the nearly identical functional similarity of DNA replication observed in all life. The universally shared features of DNA replication indicate that this biochemical system emerged prior to the emergence of LUCA. The universal features of DNA replication include:

- semiconservative replication;
- initiation at a defined origin by an origin replication complex;
- bidirectional movement of the replication fork;
- continuous (leading strand) replication for one DNA strand and discontinuous (lagging strand) replication for the other;
- use of RNA primers;
- use of nucleases, polymerases, and ligases to replace RNA primers with DNA.

In 1999, NIH researchers made an unexpected discovery. It appears that the core enzymes in the DNA replication machinery of Bacteria and Archaea/Eukaryotes (the two major trunks of the evolutionary tree of life) did not share a common evolutionary origin—because the gene sequences that code for the proteins involved in DNA replication don't trace back to a shared ancestral sequence. From an evolutionary perspective, it appears as if the process of DNA replication emerged separately in Bacteria and Archaea *after* these two evolutionary lineages diverged from the LUCA. Yet the major features of DNA replication are, in effect, identical in both lineages.[29] This finding strongly implies that DNA replication must be the way it is because constraints arising from the fabric of the universe itself dictate its central features. Given the complexity of DNA replication, it is unreasonable to think that the same biochemical system could emerge independently, if the process is dictated by a historically contingent process. One can't help but marvel at the apparent fact that the constraints that dictate the features of DNA replication yield a process that displays such exquisite molecular logic and is so ideally fit for life.

In this chapter, our objective was to address the question: Are there anthropic coincidences evident in the biochemical operations that produce proteins and nucleic acids? Because of the central importance of the nucleic acids and proteins, I predicted that we would see evidence for the biochemical anthropic principle in the biosynthesis of these two classes of biomolecules. Careful consideration of the key features of protein synthesis and DNA replication exposes a sophisticated and elegant molecular rationale that pervades these two biochemical processes. It doesn't seem that either protein synthesis or DNA replication arises from a historically contingent evolutionary progression that generated features that are happenstance—merely a frozen accident arising out of the RNA world.

Based on the analysis presented in chapters 4 and 5, it seems that if physical life is to exist, it must be built around a DNA and protein world that would be virtually identical to contemporary biochemical systems. And if life is organized around DNA and proteins, then a framework virtually identical to the central dogma of molecular biology must define the flow of biochemical information. In like manner, if life is comprised of DNA and proteins, it is reasonable to think that DNA replication must be template-directed and semiconservative, proceeding in a single direction initiated by RNA primers. In other words, terrestrial biochemistry appears to be universal. And this universality reflects a deep-seated teleology that undergirds life, arising out of the design of the universe and a biochemical anthropic principle.

In the next chapter, we continue our evaluation of the biochemical anthropic principle by turning our attention on cell membranes, an indispensable feature of life.

Chapter 7

Cell Membranes

It just might be the greatest rock song of all time.

Recorded by one of my all-time favorite rock bands, "Stairway to Heaven" appeared on Led Zeppelin's untitled fourth studio album (sometimes called *Led Zeppelin IV* or the *ZOSO* album). During the 1970s, "Stairway to Heaven" was the most requested song on FM radio stations even though it was never released as a single.

Controversy has surrounded "Stairway to Heaven." Since the song's release, several people have pointed out the close similarities between the opening section of "Stairway to Heaven" and the song "Taurus" by the rock band Spirit—including Spirit's guitarist Randy California (who wrote "Taurus"). Led Zeppelin toured with Spirit two years before "Stairway to Heaven" was recorded, raising suspicions that Jimmy Page and Robert Plant plagiarized the Spirit song. These suspicions culminated in a copyright infringement lawsuit in 2014, spearheaded by Spirit bassist Mark Andes and a trust representing California (who died in 1997). The lawsuit dragged through the courts for six years before the court ruled in Led Zeppelin's favor in March 2020.

This type of controversy wasn't new to Led Zeppelin. Over the course of the band's career, Led Zeppelin was accused repeatedly of co-opting—either in part or entirely—songs, melodies, and lyrics written and recorded by other artists. In some instances, Led Zeppelin provided the appropriate attribution, yet more often than not they didn't. Because of this practice, Led Zeppelin was sued numerous times, with the injured parties seeking songwriting credit, royalties, and damages.

Because the band drew much of its inspiration from the blues, Led Zeppelin frequently borrowed from songs written by legendary blues figures, such as

Sonny Boy Williamson, Willie Dixon, Robert Johnson, Sleepy John Estes, Bukka White, Howlin' Wolf, and Bobby Parker. Unbeknownst to me at the time, it was the blues imprint on their sound that made Led Zeppelin one of my favorite rock bands. And the highly publicized controversies about songwriting credits and Led Zeppelin's well-known "thievery" piqued my curiosity, so much so, that I bought blues records so that I could compare the Led Zeppelin songs with the original recordings. In the process, as a teenager, I was introduced to some of the early masters of the blues and some of the most important figures in the history of American music.

It wasn't long before my curiosity about songwriting credits became a lifelong obsession with the blues. I quickly became familiar with many of the early blues artists. Many of them lived quite colorful lives. Yet the blues musician I found most interesting was Blind Willie Johnson. In fact, today, he ranks among my favorite musicians of all time. Led Zeppelin recorded two songs that were first written and recorded by Blind Willie Johnson. "In My Time of Dying" appeared on their 1975 album, *Physical Graffiti*, and "Nobody's Fault but Mine" appeared on their 1976 release, *Presence*. Based on Psalm 41:3, "In My Time of Dying" is a desperate plea to Jesus Christ for mercy and acceptance, as the singer imagines his own death. "Nobody's Fault but Mine" is also a song of desperation, as the singer comes to grips with the fact that his damnation is unavoidable—and he has no one to blame but himself, with the only glimmer of hope to "save my soul tonight" found in the Bible.

Despite Blind Willie Johnson's influence on these compositions, the songwriting credit for the two tracks went to Led Zeppelin band members John Bonham, John Paul Jones, Jimmy Page, and Robert Plant, and Plant and Page, respectively. Though not crediting Johnson was an injustice, I am forever grateful that Led Zeppelin covered these two Blind Willie Johnson songs. Not only did these recordings introduce me to the life and music of Blind Willie Johnson, they planted the seeds of the gospel in my psyche, drawing my attention to my deep spiritual need for forgiveness and directing me toward the path of salvation through Jesus Christ—a journey I took years later when I converted to Christianity. I can still remember all the times I wandered around my house as a teenager singing, "Jesus gonna make up my dyin' bed." It was a prelude of things to come for me when I did, indeed, make Jesus my hope.

Hands down, my favorite Led Zeppelin song is another cover of a blues classic, "When the Levee Breaks." This song was originally written and recorded by Kansas Joe McCoy and Memphis Minnie in 1929 for Columbia Records. In this instance, Led Zeppelin included Memphis Minnie in the songwriting credits.

"When the Levee Breaks" appears as the last track on the *ZOSO* album and is widely acclaimed as one of the record's best songs, on par with "Stairway to Heaven." The song's signature is John Bonham's drumming and Robert Plant's harmonica playing. To achieve the just-right sonic effects, the harmonica tracks were overdubbed with a reverse echo and the drum tracks were recorded in the lobby of the recording studio using hanging microphones positioned at different locations in the staircase leading from the lobby to the second floor. Even though many people consider this song one of Led Zeppelin's finest accomplishments, the band played it live only a few times because it was too difficult to recreate the song onstage.

"When the Levee Breaks" describes the catastrophe that resulted from the Great Mississippi Flood of 1927. This event impacted over 27,000 square miles of the Mississippi Delta, killing several hundred people and displacing another 600,000, including over 200,000 African Americans. The flood's devastation triggered the great migration of African Americans to the northern industrial cities, with many people choosing to leave their homes rather than to return to the destruction left behind by the flooding.

"When the Levee Breaks" became an important addition to my life's soundtrack during the time I spent in graduate school and postdoctoral research. During that time, I devoted most of my waking hours to learning about the biochemistry and biophysics of cell membranes—life's levee system. I would often catch myself humming or singing this song as I spent long hours at the lab bench carrying out experiments designed to probe the interaction between a class of antimicrobial peptides called magainins and the outer and inner cell membranes isolated from strains of *Salmonella typhimurium*. Magainins exert their antimicrobial activity by causing the inner membrane to become leaky, breaking down the biochemical levee that separates the exterior environment from the cell's internal contents. By disrupting the bacterial inner membrane, the contents of the cell are lost, leaking into the environment, while materials from the cell's exterior rush into the cell's cytoplasm. For bacterial cells exposed to magainins, when the biochemical levee breaks, "Cryin' won't help you, prayin' won't do you no good."

• • • • • • • • • • • •

Cell membranes serve as life's boundaries by segregating cellular contents and processes from the exterior environment. Membranes play a critical role in cell survival. The matrix of the cell membrane is impermeable to many different types of materials. So, like a well-built levee, it keeps harmful materials from entering the cell and, at the same time, sequesters beneficial compounds in its interior. Proteins embedded in the cell's membrane regulate the traffic of materials into and out of the cell. These transport proteins ensure that the cell imports necessary nutrients and deports waste products in an orderly manner.

Complex (eukaryotic) cells also possess an *internal network* of membranes that segregate the cell's activities. The membranes of organelles (structures inside the cell that carry out specific functions) serve as key sites for photosynthesis (in plant cells) and energy production. In both cases, the selective nature of cell membranes maintains a gradient of positively charged protons across the membrane, creating an electrochemical gradient. This gradient can be used to drive energy-capturing processes when protons move through proteins embedded in the membrane, generating the chemical energy needed to drive cellular processes. (See chapter 8.)

Because of the central importance of cell membranes for living systems, it is reasonable to expect that these biomolecular complexes would evince the biochemical anthropic principle, if such a principle exists. To determine if this expectation bears out, we adopt a similar approach to the one used in previous chapters. Specifically, we will:

- Determine if a molecular rationale undergirds the structure and function of cell membranes and their component molecules.
- Determine if there are physicochemical constraints that arise from the laws of nature that dictate the properties of cell membranes.
- Compare the features of cell membranes and their components to alternative systems to determine if these biochemical systems are suited for life.

If these three criteria are met, then we can reasonably conclude that cell membranes aren't solely the product of the contingent outworking of evolutionary processes, but, instead, are largely specified by constraints that arise from the laws of nature—constraints that produce biomolecular systems with a suitable set of properties for life.

Before we begin our assessment, it is important to know a bit about the basics of cell membrane biochemistry. (For readers familiar with the biochemistry of cell membranes, please feel free to skip ahead to Cell Membranes and the Biochemical Anthropic Principle.)

An Introduction to Cell Membrane Biochemistry

Cell membranes cannot be visualized with a light microscope. Membranes are only 7.5 to 10 nanometers thick. However, these structures can be characterized by more powerful electron microscopes. In electron micrographs, cell membranes reveal a sandwich-like appearance. The inner and outer surfaces of the membrane appear dark, whereas the membrane's interior gives a light appearance.

Lipids and proteins interact to form cell membranes. Lipids are a structurally heterogeneous group of compounds that share water insolubility as their defining property. Though water insoluble, lipids readily dissolve in organic solvents. Cholesterol, triglycerides, saturated and unsaturated fats, oils, and lecithin are widely recognized examples of lipids.

Membrane Lipids

In most cases, phospholipids are the cell membrane's major lipid component. Phospholipids consist of a glycerol backbone. Two individual fatty acid molecules are linked to the first and second hydroxyl groups of glycerol through an ester linkage. One of the defining features of fatty acids is their long hydrocarbon chain. A phosphate group is bound to glycerol's third hydroxyl group. In turn, an alcohol is bound to the phosphate moiety. (See figure 7.1.)

A phospholipid's molecular shape roughly resembles a distorted balloon with two ropes tied to it (figure 7.1). As we will soon learn, there is a molecular rationale that accounts for the molecular geometry of phospholipids. Biochemists divide phospholipids into two regions that possess markedly different physical properties. The head region, comprised of the phosphate group bound to an alcohol, corresponds to the balloon. The head region (or group) readily dissolves in water. Biochemists refer to the head group as hydrophilic ("water-loving"). The phospholipid tails, made up of the hydrocarbon chains of the fatty acids, correspond to the ropes tied to the balloon. The hydrocarbon tails are insoluble in water or are hydrophobic ("water-hating").

Chemists refer to molecules such as phospholipids that possess molecular regions with distinct solubility characteristics as amphiphilic ("ambivalent in its likes"). Picture soaps and detergents.

Amphiphilicity holds great biological importance. Phospholipids' dual solubility properties play the key role in cell membrane structure. When added to water, phospholipids spontaneously organize into sheets, called bilayers, that are two molecules thick. When organized into a bilayer, phospholipid molecules align into two monolayers with the phospholipid head groups adjacent to one

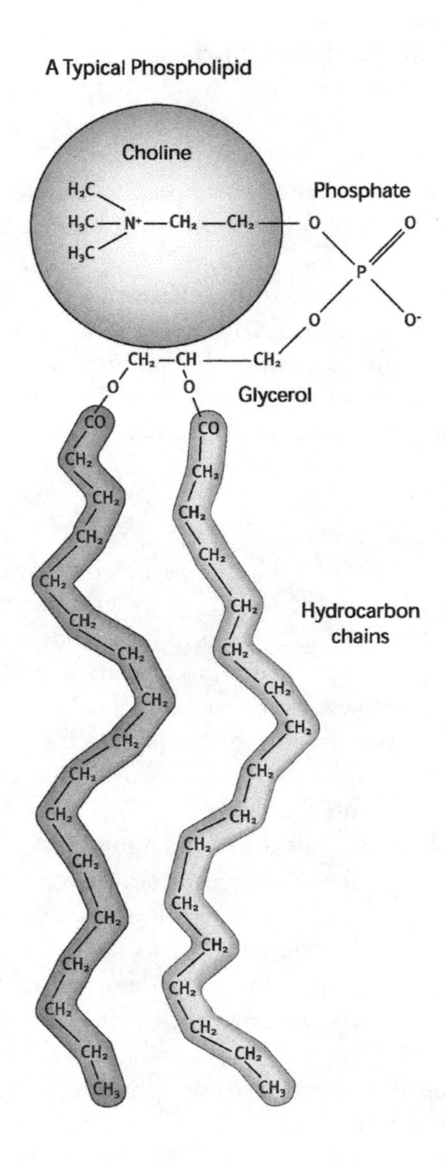

Figure 7.1: Phospholipid Structure

another and the phospholipid tails packed together closely. The monolayers, in turn, interact so the phospholipid tails of one monolayer contact the phospholipid tails of the other. This tail-to-tail arrangement ensures that water-soluble

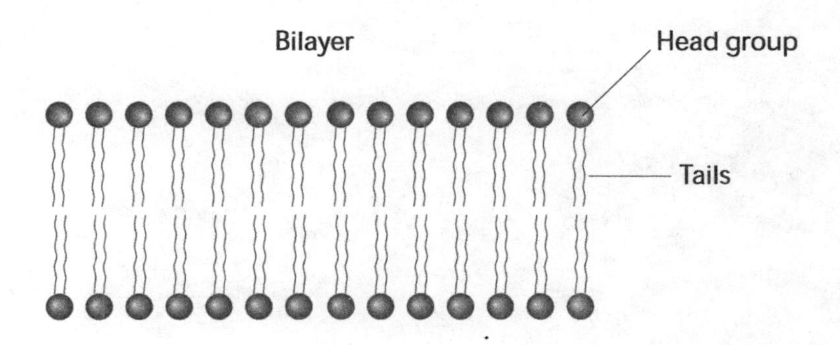

Figure 7.2: Phospholipid Bilayer Structure

head groups contact water and the water-insoluble tails sequester from water (figure 7.2). Bilayers spontaneously form when phospholipids are dispersed in water because of the hydrophobic effect. (See chapter 3.)

This arrangement of phospholipids into a bilayer structure gives cell membranes their sandwich-like appearance in electron micrographs. The electron-dense head groups forming the cell membrane's inner and outer surfaces render them dark. The phospholipid tails are less electron dense and, therefore, appear light.

Even though the phospholipids that form cell membranes possess similar physical properties, they display a wide range of chemical variability. As just noted, phospholipid head groups consist of a phosphate group bound to a glycerol backbone. The phosphate group, in turn, binds one of several possible compounds that vary in chemical and physical properties. Frequently, phospholipids are identified by their head group structure. Phospholipid head groups consist of either the amino alcohols choline (PC) or ethanolamine (PE), the amino acid serine (PS), or the polyols glycerol (PG) or inositol (PI). The fatty acid hydrocarbon chains of phospholipids also vary in length and structure. The fatty acid chains are typically long, linear hydrocarbon chains consisting of either 14, 16, or 18 carbon atoms, but can contain as few as 12 carbon atoms in length and as many as 24 carbons. Sometimes one or both of the hydrocarbon chains possess one or more permanent kinks. These kinks can occur at different locations along the chain length. Carbon-carbon double bonds (C=C) inserted into the hydrocarbon chain cause the kinks because of their cis geometry. As we will see, the head group structure, the fatty acid chain length,

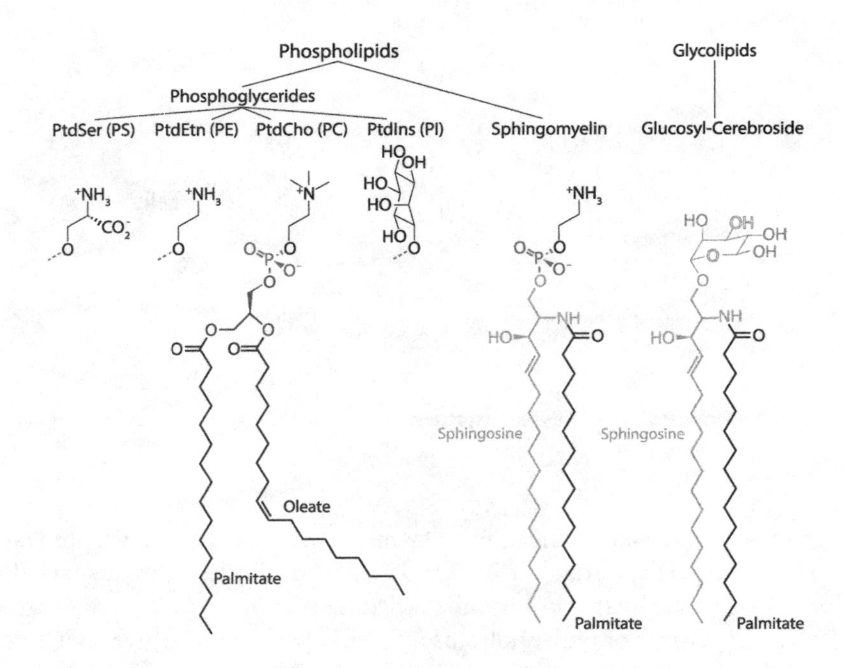

Figure 7.3: Some of the Different Phospholipid Molecules Found in Cell Membranes

and degree and type of unsaturation (a measure of the number of cis double bonds in the fatty acid hydrocarbon chains) significantly impact the biochemical and biophysical properties of the phospholipid species. (See figure 7.3.)

The cell membranes of eukaryotic organisms also consist of several other different classes of lipids (such as cholesterol, plasmalogens, sphingolipids, and glycolipids) in addition to phospholipids.

Membrane Proteins

Proteins associate with the cell membrane in a variety of ways. For example, peripheral proteins bind to the inner or outer membrane surfaces. Integral proteins embed into the cell membrane. Some integral proteins insert only slightly into the membrane interior, others penetrate nearly halfway into the membrane's core, and others span the entire membrane (figure 7.4).

Membrane proteins serve the cell in numerous capacities. Some proteins function as receptors, binding compounds that allow the cell to communicate

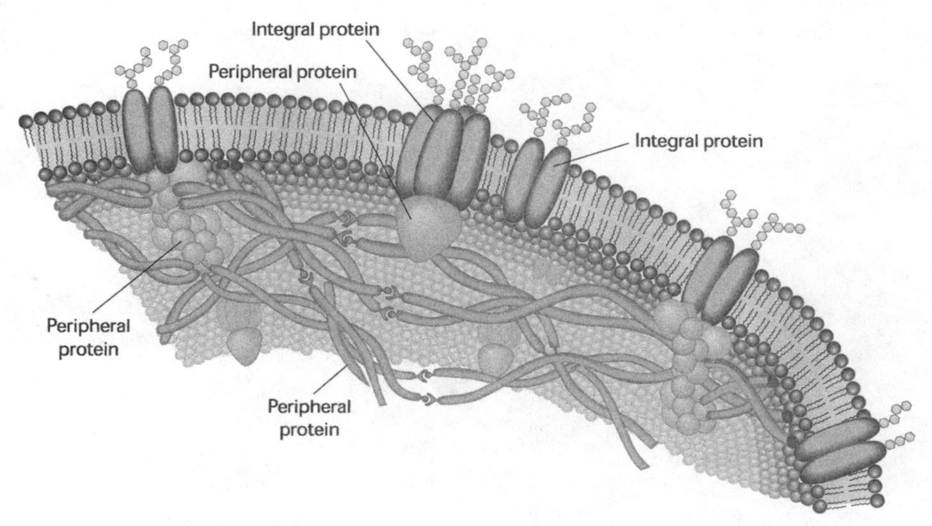

Figure 7.4: Membrane Proteins

with its external environment. Some catalyze chemical reactions at the cell's interior and exterior surfaces. Some proteins shuttle molecules across the cell membrane; others form pores and channels through the membrane. Some membrane proteins impart structural integrity to the cell membrane.

Fluid Mosaic Model

In the early 1970s, biochemists S. J. Singer and G. L. Nicolson proposed the fluid mosaic model to describe the structure of cell membranes.[1] This model depicts the bilayer as a two-dimensional fluid composed of a complex mixture of phospholipids. The bilayer serves as both a cellular barrier and a solvent for a variety of integral and peripheral membrane proteins (figure 7.5). According to the fluid mosaic model, membranes are little more than haphazard, disorganized systems with proteins and lipids freely diffusing laterally throughout the bilayer, with the bilayer itself providing the only structural organization.

An Updated Fluid Mosaic Model

Recent advances indicate that the fluid mosaic model represents an incomplete depiction of cell membranes. Biochemists now acknowledge that these biochemical systems display an exquisite hierarchical organization integral to many functions performed by cell membranes.[2]

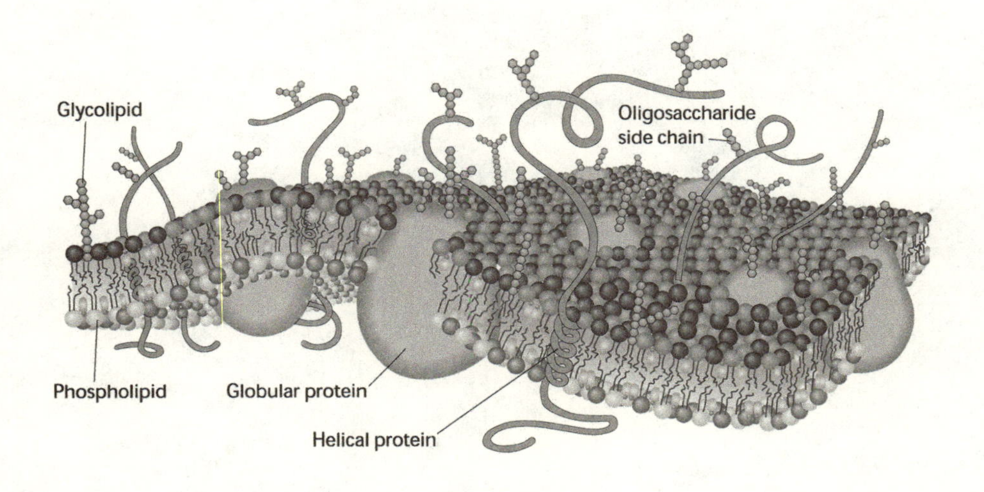

Figure 7.5: Fluid Mosaic Model

Membrane Asymmetry

Well before Singer and Nicolson proposed the fluid mosaic model, biochemists had learned that the inner and outer surfaces of cell membranes display distinct functional properties. Differences in the protein and phospholipid compositions in the inner and outer leaflets of membrane bilayers account for these differences. Because of these differing lipid and protein compositions, biochemists refer to cell membranes as asymmetric. Proteins that span the cell membrane also display a distinct orientation within the membrane.[3] This asymmetry reflects long-range order and organization in cell membranes. The asymmetry of proteins and phospholipids is established by complex biochemical processes when cell membranes are assembled and actively maintained by the cell's machinery during the membrane's lifetime.[4] (Discussion of these processes extends beyond the scope of this book.)

Because of protein asymmetry, cell membranes (1) transport materials in a single direction through cell membranes, (2) detect changes in the cell's exterior environment, (3) perform specific chemical operations inside the cell at the inner membrane surface, and (4) stabilize the cell membranes through interactions between the cytoskeletal proteins and proteins located on the interior surface of the bilayer.

The asymmetric distribution of phospholipids in the inner and outer leaflets of the cell membrane varies from membrane to membrane, but it is far from random. As a case in point, the plasma membrane that surrounds the cell typically has higher levels of PSs, PEs, and PIs in the inner leaflet and higher levels of PCs and sphingomyelins in the outer monolayer. The inner membranes of mitochondria possess greater amounts of PCs and PEs in the outer monolayer and higher levels of PIs and the unusual phospholipid cardiolipin in the inner leaflet. (Cardiolipin's structure and function are detailed later in the chapter.) The membranes of the Golgi apparatus have higher concentrations of PEs, PCs, and PSs in the membrane surface exposed to the cytoplasm and greater amounts of PI and sphingomyelins on the membrane surface in contact with the Golgi's interior.[5]

Phospholipid asymmetry has important biological consequences. Differences in phospholipid composition produce variations in the charge, permeability, and fluidity of inner and outer leaflets of the membranes. These compositional differences also make it possible for each monolayer to support and regulate the activity of proteins associated with the membrane.

Annular Lipids

Shortly after the fluid mosaic model was proposed, biochemists discovered the existence of localized regions of order and organization within the bilayers of cell membranes. These localized regions are often associated with specific membrane proteins. Biochemists first became aware of this order when they discovered that specific phospholipid species always copurify with membrane proteins when they isolated them from cell membranes. Biochemists speculated that these lipids—called annular or boundary lipids—formed a ring, about one molecular layer thick, around the integral proteins.[6]

Initially, biochemists thought that the annular lipids surrounding integral proteins behaved dynamically, with individual phospholipids associating and dissociating with specific proteins, and exchanging places with phospholipids in the "bulk" bilayer. Thanks to advances in laboratory techniques, biochemists today can directly visualize annular phospholipids in association with membrane proteins. These observations reveal that the annular phospholipids form a tightly bound layer surrounding the membrane proteins. The intimate association between the membrane protein and the phospholipids appears to be mediated by highly specific interactions between the protein and the phospholipid head groups. In many cases, the hydrocarbon chains of the lipids—aligned along the protein's axis—follow the contour of the protein's surface. In some cases, the hydrocarbon chains are straight, while other times they are kinked or bent. These types of interactions result in a uniform lipid casing that surrounds the protein. Biochemists have discovered that this close, nearly permanent association is critical for membrane stability. It allows the integral protein to fit snuggly into the bilayer without causing defects due to imperfect packing between the protein and the lipid components. These defects would make the membrane "leaky" like a poorly constructed levee—a condition detrimental to the cell.

Membrane Domains and Lipid Rafts

Today, life scientists view cell membranes as highly organized systems, comprised of several structurally and functionally discrete domains. Thus, the organization of cell membranes extends well beyond membrane asymmetry and the rings of annular lipids that surround membrane proteins.[7] As it turns out, these domains organize into supra-domains, producing a hierarchy of order and organization. Each membrane domain consists of distinct lipid and protein compositions responsible for the unique functional role of its domain.

Biochemists have discovered the existence of lipid rafts, a special type of domain in cell membranes.[8] The physical properties of the lipid rafts render them as solid-like regions of the membrane that "float" in more fluid regions of the bilayer, just like a raft at sea. Most often, lipid rafts are enriched in cholesterol and sphingomyelins. Biochemists think the structural integrity of the lipid raft is maintained through interactions between the lipid head groups that are part of the raft domain. Specific types of proteins are associated with lipid rafts, typically those involved in signal transduction. Biochemists often find high levels of protein receptors embedded in lipid rafts. These receptors bind molecules in the environment, initiating biochemical pathways that elicit specific cellular activities in response to changes in the cell's surroundings. Lipid rafts also appear to have a hand in secretion of vesicles by the Golgi apparatus.

The Compositional Fine-Tuning of Cell Membranes

One of the universal features that define all cell membranes is their unilamellar structure, meaning cell membranes are made up of a single bilayer. The

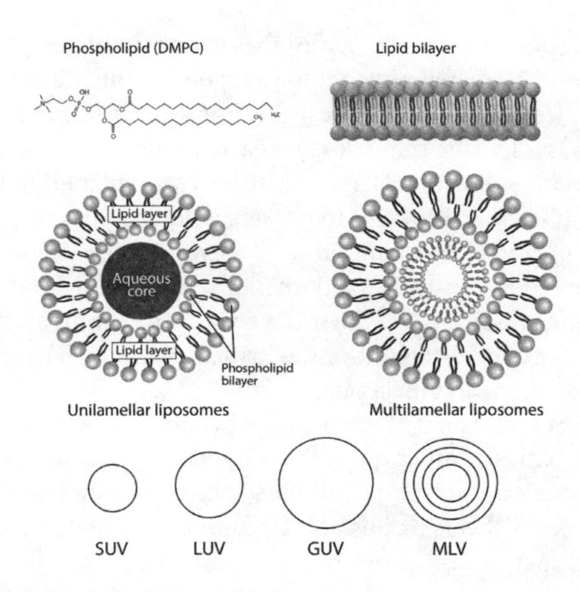

Figure 7.6: Bilayer Aggregates

structure of biological membranes stands in sharp contrast to the types of bilayers produced in the laboratory from pure phospholipids or phospholipids extracted from a biological source. Both purified and extracted phospholipids spontaneously assemble into bilayers in aqueous environments. But instead of forming unilamellar bilayers, phospholipids from these sources aggregate into stacks of bilayers (multilamellar bilayers), or they alternatively form spherical structures that consist of multiple bilayer sheets.[9] (These structures resemble an onion with each layer corresponding to one of the bilayers in the stack.) These aggregates only superficially resemble the cell membrane's structure, which, again, consists of a *single* bilayer, not bilayer stacks. (See figure 7.5.)

When dispersed in water, purified and extracted phospholipids can be forced to form structures composed of only a single bilayer, like that of a cell membrane, through laboratory manipulation. These particular aggregates arrange into hollow, spherical structures called liposomes or unilamellar vesicles.[10] (See figure 7.6.) In contrast to cell membranes, which are stable structures, liposomes exist for a limited lifetime and are considered to be a metastable phase. After a short period of time, liposomes readily fuse with one another, reverting to multilamellar sheets or vesicles.[11]

So, how is it possible for cell membranes to consist of a *single* bilayer phase when purified and extracted phospholipids spontaneously assemble into multibilayer sheets, with single-bilayer vesicles (liposomes) existing temporarily as a metastable phase? During the 1980s and early 1990s, NIH (National Institutes of Health) researcher Norman Gershfeld offered an explanation that reveals an unexpected feature of cell membranes. Gershfeld and his collaborators demonstrated that purified phospholipids will form a stable single bilayer phase, similar to the one that constitutes cell membranes, but only under a unique set of highly specific conditions.[12] (Chemists refer to phenomena that occur under a unique set of specific conditions as critical phenomena.) In other words, unilamellar bilayers are a critical phase.

As it turns out, the single bilayer phase occurs only at a specific temperature (called the critical temperature) that happens to be characteristic for each phospholipid species. Pure phospholipids spontaneously transform from either multiple bilayer sheets or unstable liposomes into stable single bilayers at their critical temperature.[13] Above or below this temperature the unilamellar phase collapses into multilamellar structures. Again, the critical temperature varies, depending on the specific chemical makeup of the phospholipid head group and fatty acid chain lengths. In the case of bilayers formed from a mixture of phospholipids, the critical temperature depends on the specific phospholipid composition.[14]

Remarkably, Gershfeld and his team learned that organisms appear to actively maintain precise phospholipid compositions in their membranes to ensure that they remain in the critical single bilayer phase, with the critical temperature corresponding to the organism's physiological temperature.[15] Gershfeld's group also discovered that animals that don't regulate their own body temperature (i.e., "cold-blooded" creatures like frogs and salamanders) and microbes actively adjust the phospholipid composition of their cell membranes so that the critical temperature matches the environmental temperature.[16] Gershfeld's team also noted that the bacterium *E. coli* adjusts its cell membrane phospholipid composition to maintain a single bilayer phase as growth temperature varies.[17] Gershfeld and his collaborators have demonstrated the physiological importance of the relationship between the phospholipid composition of cell membranes and the critical temperature of the bilayers by studying the rupturing of human red blood cells. As it turns out, these cells rupture (hemolysis) when incubated at temperatures exceeding 37°C (the normal human body temperature). Transformation of the red blood cell membrane from a single bilayer to multiple bilayer stacks accompanies the hemolysis of these cells, indicating a loss of the cell membrane's critical state.[18]

In short, Gershfeld's work indicates that cell membranes are high-precision biochemical systems that display an exquisite level of active fine-tuning of their phospholipid composition to correspond to physiological temperatures for warm-blooded creatures, environmental temperatures for cold-blooded creatures, and growth temperatures for microorganisms.

With this basic understanding of the biochemistry and biophysics of cell membranes in place, we can address the central question of the chapter. Are anthropic coincidences evident when we consider the physicochemical properties of cell membrane components?

Cell Membranes and the Biochemical Anthropic Principle

Determining if the biochemical anthropic principle manifests in the architecture and operations of cell membranes can appear to be a daunting undertaking—at least at first glance—given the complexity of these biochemical systems. One way to simplify this task is to focus on the structure and properties of its proteins and lipid components. The question then becomes: are there anthropic coincidences associated with the structure and function of membrane proteins and lipids?

We have addressed the question of whether protein biochemistry provides support for the biochemical anthropic principle. (See chapter 4.) But what about the lipid components? For the remainder of the chapter, we will focus on examining the physical and chemical properties of phospholipids—the primary lipid component of cell membranes—with the intent of determining if the lipid components of cell membranes contribute to the case for the biochemical anthropic principle. As pointed out in previous chapters, this approach is justified given the central role that phospholipids play in forming the bilayer matrix and their active role in many biochemical processes associated with cell membranes. In other words, it seems reasonable to think that if the biochemical anthropic principle has any influence on cell membrane structure and function, it would be evident in the physicochemical properties and behaviors of phospholipids.

The Physical Properties of Phospholipids

Biochemists have long understood that the physical properties of phospholipids have biological significance. These properties stem largely from their amphiphilic nature, making these molecules ideally suited for their role in forming the bilayer matrix—a structure that serves as the foundation of cell membrane architecture. As discussed, the amphiphilic properties of phospholipids arise from their water-soluble head group and water-insoluble hydrocarbon chains. Because of the hydrophobic effect, this dual solubility forces phospholipids to spontaneously aggregate into bilayers when dispersed in water.

The geometric shape of phospholipids is another physical feature that contributes to their bilayer-forming capacity. The importance of the molecular shape of phospholipids was first recognized in the 1970s by biophysicist Jacob Israelachvili and his collaborators.[19]

To understand Israelachvili's insight, it is important to review how phospholipid molecules aggregate when dispersed in water to form bilayer structures. These molecules interact laterally with one another: the hydrocarbon chains of phospholipids line up with one another to form the bilayer interior. The head groups also align to form the surface of the bilayer, which comes into direct contact with the aqueous phase.

Even though phospholipids readily form bilayers, other amphiphilic lipid molecules don't. For example, fatty acids (which consist of a water-soluble carboxylic acid head group and a single water-insoluble hydrocarbon chain) form solid spherical structures when dispersed in water, with the hydrocarbon chains forming the interior of the sphere and the carboxylic acid head groups located at the sphere's surface. (See figure 7.7.) These structures are sometimes called micelles.

Israelachvili discovered that the overall shape of amphiphilic lipid molecules influences how these molecules laterally pack together when dispersed in water. This packing, in turn, influences the overall architecture of the aggregates they form, explaining why some amphiphilic lipid molecules, such as phospholipids, will form bilayers and others won't. The differences in aggregate structures arise from the differences in molecular geometries.

To help characterize this phenomenon, Israelachvili developed the concept of the packing parameter—a numerical value that reflects the geometric structure of amphiphilic lipid molecules. The packing parameter (P) is defined as the ratio of the volume (V) occupied by the hydrocarbon chain(s) to the product of the maximum length (L) of the hydrocarbon chain(s) and the area (A) occupied by the head group at the air-water interface. That is, $P = V/AL$.

Israelachvili and his coworkers discovered that if the packing parameter is less than 0.33, the molecular geometry of the amphiphilic lipid is conical. The lateral packing of cones results in aggregates that are solid spheres. If P varies between 0.33 and 0.5, then the amphiphilic lipids adopt a wedge-shape geometry, and the aggregates they form in water take on the shape of a solid

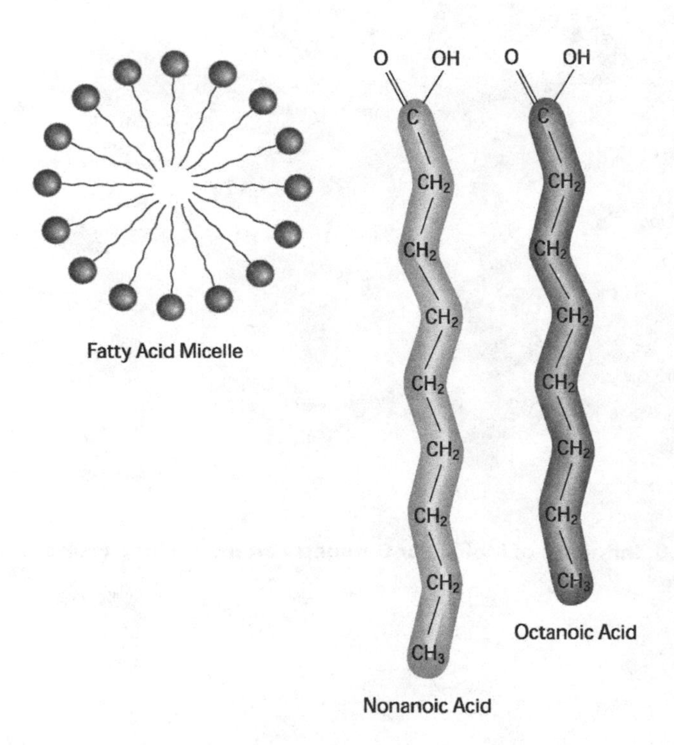

Figure 7.7: The Structure of a Typical Fatty Acid and the Micelles It Forms

cylindrical tube, with the hydrocarbon chains forming the tube's interior and the head group forming the tube's surface. If the packing parameter ranges approximately between 0.5 and 1, the amphiphilic lipids display a molecular geometry that ranges from a truncated cone to a cylinder. (See figure 7.8.) Both geometries form bilayer structures, with the truncated cone geometry leading to bilayers with some curvature, and the cylindrical geometry producing bilayers that are strictly planar. Finally, if P is greater than 1, then the amphiphilic lipid adopts the geometry of an inverted truncated cone. When these molecules pack together, they form structures called inverted micelles. In this configuration, the head groups line channels filled with water and the hydrocarbon chains are oriented toward the exterior surface of the tube. Each inverted micelle interacts with another to form an amalgam of tubes, shielding the hydrocarbon chains from the aqueous phase.

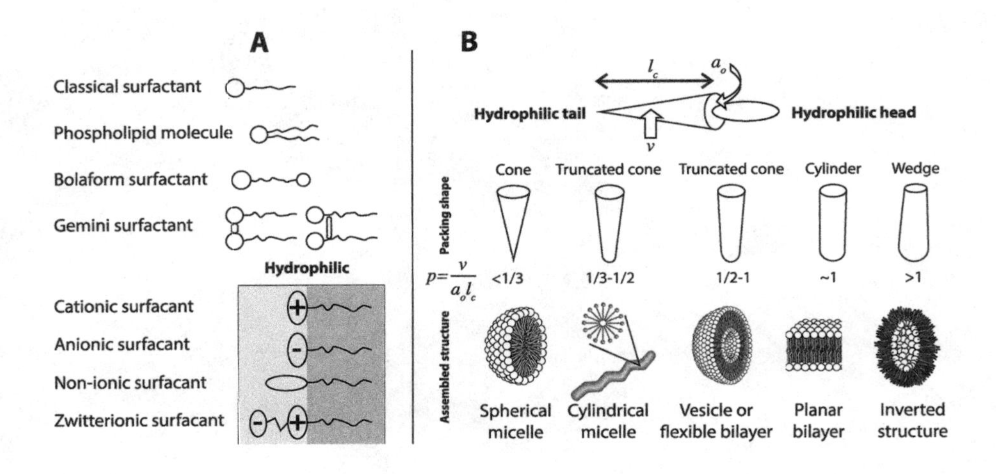

Figure 7.8: Influence of Molecular Geometry on the Packing Properties of Phospholipids

As it turns out, the packing parameter for the phospholipids that make up cell membranes falls between 0.5 and 1. It is a bit suspicious to think that the packing parameters—and, hence, the molecular geometries—for all the phospholipid species that comprise cell membranes (such as PCs, PEs, PSs, and PGs) are precisely what they need to be for cell membranes to form and for cellular life to exist. Phospholipids have the just-right molecular shapes that make them remarkably well-suited for life. The implications of this discovery are profound. It means that the bilayer-assembling properties of phospholipids—which are critical for cell membranes to form and for life to exist—are due primarily to thermodynamic and geometric considerations. Recall that these two factors constrain the folding of protein chains as well. (See chapter 4.)

The thermodynamic constraints on phospholipid structure dictate their amphiphilic properties and manifest in an aqueous environment through the hydrophobic effect (which drives phospholipids to aggregate in water) and, ultimately, the second law of thermodynamics. And, as I have just discussed, geometric considerations dictate the way phospholipids pack when they associate to form aggregates. As a corollary, this insight also means that the selection of phospholipids as the main lipid component of cell membranes most likely isn't the accidental consequence of a historically contingent evolutionary process. Instead, it appears as if the laws of nature fundamentally constrain and dictate the choice of phospholipids as the main lipid component of cell membranes because of their just-right physicochemical properties.

Are Phospholipids Unique?

But are phospholipids unique in their bilayer-forming properties? Are there reasonable alternatives that could have been used in their place? We can address this question by comparing the packing parameters of phospholipids with that of other close chemical analogs that could have, in principle, served as the matrix-forming materials of cell membranes.

Lysophospholipids

Let's begin by considering the physicochemical properties of lysophospholipids. These compounds share many of the same structural features as phospholipids. Instead of having two hydrocarbon chains linked to a glycerol backbone via ester linkages, lysophospholipids only have a single hydrocarbon chain bound to a glycerol backbone (via an ester linkage). (See figure 7.9.) Like phospholipids, lysophospholipids are amphiphilic, forming aggregates in water. But because of their molecular geometry (which results in a packing parameter that falls between 0.3 and 0.5), lysophospholipids form tubular micelles. In other words, even though lysophospholipids are simpler than phospholipids—and would be preferred as cell membrane components for that reason (because their synthesis would consume less of the cell's resources than the synthesis of phospholipids)—these compounds could never serve as the dominant lipid species in cell membranes because of their molecular geometry.

Still these biomolecules are important. Lysophospholipids serve as metabolic intermediates and function as signaling molecules within cell membranes. (Lysophospholipids are derived from phospholipids via the activity of an enzyme called phospholipase A.) When these biomolecules do occur in cell membranes, they exist temporarily as transient species and occur at low levels. Because they are minor components in cell membranes, their molecular geometry has minimal impact on the overall bilayer structure. (As an interesting aside, the toxicity of some snake venoms is due to the inclusion of phospholipase A. By converting phospholipids in cell membranes into lysophospholipids, the enzyme causes cells to rupture.)

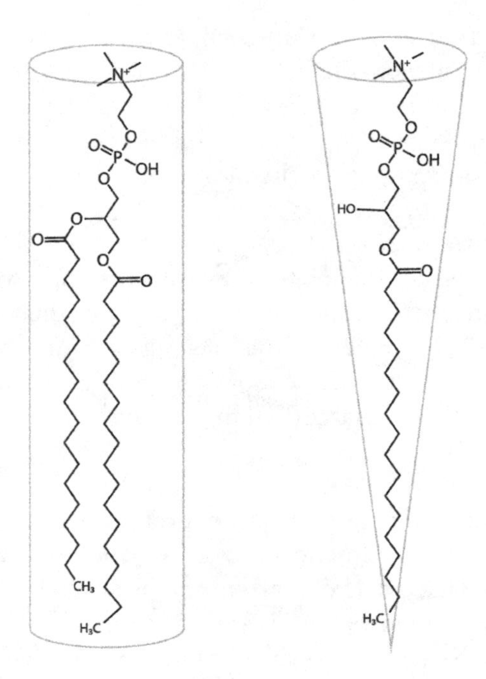

Figure 7.9: Comparison of the Structures of Phospholipids and Lysophospholipids

Fatty Acids

Work by origin-of-life investigators designed to understand the chemical evolutionary pathways that could have conceivably led to the origin of the first cell membranes can also be used to assess the uniqueness of phospholipids. Origin-of-life researchers speculate that the first membrane systems weren't initially comprised of phospholipids, but instead were primitive systems composed primarily of fatty acids (or perhaps other types of single-chain amphiphiles). Most origin-of-life researchers doubt that phospholipids would have been available on the early Earth, because of the difficulties generating these compounds under the prebiotic conditions. On the other hand, reasonable prebiotic routes exist for the formation of fatty acids and other simple amphiphilic materials. Origin-of-life investigators also point out that fatty acids, along with other lipidic materials, could have been delivered to Earth by extraterrestrial materials, such as asteroids and comets. So, despite the tendency of fatty acids (and other amphiphilic materials) to form non-bilayer spherical and tubular micelles, origin-of-life investigators have speculated that these materials could have been the raw materials that chemical evolutionary processes used to construct the first primitive bilayer systems.

Support for this idea comes from studies which demonstrate that fatty acids can form bilayers—under highly specific solution conditions (of exacting pH and temperature, for example)—when mixed with other types of long-chain compounds, such as fatty alcohols.[20] Building off this insight, origin-of-life researchers have demonstrated that pure fatty acids with hydrocarbon chains more than eight carbon atoms long can form bilayer vesicles, again, under a precise set of solution conditions.[21]

Moreover, origin-of-life investigators have observed the formation of bilayer structures from lipid-like materials extracted from the Murchison meteorite, again, under a highly specific set of solution conditions. Likewise, bilayer structures also form from extracts of simulated cometary and interstellar ice that has been irradiated with ultraviolet light.[22] This irradiation process generates organic materials, including lipid-like substances. Because extracts from the Murchison meteorite and simulated cometary ices contain aromatic hydrocarbons mixed with octanoic and nonanoic acids, the authors of these studies believe that life's first membrane systems could have been formed from these materials, with membranes containing phospholipids arriving later in evolutionary history.

For our purposes, it appears that the work of these origin-of-life researchers raises questions about the uniqueness of phospholipids with respect to their bilayer-forming capacities. In other words, experimental evidence indicates that bilayers can readily and spontaneously form from simpler lipid systems, such as fatty acids—particularly when mixed with fatty alcohols—when the right set of solution conditions exist.

This conclusion may be a bit premature. The bilayer vesicles formed from fatty acids are metastable systems that form only at relatively high fatty acid concentrations. For example, the critical micelle concentration (CMC) for nonanoic acid (which consists of 9 carbon atoms) is 95 mM (millimolar). (The CMC is the minimum concentration required by a particular amphiphile in solution for aggregate formation. At concentrations below the CMC, amphiphiles exist as individually dispersed molecules.) By comparison, the CMC for a phosphatidylcholine with two C9 fatty acid chains is 0.029 mM, over 3,000 times lower than the CMC for nonanoic acid. The CMC for a phosphatidylcholine with two C16 fatty acid chains (which is an ideal representative of the types of phospholipids typically found in cell membranes) is 0.046 nM, over 2.0×10^{15} times lower.[23] Clearly, phospholipids have a much greater propensity to adopt bilayer structures than fatty acids.

Fatty acid bilayer vesicles will collapse into spherical micelles if the temperature or pH fluctuates ever so slightly from the precise set of values needed for the bilayer vesicles to form. Also, these bilayer vesicles can only exist in pure water, being highly intolerant of even low levels of salt dissolved in solution.[24] On the other hand, phospholipids readily form stable bilayers under a wide range of temperatures, pH, and solution conditions. This feature makes them ideally suited to form the matrix of cell membranes, given the ever-changing environmental conditions experienced by living systems. Yet fatty acids could never serve as the dominant lipid component in cell membranes because they would be intolerant to fluctuations in the environment and would readily collapse into non-bilayer structures.

Fatty acid bilayers readily collapse into spherical micelles because of the conical geometry of fatty acids. The fact they form bilayers at all is surprising. As it turns out, the strict pH requirement helps explain why fatty acids form bilayers. The head group of fatty acids is a carboxylic acid functional group. Carboxylic acids can exist in one of two forms: un-ionized or ionized. Carboxylic acids react with water to form a carboxylate ion, which bears a negative charge. The extent of this reaction depends on the solution pH. When the pH of the solution is low, the reaction won't occur at all. And when the pH of the solution is high, the reaction will proceed to completion. At a specific pH (called the pK_a), the reaction proceeds to 50 percent completion. Under these conditions, half of the fatty acid species are ionized and half are un-ionized. In the un-ionized form, the fatty acid head group can interact with the head group of an ionized fatty acid through hydrogen bond interactions. These interactions result in the formation of fatty acid pairs in tight association with one another.

The overall geometry of these pairs approximates a cylindrical shape, explaining their capacity to form bilayers—at the just-right pH. The inclusion of a fatty alcohol into the fatty acid bilayer produces a similar effect, with the alcohol functional groups forming hydrogen bonds with the ionized head groups of fatty acids, again, leading to the formation of aggregates with cylindrical shapes. (It should be noted that fatty alcohols exert less of a stabilizing effect than un-ionized fatty acids.) When pH significantly deviates from the pK_a, it alters the proportion of ionized and un-ionized fatty acids, reducing the number and size of the fatty acid clusters. When this change takes place, it reduces

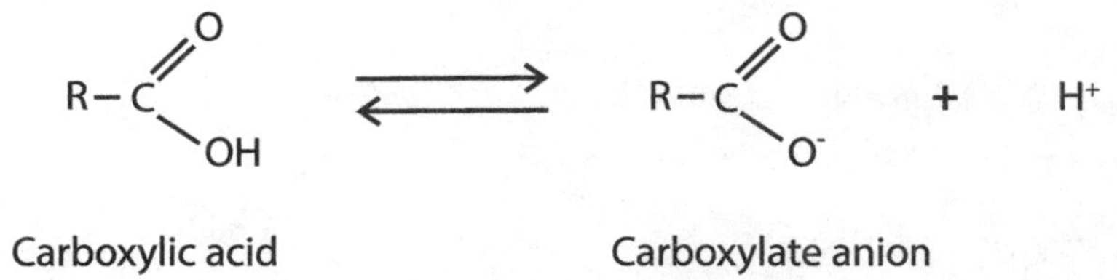

$$R{-}C(=O){-}OH \rightleftharpoons R{-}C(=O){-}O^- + H^+$$

Carboxylic acid Carboxylate anion

Figure 7.10: Ionization Reaction of the Fatty Acid Carboxylic Acid Head Group

the number of cylindrical-shaped lipid aggregates and increases the number of conical-shaped amphiphiles, leading to a collapse of the bilayer structure. This model also explains the sensitivity of fatty acid bilayer vesicles to the presence of salt in solution. The positively charged species in salt's makeup interact with the ionized head groups of the fatty acids, disrupting the aggregates.

Origin-of-life researchers have shown that, beyond fatty alcohols, the inclusion of other lipidic materials also stabilizes fatty acid vesicle bilayers. For example, the one-to-one combination of alkyl amines and fatty acids forms bilayer vesicles that are stable at pH extremes and resistant to the disrupting effects of salts dissolved in the aqueous phase.[25] Once again, this stability arises from the hydrogen-bonding interactions of the fatty acid and alkyl amine head groups. This interaction results in a lipid pair that, effectively, has a large head group and two hydrocarbon tails, yielding a cylindrical geometry for the paired lipids. The formation of this type of bilayer requires a precise 1:1 ratio of fatty acids to alkyl amines. Deviation from this ratio reduces the number of lipid dimers and increases the number of unpaired fatty acids, which are conical in shape, disrupting the stability of the bilayer. As interesting as this result may be, it most likely has no direct bearing on the evolutionary origin of cell membranes, because there is no indication that alkyl amines could form under plausible prebiotic conditions.

Origin-of-life researchers have also learned that incorporating glycerol monoacyl compounds (GMAs) into fatty acid bilayer vesicles stabilizes the bilayer structure.[26] GMAs consist of a fatty acid joined to one of the hydroxyl groups of glycerol via an ester linkage. (See figure 7.11.) Origin-of-life researchers believe that these types of compounds could have been formed under the conditions of early Earth, thereby contributing to the chemical evolution of cell membranes.[27] When GMAs are incorporated into fatty acid vesicle bilayers

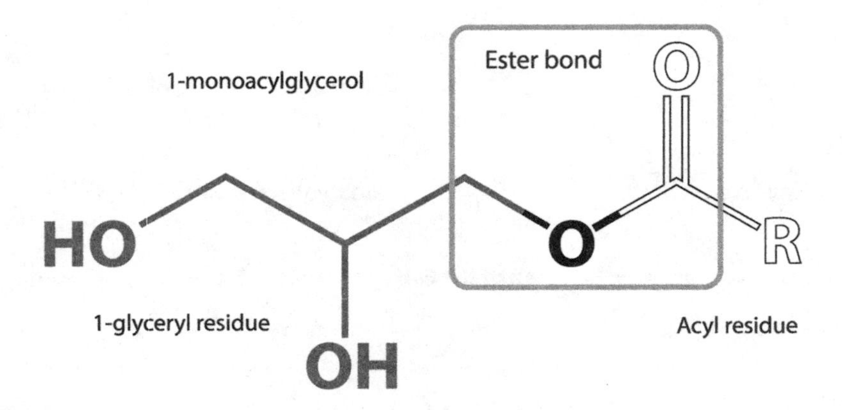

Figure 7.11: A Typical Glycerol Monoacyl Compound

(at a ratio of 1 GMA molecule for every 2 fatty acid molecules), the CMC for bilayer formation is lowered and the temperature stability is enhanced. Still, long-term incubation at elevated temperatures eventually results in bilayer collapse into spherical micelles. So, in short, these types of bilayers are still metastable systems. Unfortunately, GMAs don't seem to have any effect on the sensitivity of fatty acid vesicles to changes in pH or salt in solution.

The bottom line is that fatty acid bilayer systems, whether pure or mixed with other amphiphiles, aren't suitable for forming cell membranes, even though they can form bilayer vesicles. These systems are metastable systems that are, generally speaking, sensitive to temperature, pH, and the ionic strength of the system. Adding other amphiphiles to the fatty acid bilayers can improve the stability of the vesicles, but only when the amphiphiles have the just-right properties and are incorporated into the bilayer at the just-right concentrations.

Another shortcoming displayed by fatty acid bilayers relates to their permeability. The selective permeability of phospholipid bilayers makes them nicely suited for their role in forming the matrix of cell membranes. This limited permeability prevents unwanted water-soluble materials in the environment from entering the cell and ensures that the water-soluble cellular components are retained within the cell's interior. On the other hand, the bilayer vesicles made from fatty acids display a high degree of permeability, readily allowing the influx and efflux of a wide range of materials in and out of the vesicles. The greater permeability of fatty acid bilayers renders them useless as membrane-forming components.[28]

One other limitation of fatty acid bilayers stems from the ionization reaction that the fatty acid head group undergoes in water. (See figure 7.10) This reaction is reversible, meaning that the fatty acid head group undergoes ionization when the pH is high (and the hydrogen ion concentration is low) and becomes un-ionized when the pH is low (and the hydrogen ion concentration is high). Because of this chemical behavior, fatty acid membranes cannot establish and maintain a proton gradient across the bilayer. As we will see in the next chapter, the formation of proton gradients across cell membranes plays a central role in the energy-harvesting biochemical processes the cell uses to power its operations. When proton concentration is high on one side of the membrane, the ionized form of the fatty acid head group will react with it, forming a un-ionized carboxylic acid. In the un-ionized form, the fatty acid can flip-flop in the bilayer, moving from one monolayer to the other. In this new orientation, the fatty acid head group will be exposed to low proton concentrations, causing the fatty acid to undergo an ionization reaction. Over time, this process will prevent a proton gradient from forming (or cause an existent proton gradient to dissipate). On the other hand, phospholipid bilayers are ideally suited for forming and sustaining proton gradients across the bilayer because they are impermeable to positively charged hydrogen ions.

Lipoamino Acids and Lipopeptides

Origin-of-life researchers have also contemplated membrane systems composed of lipoamino acids and lipopeptides. Origin-of-life investigator Gordon Sproul has shown that these compounds—which consist of fatty acids linked to either amino acids or small peptides via an ester linkage—can be made under simulated conditions thought to be present on early Earth by heating fatty acids and amino acids (or small peptides) in the presence of certain dehydrating agents.[29] Sproul thinks that these compounds could have served as an important lipid component in primitive membrane systems. It is not likely that these compounds could ever be the single or dominant lipid component in cell membranes, however, because aggregates of lipoamino acids adopt micellar and other non-bilayer phases in water.[30] These results make sense considering the conical molecular geometry of these amphiphiles. Lipoamino acids and lipopeptides can form vesicle bilayers when combined with other single-chain amphiphiles that possess the right type of head group structure that leads to the formation of lipid pairs.

Mixed Lipid Systems

In effect, the work of origin-of-life researchers exemplifies a general principle discovered by chemists and material scientists who work on surfactant systems. These researchers have learned that, regardless of its chemical identity (whether synthetic or a natural product), single-chain amphiphiles can form bilayers when mixed with the just-right lipid companion.[31] For these mixtures to form bilayer vesicles, the head groups of the two different amphiphiles must interact in such a way that they form a molecular pair. When this happens, the conical geometry of each of the individual amphiphiles transforms into the cylindrical geometry of the paired molecular species. Alternatively, mixed single-chain amphiphiles can form bilayer structures if the geometry of the two amphiphiles offset, so that one adopts a conical shape and the other forms an inverted cone. As origin-of-life researchers have discovered, these types of bilayer systems require a precise ratio of the individual components (typically a 1:1 ratio) and are often metastable, being sensitive to the solution conditions.

This survey points to the uniqueness of phospholipids, highlighting how remarkably well-suited they are for forming the bilayer matrix of cell membranes. Phospholipids aren't alone in their capacity to form bilayer structures, but compared to all reasonable alternatives, they are the only amphiphilic system that readily forms bilayers at extremely low lipid concentrations. The bilayers that they form are invulnerable to changing solution conditions and environmental temperatures. Phospholipid bilayers are also impermeable to a wide range of materials, making them ideally suited as a biochemical barrier.

Non-bilayer–Forming Phospholipids

Most phospholipid species possess a cylindrical shape and, hence, readily form bilayers when they aggregate in solution. However, there are some phospholipid species that adopt an inverted conical geometry. These phospholipids form inverted micelles when they aggregate in solution.[32] Included in this list of phospholipids are some PEs. At first glance, it is surprising to realize that non-bilayer–forming phospholipids (along with other non-bilayer–forming membrane components, such as certain glycolipids and cholesterol) occur in cell membranes. However, their presence doesn't destabilize the membrane structure due to the strong bilayer-forming tendencies of most phospholipid species present within the cell membrane. As it turns out, these non-bilayer–forming phospholipids serve an important biological function in the fusion of cell membranes.[33] Biological membranes can fuse with one another. Cell membranes can also "bleb," with a portion of the membrane becoming pinched

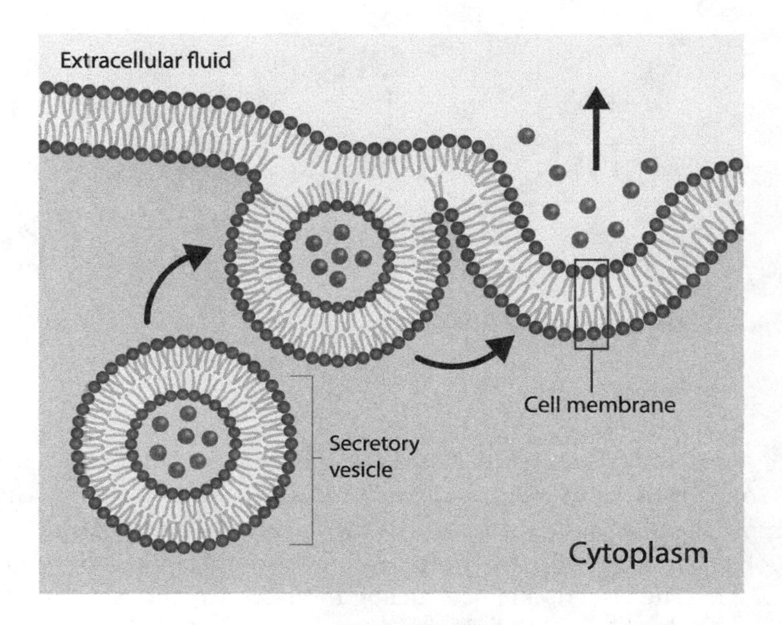

Figure 7.12: The "Blebbing" Process during Endocytosis and Exocytosis

off to form a vesicle. (See figure 7.12.) This process occurs during endocytosis and exocytosis. Membrane fusion and blebbing involve non-bilayer structures that form as transient intermediates. Biochemists believe that the non-bilayer–forming phospholipids play a critical role in this process by temporarily forming non-bilayer domains within the cell membrane. It appears that an exquisite molecular rationale exists for the inclusion of non-bilayer–forming phospholipids in cell membranes.

The Chemical Structure of Phospholipids

Phospholipids possess the just-right thermodynamic properties and geometric features that make them ideally suited to form the bilayer structure of cell membranes. But what about the chemical structure of phospholipids? Is there a molecular rationale that explains the chemical makeup of these biomolecules?

As we discovered, for phospholipids to form bilayer structures, these molecules must have a cylindrical geometry with a distinct head group region that is hydrophilic and a distinct tail region, comprised of two long hydrocarbon chains that are hydrophobic. This arrangement requires a molecular species to

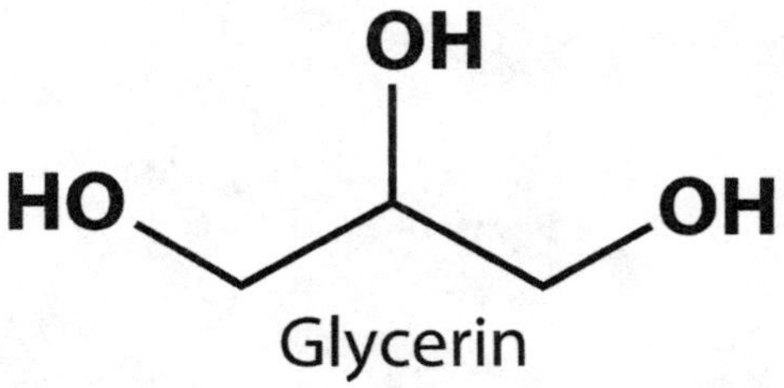

Figure 7.13: The Chemical Structure of Glycerol

serve as an interface between the head and tail regions with three attachment points, one for the head group and two for the hydrocarbon chains. The simplest molecule that can serve this role is the three-carbon compound glycerol. This polyol has three hydroxyl groups, each one bound to a carbon atom. (See figure 7.13.) The closest chemical analog to glycerol is the two-carbon diol, ethylene glycol. This molecule has only two attachment points. To make an amphiphilic molecule, the hydroxyl groups of ethylene glycol would have to bind a hydrocarbon chain and the head group. While amphiphilic, these types of compounds would have a conical geometry.

In phospholipids, the phosphate group forms bonds with the glycerol backbone and an additional substituent (either an amino alcohol, polyol, or amino acid) that gives the head group—and, hence, the phospholipid—its distinct chemical identity. The phosphate moiety is ideal for its role as a head group. As discussed in chapter 5, phosphates can form diester linkages. Recall that the two closest chemical analogs to the phosphate moiety are sulfate and arsenate. While a sulfate can form diester linkages, it would be unsuitable to form the head group. When the phosphate group takes part in a diester linkage, it still retains a negative charge. This charge is critical because it ensures that the head group interacts strongly with the aqueous phase, creating the amphiphilic character needed for bilayer formation. On the other hand, when the sulfate group participates in a diester linkage, the sulfate is neutral, making it less suitable for bilayer formation, since the head group would not interact as strongly with the aqueous phase.

It is possible that sulfate could be used as a head group by forming a single ester linkage with the glycerol backbone. In this type of bonding arrangement, the sulfate would retain its negative charge, successfully contributing to the resulting compound's amphiphilicity. Unfortunately, a mere sulfate head group would not be ideal for bilayer formation. Part of the head group size in phospholipids comes from the substituent group. Without this moiety, the head group would be too small, resulting in an inverted, truncated conical geometry.

One of the phospholipid species that occurs in cell membranes at relatively low levels is phosphatidic acid. Unlike other phospholipids, the head group of phosphatidic acid consists solely of a phosphate moiety, lacking a substituent on the phosphate head group. The absence of a substituent bound to the phosphate renders the head group relatively small. Consequently, phosphatidic acids adopt an inverted, truncated conical geometry and tend to form inverted micelles. While their tendency is not to form stable bilayers, phosphatidic acids have minimal impact on the integrity of the bilayers of cell membranes, because they occur at low levels in cell membranes, serving as metabolic intermediates and signaling molecules.

Additionally, the absence of the head group constituents would strip the sulfate analogs of their chemical distinctives. Therefore, only a single type of lipid species (based on head group structure) would exist. Head group diversity plays a critical role in the other biochemical functions performed by phospholipids, beyond their bilayer-forming role, such as their interactions with membrane proteins. (See "Why Are Cell Membranes So Complex?")

The use of fatty acids to form the hydrocarbon tails of phospholipid molecules seems intuitive. Fatty acids are the simplest molecules with a long hydrocarbon tail that will readily form a linkage with the hydroxyl groups of the glycerol backbone, thanks to the carboxylic acid head group. Fatty acids also play a central role in energy metabolism. For this reason, these compounds are readily available in the cell. Using these biomolecules as molecular components for phospholipids creates an efficiency for the cell. Instead of having to synthesize fatty acids from scratch, so that they can be used to make phospholipids, the cell's machinery can simply make use of compounds already present in the cell.

In short, there appears to be an exquisite molecular logic that undergirds the chemical features of phospholipid molecules. Every aspect of their structure plays an indispensable function in service to the role phospholipids play in forming the bilayers of cell membranes and, as we will soon discover, their role interacting with membrane proteins. This underlying molecular rationale suggests that the selection of phospholipids as the dominant membrane lipid component wasn't merely happenstance. Instead, it appears to reflect deep-seated constraints that arise out of the laws of nature. Yet a detractor might argue against this conclusion, pointing out that a different set of phospholipids is used to construct cell membranes in the Archaea.

Archaeal Phospholipids and the Lipid Divide

Biologists have defined three domains of life: Eukarya, Bacteria, and Archaea. The phospholipid species described and discussed up to this point in the chapter are found exclusively in the Eukarya and Bacteria domains. The phospholipids found in the third domain of life, Archaea, are chemically distinct. This difference suggests that perhaps the phospholipid constituents of cell membranes may not be as uniquely fit as we thought. If so, it makes it more reasonable to view the lipid composition of cell membranes as the consequence of a historically contingent evolutionary pathway. This would deal a blow to the idea that cell membranes point to the existence of a biochemical anthropic principle.

Yet a careful consideration of the structural and physical properties of archaeal phospholipids indicates that it may be a bit premature to abandon the case for the biochemical anthropic principle. As we will soon discover, there does appear to be a rationale for the differences in phospholipid structure of members of Archaea compared to the structure of phospholipids found in Bacteria and Eukarya.

Life scientists refer to the differences in the structure of phospholipids found in Archaea compared to those found in Bacteria and Eukarya as the lipid divide.[34] All three domains of life make use of phospholipids as the dominant lipid component of their cell membranes. Accordingly, the phospholipids found in all three domains possess a phosphate head group bound to an alcohol substituent. In Bacteria, ethanolamine and glycerol are the most prominent phospholipid head groups. In Eukarya, the most prominent head groups include choline, ethanolamine, glycerol, serine, and inositol as the head group substituents. In Archaea, the head group constituents consist of ethanolamine, glycerol, serine, and inositol—but not choline. The phospholipids found in all three domains of life also make use of glycerol as the interfacial backbone of the molecule. The similarity ends there.

To appreciate one of the most significant differences between the two classes of phospholipids, I need to introduce the concept of chirality. It turns out some molecules can exist in two forms that are mirror images of each other. These two versions of the same molecule result when four different chemical constituents bind to a central carbon atom. (The central carbon atom is called the chiral carbon.) These chemical groups can orient in space in one of two

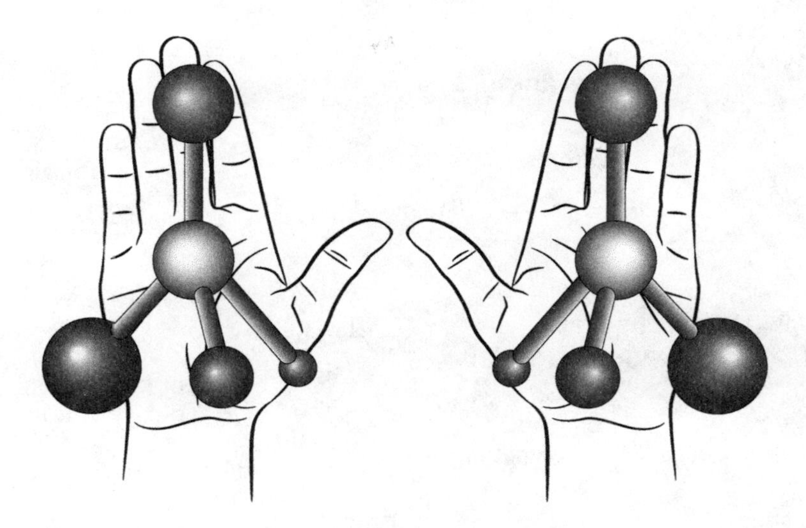

Figure 7.14: Chirality

possible arrangements that turn out to be mirror reflections of each other. As mirror images, these compounds cannot be overlaid on one another so that all the chemical groups coincide in space. Because they can't be superposed, these molecular mirror images (called enantiomers) are distinct chemical entities. (See figure 7.14.) Biochemists use the d- and l- label as one way to distinguish between the two enantiomers.

Chirality is a very important feature of many biomolecules. For example, most of the canonical amino acids, and the sugars deoxyribose and ribose, are chiral compounds. In fact, the amino acids that comprise proteins and the sugars that are the constituents of DNA and RNA have uniform chirality, a condition biochemists call homochirality. In other words, all the amino acids in proteins have the same chirality. And all the sugars in DNA and RNA have identical chirality as well.

Homochirality is a strict requirement for life. Chirality dictates the three-dimensional positioning of chemical groups in space. And the spatial location of the chemical moieties (equal parts) plays an essential role in the interactions that stabilize the three-dimensional structure of proteins and the nucleic acids. (See chapters 4 and 5.) As a case in point, for some proteins the incorporation of even one amino acid of the opposite mirror image into its backbone will disrupt the protein's capacity to form a stable three-dimensional structure and,

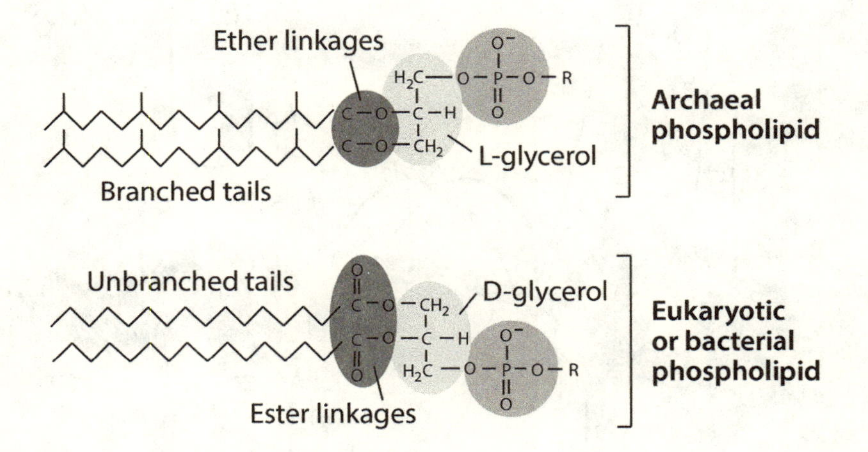

Figure 7.15: Differences in Chemical Structures of Phospholipids from Archaea, Bacteria, and Eukarya

hence, its function.

As it turns out, when glycerol has two different chemical constituents bound to the hydroxyl groups of its first and third carbon atoms, the second carbon atom becomes a chiral center. For the phospholipids found in Eukarya and Bacteria, the backbone is made up of d-glycerol, which has a phosphate moiety bound to the glycerol on the third carbon atom (called the *sn*-3 position). The two fatty acids are bound to the d-glycerol backbone at the *sn*-1 and *sn*-2 positions. In contrast, archaeal phospholipids are constructed around a l-glycerol backbone. The phosphate moiety is attached to the *sn*-1 position of glycerol. Two hydrocarbon chains are bound to the *sn*-2 and *sn*-3 positions of l-glycerol.

The hydrocarbon chains also differ for these two classes of phospholipids. Instead of consisting of fatty acids linked to the glycerol backbone via an ester linkage, archaeal phospholipids make use of hydrocarbons called isoprenoids linked to the glycerol backbone via an ether linkage. (See figure 7.15.) Isoprenoids consist of a polyprenyl chain decorated with regularly repeating methyl branches. These hydrocarbon chains also incorporate cyclopentyl substituents. The hydrocarbon chains of fatty acids in the phospholipids from Bacteria and Eukarya range from C12 to C22 in length, with the most common being the C14, C16, and C18 species. On the other hand, the hydrocarbon

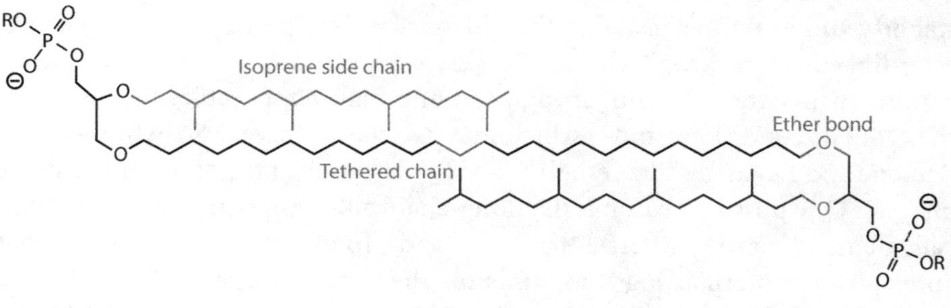

Figure 7.16: An Example of a Tetraether Bipolar Phospholipid

chains of the archaeal phospholipids typically span 20 to 40 carbon atoms in length.

In addition to these types of phospholipids, some members of Archaea contain highly unusual phospholipids described as tetraether bipolar phospholipids. (See figure 7.16.)[35] These phospholipids are organized around a "di-isoprenyl chain" that culminates with an ether group at both ends. Each of the ether groups binds to a separate l-glycerol molecule at the *sn*-2 position. Each of the l-glycerol molecules, in turn, binds a phosphate head group in the *sn*-1 position and an isoprenyl ether at the *sn*-3 position. When incorporated into a bilayer, one head group locates to the inner bilayer surface and the other locates to the outer bilayer surfaces. The hydrocarbon chains bound to the *sn*-3 position contribute to the inner and outer monolayers.The di-isoprenyl chain bound to the *sn*-2 positions of the two glycerol backbones spans the length of the bilayer interior. In effect, the di-isoprenyl chain covalently links together the inner and outer monolayers. For this reason, pure bilayers made from tetraether bipolar lipids are unusually stable.

The unusual stability of these types of bilayers provides a rationale for why archaeal phospholipids differ from phospholipids found in Bacteria and Eukarya. Most Archaea live in extreme environments (high temperatures, high acidity, high alkalinity, high salinity, etc.). And archaeal phospholipids are ideally suited for these locales—much more so than the phospholipids found in Bacteria and Eukarya. For example, ether linkages are more chemically stable than ester linkages, explaining why ether-linked hydrocarbon chains are found in the Archaea. Laboratory studies demonstrate that bilayers made up of pure phospholipids containing ether-linked hydrocarbon chains display greater stability and lower permeability than bilayers made up of phospholipids with ester-linked hydrocarbon chains. Bilayers comprised of ether-linked hydrocarbon chains are stable under high-salt and high-temperature conditions—the types of conditions under which many Archaea thrive.[36] So, why wouldn't Bacteria and Eukarya utilize ether-linked lipids as well, if they form more stable bilayers? One possible reason is that long-chain alkyl ethers are more metabolically demanding to synthesize than fatty acids. In other words, it appears as if an exquisite molecular logic accounts for the differences in the structures of phospholipids in Archaea and Bacteria and Eukarya. For this reason, it appears as if the phospholipid compositions of Eukarya, Bacteria, and Archaea aren't happenstance.

Why Are Cell Membranes So Complex?

By this point, it should be clear that compositional complexity stands as one of the characteristics of cell membranes. This leads us to ask: Is this complexity necessary? Or is it merely the consequence of a historically contingent evolutionary history? According to the latter view, over vast periods of time, a variety of different phospholipid molecules found their way into membranes as metabolic pathways for the production of the various phospholipids emerged and diversified under the auspices of natural selection. If so, it challenges the idea that cell membranes manifest a biochemical anthropic principle. On the other hand, it is possible that a rationale exists for the complex compositional makeup of cell membranes. If this is the case, it means that phospholipids must assume roles beyond bilayer formation. As biochemists P. R. Cullis and B. de Kruijff point out:

> The fact remains however, that a single phospholipid species such as phosphatidylcholine could satisfy such structural requirements. In this context, the observation that a typical mammalian cell membrane contains one hundred or more distinctly different lipids implicitly suggests that lipids play other functional roles.[37]

Biochemists now recognize that the seemingly chaotic mix of phospholipids in cell membranes *is* necessary and points to a deep rationale that underlies the membrane's composition.[38]

Membrane Fluidity

A large variety of fatty acids (differing in chain lengths and degree of unsaturation) find their way into phospholipids. Biochemists now have some insight into the rationale for this chemical diversity. The physical properties of cell membranes are impacted by the variable length and degree of unsaturation of the phospholipids' hydrocarbon chains. A preponderance of phospholipids with short hydrocarbon chains and/or hydrocarbon chains with kinks (caused by unsaturation) cause the bilayer's interior to become fluid and liquid-like. On the other hand, if bilayers are comprised of phospholipids with longer hydrocarbon chains and chains that are straight, lacking sites of unsaturation, the cell membranes possess solid-like interiors.[39]

The cell membrane's fluidity has important biological consequences. As a case in point, the bilayer's physical state regulates the function of integral proteins. Regional differences in phospholipid composition create locales within the bilayer's interior with varying degrees of fluidity, with some areas solid-like and others more liquid-like. These fluidity differences help segregate the cell membrane's components into functionally distinct domains within the bilayer. In short, phospholipids with a wide range of hydrocarbon chain lengths and shapes (straight or kinked) make it possible for the cell to precisely adjust bilayer fluidity. Without a wide range of chemically distinct phospholipid species, it becomes difficult to envision how life could achieve both the precise regulation of membrane protein activity and the creation of functionally distinct domains.

Protein Interactions

Phospholipids do more than form bilayers. They play a critical role in controlling the activity of proteins associated with the cell membranes and, in some instances, those located in the cytoplasm.[40] This control extends beyond simply dictating bilayer fluidity. Phospholipids that serve as annular lipids regulate protein function through direct and highly specific interactions, usually involving their head groups.

As a case in point, phospholipids with glycerol and serine head group substituents bind to target proteins through head group-mediated interactions. These two phospholipids are both negatively charged and have distinct chemical features. Both properties dictate their interactions with membrane proteins. In bacteria, PGs activate some of the proteins involved in replicating DNA, in assembling the outer membrane, and in moving proteins across cell membranes. In eukaryotic cells, PSs play a central role in activating proteins

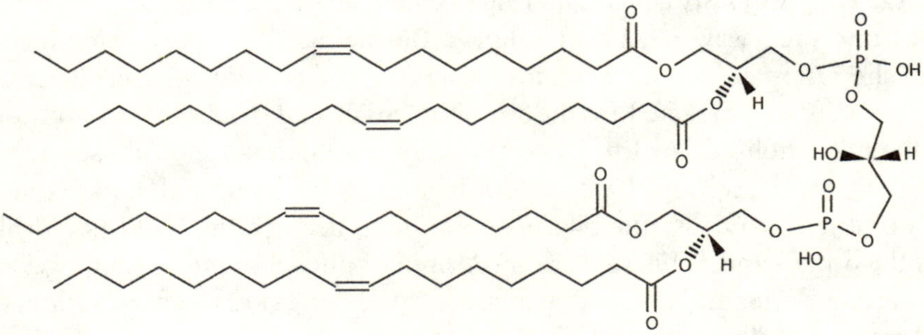

Cardiolipin

Figure 7.17: The Chemical Structure of Cardiolipin

that are part of the cell signaling pathways in the cytoplasm. These pathways alter the cell's metabolism in response to changes happening outside of the cell.

Phospholipids with choline and ethanolamine as part of their head groups are neutral in charge. Interestingly, these two types of phospholipids have regions that are both negatively and positively charged. (The charges cancel each other to yield overall electrical neutrality.) PCs and PEs are the major phospholipid components of cell membranes. Even though PCs and PEs bear a neutral charge, their chemical structures differ sufficiently so that these two phospholipids play distinct roles in cell membranes. The primary role of PCs is bilayer formation. PEs also serve in this capacity, particularly in Bacteria and Archaea, which lack phosphatidylcholines. As pointed out, phosphatidylethanolamine-rich regions in cell membranes can adopt non-bilayer structures. These non-bilayer phases regulate the activities of some proteins and play a central role in cell division and membrane fusion events. PEs also trigger the activity of membrane-embedded proteins that shuttle materials across cell membranes.

The unusual phospholipid cardiolipin occurs in bacterial membranes and in the inner membranes of mitochondria. Cardiolipin forms when two phosphatidylglycerol molecules react, producing a diphosphatidylglycerol (aka cardiolipin). In effect, cardiolipin can be envisioned as a molecule formed when two phosphatidic acid molecules share the same glycerol head group with a phosphatidic acid bound to the hydroxyl groups of the first and third carbon atoms of glycerol, respectively.

Biochemists learned that cardiolipin—which constitutes close to 20 percent of the phospholipid component of the inner membrane of mitochondria—forms close associations with several proteins found in this membrane. These proteins play a role in harvesting energy for the cell to use. Compared to other lipid components found in the inner membrane, cardiolipin appears to associate preferentially with these proteins. Evidence indicates that cardiolipin helps to stabilize the structures of these proteins and serves to organize the proteins into larger functional complexes within the membrane.[41]

Of interest are the interactions between cardiolipin and F_1-F_0 ATPase. Embedded within the inner membrane of mitochondria, this complex is a biomolecular rotary motor that produces the compound ATP—an energy storage material the cell's machinery uses to power its operations. Instead of binding permanently to the surface of the protein complex, cardiolipin interacts with F_1-F_0 ATPase dynamically. Researchers think this dynamic association and the unusual chemical structure of cardiolipin (which gives it the flexibility to interact with a protein surface) are critical for its role within the mitochondrial inner membrane. It turns out cardiolipin not only stabilizes the F_1-F_0 ATPase complex (as it does for other inner membrane proteins), but it also lubricates the protein's rotor, allowing it to turn in the viscous cell membrane environment. Also, its unique structure helps move protons through the F_1-F_0 ATPase motor, providing the electrical power to operate this biochemical motor.[42]

This brief survey of phospholipid function demonstrates that phospholipids are intimately involved in a legion of biochemical processes. As biochemist William Dohan notes:

> The wide range of processes in which specific involvement of phospholipids have been documented explains the need for diversity in phospholipid structure and why there are so many membrane lipids.[43]

In other words, cell membrane complexity doesn't appear to be solely the outworking of unguided historically contingent evolutionary processes. Instead, it appears to reflect a deep-seated molecular rationale, with each phospholipid serving a necessary role in forming the cell membrane matrix, contributing to domain structure, and mediating the function of membrane proteins.

Do Cell Membranes Reflect the Biochemical Anthropic Principle?

The goal of this chapter is to determine whether cell membranes provide evidence for the biochemical anthropic principle. Because of the complexity of cell membranes, my assessment adopted an approach that focuses on the structures and properties of the components that form cell membranes: proteins and lipids. Chapter 4 presented evidence that proteins evince the biochemical anthropic principle. So in this chapter I examined phospholipids—the dominant lipid component of cell membranes. Given their importance for living systems and given the abundance of phospholipids in cell membranes, I predicted that phospholipids should, too, offer added support for the existence of the biochemical anthropic principle.

Making use of the same criteria employed in previous chapters, it can be determined that:

- An exquisite molecular logic does indeed exist and accounts for the chemical structure of the phospholipids found in Bacteria and Eukarya. Likewise, a molecular rationale explains the differences in the structure of phospholipids found in Archaea, compared to Bacteria and Eukarya. A molecular rationale also accounts for the extensive chemical diversity of phospholipids found in cell membranes.
- Physicochemical constraints arise from the laws of nature that dictate the properties of phospholipids. To form bilayers in water, phospholipids must form cylindrically shaped molecules with a highly hydrophilic head region and a strongly hydrophobic tail region. This amphiphilicity drives phospholipids to form aggregates in water, due to the hydrophobic effect, which ultimately arises out of the second law of thermodynamics. The geometric shape of the phospholipid influences the way individual molecules pack, allowing bilayers to form. The geometric shape is described by the concept of a packing parameter, developed by Jacob Israelachvili. It is remarkable that phospholipids found in membranes possess the just-right packing parameter to form bilayers.
- Phospholipids are unusually, if not uniquely, suited for life. Other amphiphiles that would be conceivable alternatives to phospholipids—such as lysophospholipids, fatty acids, and lipoamino acids and lipopeptides—don't naturally form bilayer structures in water. Under the right set of conditions they can form bilayers. However, the bilayers they form are metastable, readily collapsing into non-bilayer phases

> with even the slightest fluctuations in environment and solution conditions. In other words, phospholipids appear to be unusually well-suited to form bilayers.

In effect, phospholipids satisfy all three criteria for the biochemical anthropic principle, as expected. The next chapter will continue the quest to determine whether the biochemical anthropic principle is real by focusing on the cell's energy-harvesting pathways. These pathways are an indispensable necessity for life. If the biochemical anthropic principle is valid, these systems should evince it.

Chapter 8

Energy-Harvesting Pathways

I'm obsessed with the blues. It isn't just the musical sounds and style of the blues that account for my love for the genre. It's my fascination with the colorful characters that have populated the history of the blues—along with the riveting personalities of the blues scene today.

If any blues musician holds enduring interest for me, it is the late Johnny Winter (1944–2014). Born in Beaumont, Texas, he was one of the first blues rock guitar virtuosos, pioneering the blues power trio with his 1968 debut album *The Progressive Blues Experiment*.

In 1968, Winter signed with Columbia Records, receiving $600,000—the largest ever advance paid by the record industry to any musical artist at that time. Winter released a flurry of albums in the late 1960s and 1970s, beginning with his 1969 blues-oriented offerings *Johnny Winter* and *Second Winter*.

Winter had an arresting stage presence. Both he and his brother Edgar Winter were born with albinism. Winter's pale skin and long white hair highlighted his flamboyant and flashy appearance on stage. Winter sang with a distinct blues howl that accompanied his high-octane guitar playing. He played the guitar fast and furiously.

I witnessed the power of his performances at the Canyon music venue in Agoura Hills, California, a few years before he died. It was a bucket list experience for me. At that time, Winter was in his late sixties. The life he led as a blues and rock musician had caught up with him. He looked extremely frail and feeble, in obvious poor health. So much so, he had to be helped onto the stage and carefully escorted to a chair placed front and center. He couldn't even carry his own guitar. The aging process had stripped Winter of his flair and flamboyance—but it didn't quench his spirit. The scene is forever etched in my memory. Johnny Winter sat down. A crew member handed him his guitar. And suddenly sonic fireworks ensued. Music just exploded from his guitar with the same ferocity as in his younger years. The power of his playing that night simply didn't comport with Winter's diminutive appearance on the stage.

For many people, Winter's high-intensity, hyperactive approach to playing the blues stands as his most innovative contribution to the genre. But perhaps his most significant contribution was his work in the late 1970s as a record producer. In 1974, Johnny Winter invited Muddy Waters to record for Blue Sky Records. It resulted in four commercially successful and critically acclaimed albums (*Hard Again*, *I'm Ready*, *King Bee*, and the live album *Muddy "Mississippi" Waters Live*), three of which won Grammys. This success gave a late-life resurgence to the career of the legendary bluesman. Muddy Waters was well on his way to becoming an afterthought in the early 1970s, recording and releasing a series of albums that were uninspired at best. (In blues parlance, Waters had lost his mojo.)

I am almost always in the mood to hear a Johnny Winter song. No matter the occasion. If I am feeling upbeat, his music helps me celebrate. If I am in a funk, the energy and intensity of his guitar playing picks me up.

I find Johnny Winter's life in music nothing short of inspirational. Instead of his albinism holding him back, Winter embraced it and made it part of his onstage persona. Against all odds, he overcame very real challenges and limitations to achieve musical success. As journalist David Marchese writes in a tribute to Johnny Winter:

> It's the barest facts that remain the most inspiring. Johnny Winter, from little Beaumont, Texas, afflicted with albinism and 20/400 eyesight in one eye and 20/600 in the other, made an iconic life for himself by playing the blues. What are the odds of that story coming true? What levels of self-belief, resilience and talent did it take to transform those biographical details . . . into the stuff of a legendary career? Winter was . . . a testament to the idea that with a lot of skill and dedication and more than a little luck, music can open any door.[1]

Whenever I listen to a Johnny Winter song, I can't help but marvel at the gritty tenacity that he displayed throughout his life. Winter's example has long motivated me not to squander the gifts I have been given. So, whenever I begin to feel discouraged or treated unfairly by life, I always come back to this point:

If Johnny Winter could achieve success, overcoming the physical limitations of his albinism, then what is my excuse?

In recent years, Winter's success in music has served as fresh inspiration for me. I lost the sight in one of my eyes due to an archery accident. It impaired my vision to a greater degree than I thought it would immediately after the injury. After a short bout of feeling sorry for myself, I decided to embrace my vision loss as a challenge, not a limitation. I pressed on.

I first learned about Johnny Winter as a teenager thanks to the music of his brother Edgar Winter. The two Winter brothers frequently collaborated, appearing on one another's albums and performing together on stage. Unlike Johnny, who focused his efforts on the guitar, Edgar was a multi-instrumentalist who played the guitar, keyboards, and saxophone. The music written and recorded by Edgar Winter and his bands spanned several genres including blues, rock, jazz and pop. Winter's greatest success came in the 1970s with his band White Trash and, later, with the Edgar Winter Group. Their 1972 release, *They Only Come Out at Night*, is one of my favorite rock albums. The two singles from the album—"Free Ride" and the rock instrumental "Frankenstein"—are among the most recognizable rock songs of all time, still receiving radio airplay on classic rock stations and serving as the soundtrack for commercials and video games. These two songs often find their way into the rotation of songs in most of my playlists. Yet the track I like the best on the album is the obscure tune "Round & Round." A country rocker with pop sensibilities, this song describes the whirlwind of confusion—and frustration—that often surrounds the relationship between two people in love. And somehow, their love for one another manages to grow, even amid their struggles.

So round and round and round she goes
And where she stops nobody knows
Love is the feelin' that grows that grows
And where it stops nobody knows

This song never fails to evoke in me a sense of hope, buoying me during those times that I experience confusion and frustration. Whenever I listen to this song, I know that confusion and frustration are part of life, and somehow things will get better—even if I can't see how.

This song also harbors meaning for me in another way. I often found myself humming or singing it during graduate school, particularly when studying intermediary metabolism in my biochemistry course work.

• • • • • • • • • • • •

Biochemists use the term *metabolism* to reference the myriad chemical reactions that occur inside the cell. Without these chemical processes life simply would not be possible. Living systems rely on metabolic activity to extract energy from the environment and construct life's components. Through these processes, organisms grow, reproduce, maintain biological structures, and respond to changes in the environment. As we saw in chapter 6, metabolic reactions include DNA replication and the production and breakdown of proteins and RNA molecules. They also include the assembly of cell membranes and cell walls.

Metabolism also refers to the chemical interconversion of small molecules. A significant number of metabolic reactions produce subunit molecules the cell's machinery utilizes as building blocks to assemble proteins, DNA, RNA, and cell membrane bilayers. On the other hand, some metabolic activities break down compounds, such as glucose and fats, into smaller molecules. This breakdown process provides energy for the cell's operations. Some metabolic activities prepare materials the cell no longer needs for elimination as cellular waste. Other reactions detoxify materials harmful to the cell.

Within the cell's interior, metabolic processes are often organized into pathways comprised of a sequential series of chemical reactions. These reactions transform a starting compound into a final product via a series of small, stepwise chemical transformations. Biochemists refer to each molecular species in the sequence as a *metabolic intermediate*. For this reason, these pathways are collectively grouped under the umbrella of *intermediary metabolism*.

Each step in a metabolic route is mediated by a protein (called an enzyme) that assists in the chemical transformation. These pathways can be linear, branched, or circular. An example of a linear pathway is glycolysis, and an example of a circular pathway is the Krebs cycle (or the tricarboxylic acid cycle). See figures 8.1A and 8.1B.

Whenever I would sit down to study the Krebs cycle, I couldn't help but recall the song "Round & Round" as I envisioned the intermediates in this metabolic pathway going around and around, hoping to know where they would stop. Because if I didn't, it wouldn't bode well for my performance in class.

To say that intermediary metabolism is complex is an understatement. It can be bewilderingly complex. Part of this complexity arises from the fact that the chemical components that form a segment of a particular metabolic

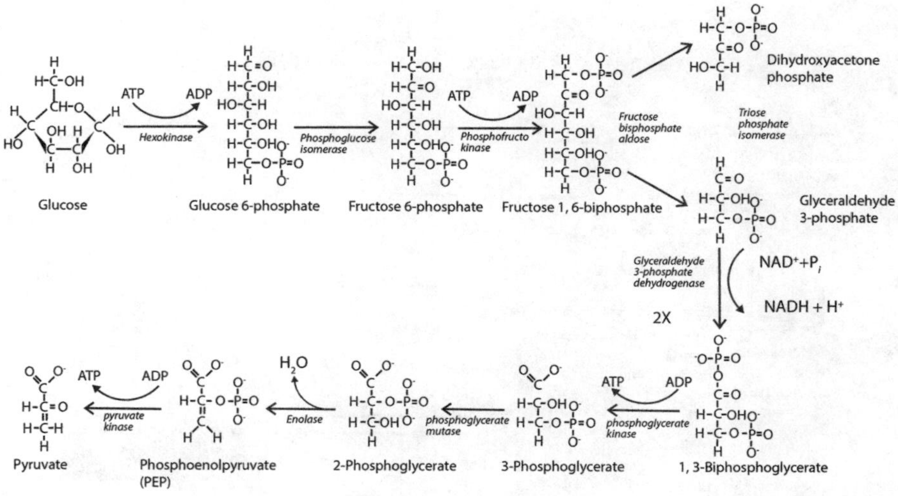

Figure 8.1A: Linear Metabolic Route: The Glycolytic Pathway

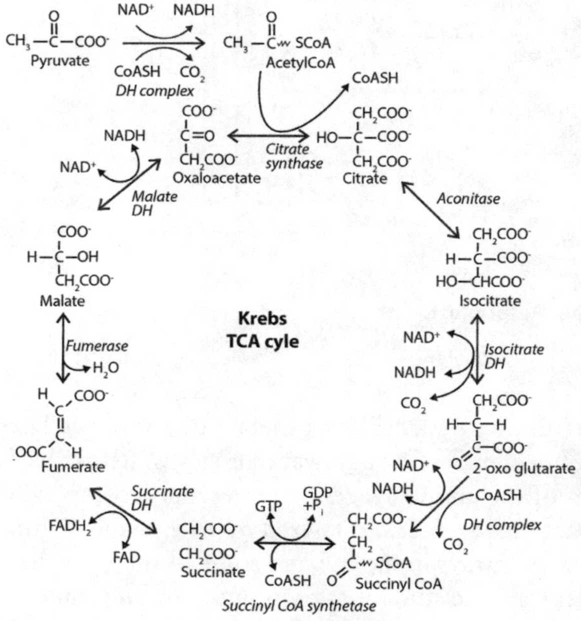

Figure 8.1B: Circular Metabolic Route: The Krebs Cycle

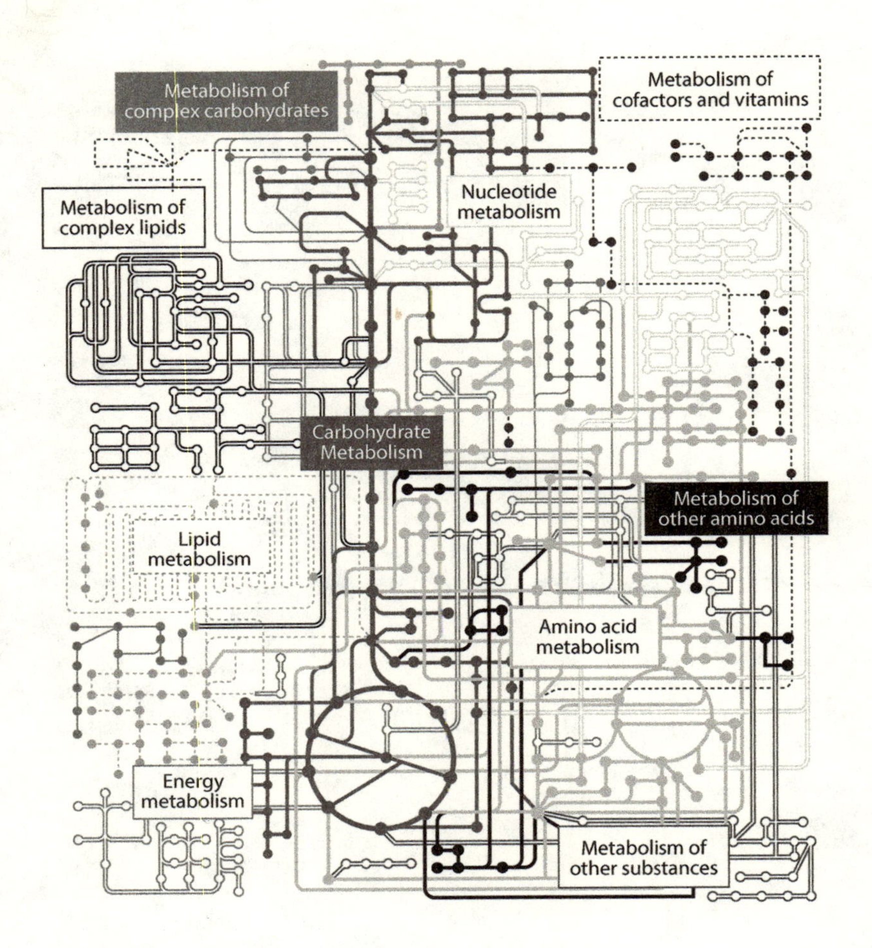

Figure 8.2: The Metabolic Chart

sequence sometimes take part in other metabolic pathways. These shared compounds interconnect metabolic pathways into a vast network of chemical reactions. The sum of metabolic processes represents a complex, reticulated web of chemical reactions, each one catalyzed by an enzyme. Figure 8.2 shows the major metabolic pathways in the cell and some of the ways they interconnect.

As an undergraduate student attempting to master intermediary metabolism, I often felt frustrated as I worked to commit to memory all the chemical intermediates, enzymes, points where the pathways connected, etc. At times, it felt as if there was no rhyme or reason to intermediary metabolism. I knew people who gave up on biochemistry because they were unwilling to do the hard work that went along with trying to grasp intermediary metabolism. However, I found the complexity alluring. I loved to immerse myself in the seemingly never-ending intricacies of metabolic pathways as I tried to make sense of them. Still, I wished there were some opportunities for clarity; I hoped for some means to make sense of the swirling confusion surrounding all the differing metabolic routes.

I found that clarity in graduate school. I learned that, despite the complexity and chemical diversity of the metabolic pathways in the cell, a set of principles does indeed exist that dictates the architecture and operation of metabolic routes. Armed with an understanding of these principles, I soon realized that I didn't have to memorize each of the metabolic pathways. Instead, I could employ these principles to write out each of the metabolic routes in the cell and identify the places where the metabolic pathways connect to one another.

The fact that it is possible to tame the unwieldy complexity of intermediary metabolism through a set of principles suggests that a molecular rationale undergirds these biochemical systems. If so, these processes may not be best explained as the haphazard product of a contingent evolutionary history. Instead, the structure and function of these systems may well be fundamentally dictated and constrained by the laws of nature. If true, it raises the distinct possibility that the intermediary metabolic routes in the cell evince the biochemical anthropic principle.

Determining if the biochemical anthropic principle is evident in the architecture and operations of metabolic systems isn't for the faint of heart. It appears to be a nearly impossible undertaking, given the extensive complexity of these biochemical systems. To make this part of the project tractable, I chose to focus primary attention on three textbook examples of metabolic pathways that are central to energy harvesting for the cell: (1) glycolysis, (2) the Krebs cycle (TCA), and (3) the electron transport chain (ETC). These pathways are nearly universal among living organisms. From my viewpoint, focusing on these three pathways is justified by the central importance of energy-harvesting metabolic routes to all life on Earth. If the biochemical anthropic principle is a valid idea, I would expect these three pathways to be characterized by anthropic coincidences.

To determine if this expectation bears out, I will adopt a similar approach to the one used in previous chapters. Specifically, I will:

- Determine if a molecular rationale undergirds the structure and function of energy-harvesting pathways.
- Determine if there are physicochemical constraints that arise from the laws of nature that dictate the properties of these metabolic routes.
- Compare the features of metabolic pathways to alternative systems to determine if these biochemical operations are unduly suited for life.

If met, then we can reasonably conclude that energy metabolism isn't necessarily the contingent outworking of evolutionary processes, but instead is primarily specified by constraints that arise from the laws of nature—constraints that produce biomolecular systems with a suitable set of properties for life.

Glycolysis

The glycolytic pathway appears in all three domains of life (Archaea, Bacteria, and Eukarya) and could rightly be viewed as a universal biochemical system. Glycolysis is not only one of the chief energy-harvesting pathways in the cell, it's also grouped with a set of metabolic pathways that take part in central carbon metabolism. These pathways (which also include gluconeogenesis, the pentose phosphate pathway, and the Krebs cycle) play the primary role in converting foodstuff into the building-block materials used by the cell's machinery to form the biomolecules, such as proteins, nucleic acids, and phospholipids.

Glycolysis involves the breakdown of the six-carbon sugar glucose into two molecules of the three-carbon compound pyruvate. (During the breakdown process, the energy released from the broken chemical bonds is harvested to power cellular operations.) Pyruvate can be transformed into the Krebs cycle intermediate oxaloacetate through the addition of carbon dioxide. It also is transformed into the compound acetyl-CoA, which is the primary entry point into the Krebs cycle. It can also be converted into the amino acid alanine. Several of the metabolic intermediates in the glycolytic pathway also take part in the pentose phosphate pathway. One of the functions of this pathway is to generate the five-carbon sugars ribose and deoxyribose used by the cell's machinery to make RNA and DNA, respectively. The Krebs cycle also plays a role in harvesting energy for the cell. As with glycolysis, the intermediates of the Krebs cycle are also used to generate building-block materials, such as fatty acids and amino acids.

The glycolytic pathway formally begins with the generation of glucose 6-phosphate from glucose. The enzyme hexokinase mediates this reaction, transferring a phosphate group from adenosine triphosphate (ATP) to the hydroxyl group of glucose located in the 6 position. This transfer involves cleaving a high-energy chemical bond in the ATP molecule. This process releases energy stored in the chemical bond. The liberated energy then powers the formation of the phosphoester linkage that joins the phosphate moiety to glucose. Because of this reaction, the first step in the glycolytic pathway consumes energy.

In the second step in the glycolytic pathway, the enzyme phosphoglucose isomerase converts glucose 6-phosphate into the six-carbon sugar fructose 6-phosphate. This step is critical because glucose isn't easily cleaved, but the structure of fructose is much more amenable to breakdown.

The third step of the pathway is mediated by the enzyme phosphofructokinase, which transfers another phosphate moiety from ATP to fructose 6-phosphate, yielding fructose 1,6-bisphosphate. As with the first step of glycolysis, this reaction consumes energy. Biochemists sometimes refer to the first and third steps as the pump-priming reactions because, to extract energy from glucose, the cell needs to utilize some of its energy reserves.

Biochemists have learned that the third step of the glycolytic pathway is the rate-limiting step, because the rate of this transformation controls the rate of glycolysis. To put it another way, the cell regulates the rate of glycolysis by regulating the activity of phosphofructokinase. This enzyme could be understood as both an on-off switch and a volume control knob. When the cell's energy stores are full, phosphofructokinase's activity is dialed down. When the energy is depleted, the enzyme's activity is turned up.

There are good reasons why this step is the regulatory step of glycolysis. For all intents and purposes, the generation of fructose 1,6-bisphosphate is irreversible because it involves the transfer of a phosphate from ATP to fructose 6-phosphate. The high amounts of energy released in this reaction ensures that once fructose 1,6-bisphosphate forms it can't be readily converted back to fructose 6-phosphate. When fructose 1,6-bisphosphate forms, the glycolytic pathway will proceed unhindered. If the cell has plenty of energy reserves, it doesn't make sense for the glycolytic pathway to proceed. And it makes less sense for the cell to use up ATP unnecessarily in generating fructose 1,6-bisphosphate if it doesn't need more fuel molecules.

In principle, hexokinase could also serve as a regulatory enzyme for the glycolytic pathway because the conversion of glucose to glucose 6-phosphate also requires energy. This reaction is also irreversible. So, why hasn't hexokinase been selected as the pacemaking enzyme for glycolysis? The reason has to do with the different fates for glucose 6-phosphate and fructose 1,6-bisphosphate. The latter sugar only serves as a metabolic intermediate for glycolysis. On the other hand, glucose 6-phosphate plays a role in another metabolic process, namely the synthesis of glycogen. So, the first irreversible step in glycolysis exclusive to the pathway is the conversion of fructose 6-phosphate to fructose 1,6-bisphosphate. For this reason, it makes perfect sense that the primary control of glycolysis resides with phosphofructokinase.

Once the pump-priming steps reach completion, glycolysis enters the second stage. Here, fructose 1,6-bisphosphate undergoes a cleavage reaction—mediated by the enzyme aldolase—and forms two three-carbon sugars (glyceraldehyde 3-phosphate and dihydroxyacetone phosphate). These two compounds can be converted into one another with the enzyme triose phosphate isomerase mediating the interconversion. The next reaction involves the oxidation of glyceraldehyde 3-phosphate to 3-phosphoglycerate. Oxidation reactions liberate energy. The released energy drives the chemical reduction of nicotinamide adenine dinucleotide (NADH). This compound also stores energy that the cell can use later. The released energy also supports the addition of phosphate to the carboxylic acid in the 1 position, resulting in the final compound 1,3-bisphosphoglycerate. The phosphate attached to the carboxylic acid moiety is a high-energy compound (called a mixed anhydride). Because of the high-energy nature of the mixed anhydride, the phosphate group can be readily transferred to ADP (adenosine triphosphate) to form ATP. This reaction is mediated by the enzyme phosphoglycerate kinase, which converts 1,3-bisphosphoglycerate to 3-phosphoglycerate while generating ATP. Next, the enzyme phosphoglycerate mutase transforms 3-phosphoglycerate to 2-phosphoglycerate. This reaction is followed by another reaction catalyzed by the enzyme enolase, which converts 2-phosphoglycerate into phosphoenolpyruvate. This compound is then converted into pyruvate by the enzyme pyruvate kinase, which also generates a molecule of ATP.

The net output of the glycolytic pathways is the conversion of glucose into two molecules of pyruvate. This transformation requires the consumption of two ATP molecules and in return yields four molecules of ATP, along with two molecules of NADH. The excess of ATP molecules and the two NADH molecules represent the harvested energy from the glycolytic pathways.

Glycolysis and the Biochemical Anthropic Principle

So is there a rationale that undergirds the glycolytic pathway? Given the complexity of glycolysis, it doesn't seem unreasonable to think that this metabolic process reflects the outworking of a historically contingent evolutionary process. This view would make sense—if there were no rhyme or reason to glycolysis. But, as I already pointed out, a rationale does indeed exist for two features of this pathway, specifically (1) the conversion of glucose 6-phosphate to fructose 6-phosphate, which allows for the cleavage of the six-carbon sugar into two three-carbon sugars, and (2) the site of regulation (phosphofructokinase). Are these two steps the only ones that display a molecular logic? Or are there others? We can begin the process of addressing these questions by asking another question. Why is glucose the starting point for the glycolytic pathway? Why not another sugar?

Why Glucose?

To be useful as a fuel source, a molecule must possess several properties. First, it must be chemically reduced. That is, it must have chemical energy stored in its bonds. It must also be chemically reactive. If chemically inert, it makes it harder to initiate the metabolic processes required to extract energy from its chemical bonds. Because life exists in an aqueous milieu, the fuel molecule also needs to be water soluble. (See chapter 3, pages 68–70.)

Sugars meet these three criteria. Biochemists have determined the energy stored in carbon-containing functional groups. The greatest amount of energy is harbored in methylene (CH_2), followed by hydroxycarbons (CHOH), followed by carbonyls (CHO), and finally by carboxylic acids (COOH). Because the structures of most sugars consist of one or more hydroxycarbons, they are an ideal source of chemical energy. The added advantage of having oxygen incorporated into their molecular structures means sugars can be readily converted into carbon dioxide as a waste product.

But there are so many sugars, so why glucose, specifically? A research team from Germany discovered one reason glucose is used in the glycolytic pathway. It is not possible to generate more than one ATP molecule per sugar if the number of carbon atoms in the sugar is less than six.[2] This constraint explains why sugars made up of four and five carbons aren't used as the primary fuel source in cells. Still, myriad six-carbon sugars exist in nature. Two six-carbon sugars that can be used as an energy source are fructose and galactose. Yet these sugars (and most others) don't have a dedicated metabolic pathway devoted to their breakdown. Instead, these sugars get funneled into the glycolytic pathway. For example, galactose can be used as an energy source by entering the glycolytic pathway. The enzyme galactokinase transfers a phosphate group from ATP to galactose to form galactose-1-phosphate. After this reaction, a series of metabolic steps converts galactose 1-phosphate into glucose 6-phosphate. Fructose can be directly converted into fructose 6-phosphate or it can alternatively be converted into dihydroxyacetone phosphate or glyceraldehyde 3-phosphate.

So why does only one dedicated pathway for the breakdown of glucose exist? Why aren't there pathways devoted to the breakdown of other sugars—at least, pathways found universally throughout nature? As noted, carbonyl groups (either aldehydes or ketones) make sugars chemically reactive compounds. This reactivity makes them highly useful as fuel molecules, but if sugars are too reactive, it can be detrimental to the cell. The aldehyde and ketone functional groups of sugars will react readily with proteins, which compromises protein function. So for a molecule to be useful as a fuel source, its chemical reactivity must be balanced with chemical stability. Glucose fits the bill. It is one of the most stable sugar molecules thanks to a combination of several structural features.[3] (Because of its stability, glucose is also one of the most abundant sugars in nature.)

What is the source of glucose's unusual stability? Sugars can exist either as linear or cyclic forms. In the linear form, sugars undergo a cyclization reaction with one of their hydroxyl groups reacting with either the aldehyde or ketone functional group that is also part of their structure. In the cyclic form, sugars are much less reactive because the aldehyde or ketone are transformed into hemiacetal or hemiketal functional groups, respectively. Glucose favors the cyclic form because when it becomes cyclized, it results in a six-membered ring structure. And it turns out that, for organic compounds, six-membered rings are more stable than any other ringed structure, whether it be four-membered, five-membered, or seven-membered.

The geometry adopted by the six-membered ring of glucose also contributes to its stability. The six-membered rings formed by organic compounds can exist in one of two conformations: chair or boat. The chair conformation is more stable than the boat form. Because of its structural features, in the cyclic form, glucose adopts the chair conformation. Lastly, the hydroxyl groups extending from the ring when glucose adopts the cyclic form are in the equatorial position as opposed to the axial position. As you might suspect, the equatorial position affords greater stability than the axial position.

In other words, glucose possesses the just-right set of chemical properties that make it ideally suited to serve as the chief fuel molecule for the cell. And for that reason, an exquisite molecular rationale undergirds the near-universal use of glucose as a fuel source. It also explains the existence of a dedicated pathway

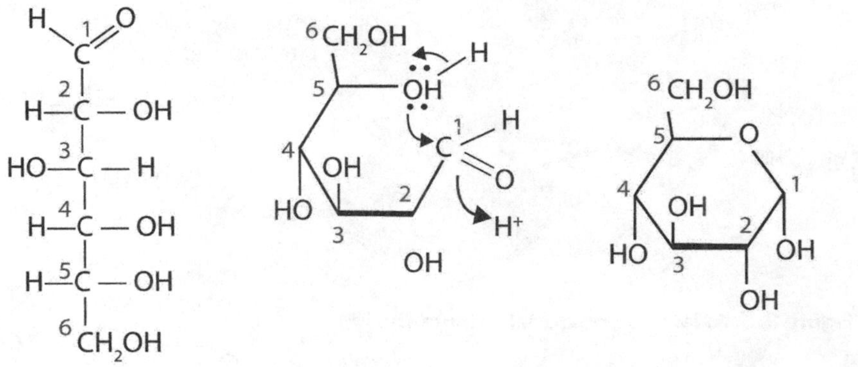

(a) Fischer projection (b) Three-dimensional representation (c) Cyclic monosaccharide

Figure 8.3: The Linear and Cyclical Forms of Glucose

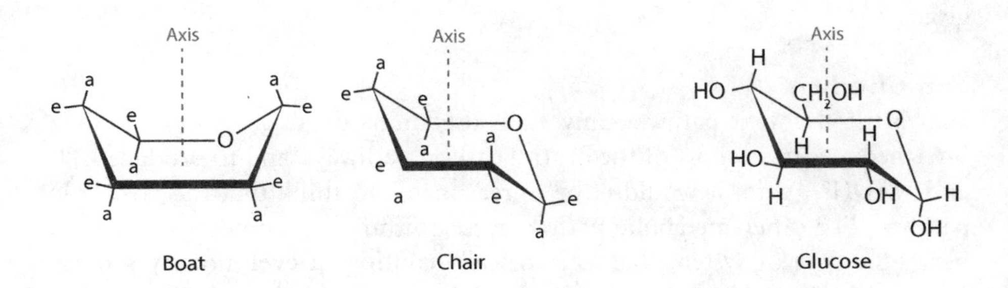

Figure 8.4: Chair and Boat Configurations

for the breakdown of glucose and the dearth of metabolic pathways devoted to the breakdown of other sugars that could potentially serve as fuel molecules. These other molecules lack the balance between stability and reactivity, displaying a greater tendency to react. This tendency brings real consequences. Highly reactive sugars can cause damage to cellular components. Fructose and the three-carbon sugars dihydroxyacetone and glyceraldehyde also are highly stable sugars, though not as stable as glucose.[4] Again, their properties make

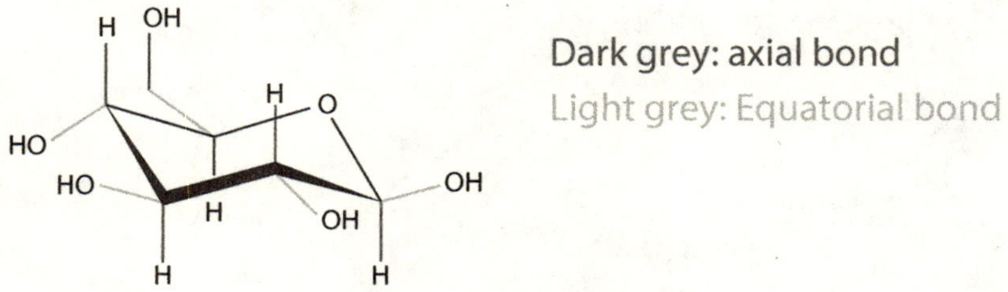

Dark grey: axial bond
Light grey: Equatorial bond

Figure 8.5: Axial and Equatorial Orientations

them ideally suited as intermediates in the glycolytic pathway (and other metabolic reactions in central carbon metabolism), but poor choices as one of the primary foodstuffs for the cell.

Of course, this leads us to the next set of questions. Why is the glycolytic pathway structured in the way it is? Is there a rationale that undergirds glycolysis?

Why Glycolysis?

Why is the glycolytic pathway universally used in nature to generate metabolic intermediates in support of the central carbon pathways, and to produce ATP and NADH? Again, it wouldn't be unreasonable to think that the glycolytic pathway, like other metabolic pathways, is arbitrary and unnecessarily complex—the type of system that a historically contingent evolutionary process would yield. But appearances can be deceiving, as the adage goes. Biochemists studying glycolysis have discovered an exquisite molecular rationale that characterizes nearly every feature of glycolysis.

If glucose is cleaved into two glyceraldehyde molecules, thermodynamically speaking, enough energy would be released to drive the formation of two ATP molecules. This energy release comes from the conversion of two hydroxycarbons (CHOH) into two aldehyde (CHO) functional groups. So, why is glucose 6-phosphate converted into fructose 6-phosphate before the cleavage reaction takes place? Biochemists have learned that the cleavage of glucose into two three-carbon sugars is not feasible mechanistically, given the chemical structure of glucose. On the other hand, a mechanism exists for the cleavage of fructose into two three-carbon sugars.[5]

Mechanistic restrictions also account for the structure of the last steps of the glycolytic pathway. Biochemists have learned that the only mechanistically realistic way that 3-phosphoglycerate can be converted to pyruvate and ATP is via the formation of phosphoenolpyruvate. The formation of this glycolytic intermediate necessitates the formation of 2-phosphoglycerate.[6]

Biochemists have discovered another reason why glucose isn't converted directly into two molecules of glyceraldehyde in order to generate two ATP molecules.[7] If this conversion took place, virtually all the energy released by the cleavage reaction would have to be employed to generate the two ATP molecules. Very little chemical energy would be released as heat. To be clear, the generation of heat when a chemical bond breaks is wasteful, but the release of heat is also necessary to push chemical reactions in the forward direction. Heat liberation is the primary motive force for chemical processes. This need explains why glucose and fructose 6-phosphate are phosphorylated (when a phosphate is added to the compound). The energy used to form the two phosphorylated sugars makes it possible to generate heat-liberating reactions later in the process.[8]

A team of German researchers demonstrated that a rationale exists for the location of the phosphorylation steps at the beginning of the glycolytic pathway. Placing the energy-consuming steps at the start of the pathway leads to a highly optimized flux of ATP production. Placing the ATP-consuming steps at another location in the pathway would result in a less-than-ideal rate of ATP generation.[9]

There is another reason for locating the phosphorylation steps at the beginning of the glycolytic pathway. It ensures that all pathway intermediates are phosphorylated.[10] The phosphorylated intermediates have a greater affinity for enzymes, ensuring the efficiency of the glycolytic pathway. Unless actively transported across cell membranes by a protein, phosphorylated intermediates can't diffuse across phospholipid bilayers. This physicochemical property prevents valuable intermediates in the glycolytic pathway from leaking out of the cell. Unmodified sugars have both sufficient water solubility and a low enough polarity that they can dissolve into and out of cell membranes. This set of properties allows them to diffuse across cell membranes, leaking into the extracellular space.[11]

In short, an exquisite molecular rationale is on full display in the design of the glycolytic pathway. Glycolysis doesn't appear to be a haphazard construct of a historically contingent evolutionary process. In fact, the glycolytic pathway appears to be highly optimized for maximum flux, maximum ATP production, and simplicity. The researchers determined that the glycolytic pathway proceeds with the minimum number of steps.[12] This allows the cell to respond as rapidly as possible when its energy status begins to bottom out. It also keeps the diversity of the intermediate pool as low as possible.

The glycolytic pathway appears to be well-suited for its metabolic role in the cell. But is it *uniquely* suited for this role? Are there conceivable and reasonable alternatives to glycolysis?

Is the Glycolytic Pathway Unique?

In an attempt to understand the evolutionary origin of glycolysis and gain insight into the very nature of the evolutionary process, several research groups have sought to understand how the glycolytic pathway compares to alternative pathways that could have been used to break down glucose and harvest ATP.[13] In one study, a team of German researchers created a library of a vast number of hypothetical—but biochemically feasible—pathways that would break down glucose, generating ATP in the process. Starting with this random population of pathways, the researchers selected those with the highest ATP flux. Then the researchers altered the selected pathways to generate a new set that, in turn, they evaluated for ATP flux. Through this iterative process, the researchers discovered a number of optimized pathways that share the properties of the glycolytic pathway, including the positioning of ATP-consuming steps near the pathway start and the placement of ATP-producing and NADH-producing pathways near the end of the metabolic route. This result indicates that, while the glycolytic pathway may not be unique, it does stand apart compared to a vast number of alternatives and it possesses the features of a highly optimized metabolic route.[14]

More recently, a research team from the UK used an algorithm to generate all biochemically feasible pathways for the lower reactions of glycolysis. (These are the reactions that convert glyceraldehyde 3-phosphate into pyruvate.) They discovered around 6,500 alternative pathways. They also discovered that the maximum yield of any reasonable analog to the glycolytic pathway is two ATP molecules. Any pathway that produced three ATP molecules or more was thermodynamically infeasible. Of the feasible pathways, only 1,800 generated two ATP molecules. All others generated one ATP or less. Of these 1,800 pathways, the researchers learned that, under the physiological conditions of the cell, none of them outperformed the glycolytic pathway in terms of (1) rate of ATP production, (2) minimal number of steps, and (3) metabolic cost of producing the enzymes needed to catalyze the steps of the metabolic routes. The team did discover that, under a different set of alternative physiological conditions to those found in the cell, some of these alternative pathways outperformed the glycolytic pathway.[15] So, it does indeed seem that the lower part of the glycolytic pathway is highly optimized, but it doesn't appear to be unique.

Up to this point, the discussion of glycolysis has focused on the "textbook" description of the pathway. But there are other metabolic pathways for the breakdown of glucose found in nature. Biochemists refer to the textbook version of glycolysis as the Embden-Meyerhof-Parnas pathway (EMP), named for the three biochemists who solved the pathway. The EMP pathway is nearly ubiquitous among eukaryotic organisms. It is also found among Bacteria and Archaea. However, other "glycolytic" pathways exist, particularly among Bacteria and Archaea. Careful analysis of these alternative pathways reveals that these routes have been optimized for features other than ATP flux through the pathway, as well as the specific conditions in which the organisms find themselves.[16] One of the most widely occurring alternatives to the EMP pathway is the Entner-Doudoroff (ED) pathway, found in numerous Bacteria and Archaea.

The Entner-Doudoroff Pathway

Like the EMP pathway, the ED pathway begins with glucose and ends with the production of two pyruvate molecules. Unlike the EMP pathway, which generates two ATP molecules, the ED route yields only a single molecule of ATP. The EMP pathway also generates two molecules of NADH. The ED pathway produces a molecule of NADH and a molecule of NADPH.

Like the EMP pathway, the ED pathway begins with the work of the enzyme hexokinase, which converts glucose to glucose 6-phosphate. The next step of the ED pathway involves the activity of the enzyme glucose 6-phosphate dehydrogenase, which converts glucose 6-phosphate into 6-phosphogluconolactone and NADPH. Following this step, the enzyme hydrolase converts 6-phosphogluconolactone into 6-phosphogluconic acid. This compound is converted into 2-keto-3-deoxy-6-phosphogluconate (KDPG) by the enzyme 6-phosphogluconate dehydratase. Next, the enzyme KDPG aldolase cleaves 2-keto-3-deoxy-6-phosphogluconate into pyruvate and glyceraldehyde 3-phosphate. The latter compound follows the same pathway found in the lower part of the EMP pathway to generate pyruvate plus a single molecule of ATP and a single molecule of NADH.

If the ED pathway produces only a single molecule of ATP, why is it so widespread among prokaryotic organisms? In 2013, a team of biochemists from

Israel offered an answer to that question.[17] They demonstrated that the choice between the EMP and ED pathways in microbes reflects a trade-off between the need for maximum ATP production and the biochemical cost of producing the enzymes needed for the glycolytic pathways. Because the ED pathway only produces a single molecule of ATP, the remaining energy liberated from the cleavage of the glucose molecule is released as heat. As discussed earlier, this heat serves as a driving force for the ED pathway, propelling it forward. While the EMP pathway brings a greater yield of ATP, the heat released during the pathway is much less than the ED pathway. This means that the EMP pathway lacks the same forward momentum. To work around this lack of momentum, cells that employ the EMP pathway need to produce higher levels of enzymes to process the same amount of glucose in the same period as the ED metabolic route. So the capacity to generate a higher ATP yield comes with a higher cost in terms of the resources devoted to enzyme production.

This is a particularly pressing problem for microorganisms because protein synthesis controls the growth rate. So the ED pathway offers a benefit to microorganisms, particularly aerobic ones in which pyruvate is further metabolized into carbon dioxide and water by the Krebs cycle and the electron transport chain (ETC). This additional processing generates a large amount of ATP, rendering the need for an additional ATP molecule from glycolysis superfluous. In other words, aerobic organisms benefit from utilizing the ED pathway because it reduces the demand on the cell's machinery to synthesize proteins, while having minimal impact on overall ATP production. On the other hand, anaerobic microorganisms rely solely on the glycolytic pathway as an ATP source. If these organisms relied on the ED pathway, their ATP production would drop by 50 percent. So, for anaerobic microbes, the cost of producing extra enzymes in support of the glycolytic pathway is well worth it.

It does appear that an exquisite molecular rationale undergirds the design of every step of the glycolytic pathway. The most prevalent form of glycolysis, the EMP pathway, appears to be highly optimized. It outperforms thousands of conceivable alternative pathways based on ATP yield, rate of ATP production, pathway simplicity, and response time. Alternative "glycolytic" pathways do exist, but these pathways would only be preferred under physiological conditions that are not typically found in nature. Alternative pathways found in nature are less optimal than the EMP pathway when it comes to ATP yield. In effect, these pathways appear to be optimal for other considerations that arise from the specific needs of the organisms.

It appears that the glycolytic pathway meets two of the three criteria for the anthropic principle. What about the third criterion? Is the structure of the glycolytic pathway fundamentally dictated by the laws of nature? There is some provocative evidence that this is the case. I will address this evidence at the same time I address the same question for the Krebs cycle (see pages 281–283). At this juncture, I am going to shift the focus to the Krebs cycle. Does this metabolic route meet the criteria for the anthropic principle?

The Krebs Cycle

The Krebs cycle (sometimes called the tricarboxylic acid cycle or the citric acid cycle) oxidizes the pyruvate generated from the glycolytic pathway into carbon dioxide and water. This combustion process liberates a significant amount of energy, which is then captured and used to drive the formation of NADH and $FADH_2$ (flavin adenine dinucleotide). Once formed, these two molecules feed high-energy electrons into the ETC, which uses the electrons to form a proton gradient across the mitochondrial inner membrane. This proton gradient drives the formation of ATP molecules through a process called oxidative phosphorylation. (See "The Electron Transport Chain.") This process requires molecular oxygen and generates over 90 percent of the ATP used by aerobic organisms.

The Krebs cycle is also part of central carbon metabolism. Many of its intermediates serve as the precursors used to generate building-block compounds, such as glucose, fatty acids, amino acids, nucleobases, and the porphyrin ring (used as the cofactor for the hemoglobin molecule.) In fact, the Krebs cycle functions as the key metabolic hub for intermediary metabolism, integrating many of the individual metabolic routes in the cell.

The reactions of the Krebs cycle all take place in the interior of mitochondria, known as the matrix. In contrast, the glycolytic pathway takes place in the cell's cytoplasm. For these two metabolic processes to become linked, the pyruvate produced via glycolysis in the cytoplasm must be moved into the mitochondria. A membrane-embedded transport protein mediates this process. Once in the mitochondrial interior, the enzyme pyruvate dehydrogenase catalyzes the conversion of pyruvate into a molecule of carbon dioxide and a molecule of acetyl-CoA. The energy released from this reaction is also used to drive the formation of the energy storage compound NADH. Acetyl-CoA is a high-energy compound in its own right. This chemical energy is stored in the thioester linkage that joins the acetyl unit to the CoA moiety.

As noted, one of the key functions of the Krebs cycle is to extract as much chemical energy from glucose as possible by completely oxidizing pyruvate

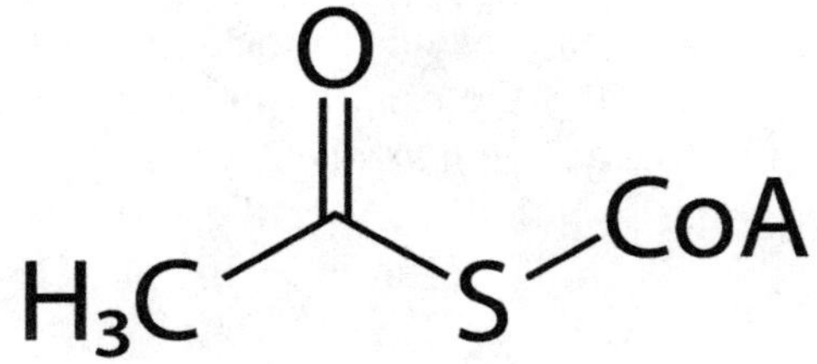

$$H_3C-C(=O)-S-CoA$$

Figure 8.6: A Partial Chemical Structure of Acetyl-CoA

into carbon dioxide and water. The Krebs cycle generates one ATP molecule per turn. (This molecule is formed indirectly from the generation of GTP, or guanosine triphosphate, during the cycle. Once it forms, GTP transfers a phosphate group to ADP to form ATP.) Most of the energy captured during a single turn of the Krebs cycle is used to form three molecules of NADH and one molecule of $FADH_2$. A series of reactions achieves this energy capture. The process begins when the acetyl group of the acetyl-CoA molecule is transferred to the four-carbon compound oxaloacetate to form the six-carbon compound citric acid. (Citric acid possesses three carboxylic acid residues in its structure. Hence, the Krebs cycle is sometimes called the tricarboxylic acid cycle.) In turn, citric acid undergoes a sequential series of reactions where two molecules of carbon dioxide are lost successively to generate first a five-carbon compound and then a four-carbon compound. The four-carbon compound undergoes a series of reactions that remove hydrogen atoms to regenerate oxaloacetate. Each decarboxylation reaction and each loss of hydrogen atom pairs liberate energy. Biochemists refer to these types of chemical transformations as oxidation reactions. Again, this liberated energy drives the formation of NADH and $FADH_2$.

The enzyme citrate synthase catalyzes citrate formation. The hydrolysis of the thioester linkage, which joins the acetyl group to CoA, drives this energy-consuming reaction. Next, the enzyme aconitase converts citrate into isocitrate. This six-carbon compound undergoes an energy-liberating reaction catalyzed by the enzyme isocitrate dehydrogenase. This reaction oxidizes isocitrate and eliminates carbon dioxide from its structure to generate alpha-ketoglutarate. Biochemists refer to this type of reaction as an oxidative decarboxylation. The liberated energy from the reaction drives the formation of NADH.

The next step in the Krebs cycle resembles the reaction that converts pyruvate into acetyl-CoA. In this case, the enzyme alpha-ketoglutarate dehydrogenase drives the oxidative decarboxylation of alpha-ketoglutarate to form succinyl-CoA. The four-carbon succinyl group is linked to CoA via a thioester linkage. As with the previous oxidative decarboxylation step, the liberated energy also powers the formation of a molecule of NADH. Next, succinyl-CoA synthetase generates the four-carbon compound succinate. The formation of GTP captures the energy liberated via the cleavage of the succinyl-CoA's thioester bond.

With the next set of reactions, the Krebs cycle enters its second phase, which involves the successive removal of hydrogen atom pairs from succinate to regenerate oxaloacetate. This process involves the conversion of two adjacent methylene groups in succinate into a double-bonded structure. Water is then added to the double bond to generate an alcohol moiety. Then the alcohol is oxidized to a ketone. Specifically, the enzyme succinate dehydrogenase converts succinate to fumarate. The energy generated by this reaction forms $FADH_2$. Succinate dehydrogenase is embedded in the mitochondrial inner membrane. In contrast, all the other enzymes reside in the mitochondrial matrix. Because of its location, succinate dehydrogenase participates directly in the electron transport chain. (See "The Electron Transport Chain.") Fumarase converts fumarate into malate. Malate dehydrogenase converts malate into oxaloacetate. Again, the liberated energy drives the formation of NADH.

The Krebs cycle also serves as an important source of precursors for the production of building-block materials. Citrate is used to make fatty acids and sterols. Alpha-ketoglutarate is used to make the amino acid glutamate, and this amino acid then takes part in pathways that build other amino acids and the purine nucleobases. Succinyl-CoA can be used to make heme and other porphyrins. Oxaloacetate can be used to make the amino acid aspartate, which takes part in metabolic routes that generate other amino acids and the pyrimidine nucleobases.

When the Krebs cycle provides precursors to the various biosynthetic pathways, the pool of intermediates that constitute the Krebs cycle are siphoned off. This loss makes it necessary to replenish the Krebs cycle components. This need is requited in part by the enzyme pyruvate carboxylase, which utilizes energy from the hydrolysis of ATP to add a carbon dioxide unit to pyruvate to form oxaloacetate.

The Krebs Cycle and the Biochemical Anthropic Principle

Do we find evidence for the biochemical anthropic principle in the design of the Krebs cycle? Is there a rationale for its design? Is the Krebs cycle unique or are there conceivable alternatives to this metabolic route? Are there physicochemical constraints that dictate the design of the pathway?

There does indeed appear to be a molecular rationale that undergirds the architectural features of the Krebs cycle. As I discussed, placement of energy-consuming steps at the front of the glycolytic pathway and energy-producing steps at the end allows for maximum efficiency and flux when it comes to ATP production. The same rationale explains why the generation of citrate from oxaloacetate and acetyl-CoA occurs at the beginning of the Krebs cycle and why the energy-harvesting pathways occur during subsequent steps.[18] The use of tri- and dicarboxylic acids at the start of the pathway also makes chemical sense. The carboxylic acids make excellent leaving groups for the oxidative decarboxylation reactions, in which the carboxylic acid moiety is cleaved from the molecule and transformed into carbon dioxide. Because carboxylic acids are one step away from being fully oxidized, they readily form carbon dioxide. The carboxylic acid intermediates are ideal components of the Krebs cycle for another reason. The carboxylic acid functional groups are ionized at physiological pH. As a result, they bear a negative charge. This charge makes it highly unlikely that the Krebs cycle intermediates will traverse the mitochondrial inner membrane and leak into the cytoplasm and, in turn, out of the cell.

Although our primary interest in the Krebs cycle relates to its role in energy metabolism, its role in central carbon metabolism can't be ignored when we seek to understand the pathway's molecular logic. As a case in point, a research team from Israel sought to identify organizing principles that could account for the design of the pathways that comprise central carbon metabolism, including the Krebs cycle.[19] They discovered that for the metabolic routes that comprise the central carbon pathways, 12 intermediates exist that serve as exit points for metabolites to leave these pathways. The intermediates at these exit points function as precursors used by other metabolic pathways to construct the building-block materials that contribute to the organism's body mass. Seven of these precursors are found in glycolysis (the EMP pathway) and four occur in the Krebs cycle. (The other intermediate is found in one of the other pathways that comprise central carbon metabolism.) The researchers learned that the central carbon pathways can be reformulated, but the alternative pathways must include these 12 precursor compounds. If they don't, then there is no way to generate the compounds that make up the cell's biomass. The other intermediates in these pathways can be bypassed. The researchers discovered that these so-called nonprecursor compounds do, however, form sub-pathways that constitute the minimum number of steps between the 12 precursors. This set of minimal paths allows the cell to get by with the fewest possible enzymes to build its biomass.

Based on this cursory survey, it appears as if a molecular rationale does undergird the structure of the Krebs cycle. But is the Krebs cycle uniquely suited for its role as an energy-harvesting pathway? To ask this question in a different way: are there alternatives to the Krebs cycle? A team of researchers from Spain addressed this question.[20] They wanted to determine if alternatives to the Krebs cycle exist and, if so, do these alternative pathways perform better or worse than the Krebs cycle? These researchers identified a vast number of theoretically possible and biochemically feasible metabolic pathways that could oxidize acetate to carbon dioxide and at the same time harvest energy through the formation of NADH. They discovered hypothetical pathways that are linear, cyclic, and a combination of both. By comparing these pathways to the Krebs cycle, they learned that the Krebs cycle stands as the best of all the possible alternatives, requiring the least number of steps while yielding the greatest amount of NADH molecules.

This insight leads us to the next criterion in our quest to determine if the biochemical anthropic principle applies to the Krebs cycle. Are there physicochemical constraints arising out of the laws of nature that dictate the design of the Krebs cycle? What about glycolysis? (Remember, we postponed addressing this question. Now is the time to address it.) Given the complexity of these two metabolic routes, it would be easy to dismiss this possibility. I am sure that for some, it might seem that evolutionary processes cobbled these pathways together—yet that doesn't appear to be the case. It is possible to identify several significant physicochemical constraints that appear to dictate the design of these two pathways. As noted, metabolic intermediates need to have a low lipophilicity. This feature makes it unlikely that metabolites will leak from the mitochondrial interior (in the case of the Krebs cycle) or from the cell by diffusing across cell membranes. This requirement helps explain why the glycolytic intermediates are phosphorylated and why the Krebs cycle intermediates consist of tri- and dicarboxylic acids. Metabolic intermediates also need the capacity to form highly specific interactions with the active sites of enzymes. Phosphorylation of metabolic intermediates and the incorporation of carboxylic acid functional groups into the structure of the intermediates help achieve this requirement. Lastly, metabolic intermediates need to be chemically stable and relatively unreactive. If they were unstable and highly reactive, metabolic intermediates would take part in unwanted side reactions, thereby depleting the pathway's molecular constituents while "gumming up" the cell's operations.[21]

Other physicochemical constraints dictate the structural features of metabolites. A research team from the Weizmann Institute of Science in Israel analyzed data sets of intermediary metabolite concentrations from bacteria, yeast, and humans to characterize the physicochemical features of intermediary metabolites that influence their concentrations in the cell. Each data set consisted of simultaneously measured concentrations of anywhere from 30 to 100 metabolites, representing 20 different sets of conditions.[22] By interrogating this data set, they discovered that four physicochemical features influence metabolite levels: lipophilicity, solubility, number of charged atoms, and the amount of the molecular surface area that is nonpolar. Metabolic intermediates with low lipophilicity and high solubility in water occur at higher concentrations in the cell, as might be expected based on our earlier discussion. This combination of properties prevents the metabolic intermediates from leaking out of the cell by passing through the phospholipid bilayers of cell membranes. Researchers discovered that possessing either positively or negatively charged atoms or being phosphorylated was particularly important. These structural features help ensure that high enough levels of amino acids and nucleotides are available to support the production of proteins and the nucleic acids. This feature is particularly important for nucleotides (because of the hydrophobic nucleobases) and amino acids that possess hydrophobic R groups. The charged residues ensure that these materials are soluble, making them readily available for the cell's biosynthetic machinery. The Weizmann researchers learned that minimizing the nonpolar surface area of metabolites is critical. In fact, it represents one of the most important physicochemical constraints for metabolites. It prevents metabolic intermediates from engaging in unproductive nonspecific binding to proteins through hydrophobic interactions. This nonspecific binding would run the risk of compromising protein function and removing metabolites from the pool of metabolic intermediates in the cell. Minimizing the nonpolar surface area of metabolites also prevents them from forming unwanted molecular aggregates in the cell, mediated by hydrophobic interactions. These aggregates are harmful to the cell because they interfere with critical cellular processes.

In many respects, the importance of nonpolar surface area for metabolites shouldn't surprise us. We have already seen that protein folds can be explained as a consequence of the hydrophobicity profile of the amino acid sequence that makes up the protein and by the geometry of the protein chain (namely, its diameter and the distance of intrachain interactions). Likewise, we learned that the structure of cell membrane bilayers can be accounted for by (1) the hydrophobicity of the fatty acid chains and (2) the ratio of the area occupied by the hydrophilic head group and the volume and length of the hydrocarbon chains. And now we learn that the same type of physicochemical constraints (molecular geometry and hydrophobicity) explain the concentration and types of compounds that can reasonably serve as metabolic intermediates, with the hydrophobic surface area of the metabolite serving as one of the most important factors toward this end.

It is remarkable that these physicochemical constraints yield a set of metabolic intermediates that can be used to construct two highly optimized metabolic pathways (glycolysis and the Krebs cycle), which also display an exquisite molecular rationale in their design. These pathways possess the just-right set of properties that make it possible for them to assume an integral role in central carbon metabolism and efficiently harvest energy for cell use. One can't but marvel at how fortuitous it all is.

Prebiotic Chemistry and the Design of the Krebs Cycle and Glycolysis

Further evidence that physicochemical constraints have predetermined the design of the Krebs cycle and the glycolytic pathway comes from work carried out by origin-of-life investigators. Because the Krebs cycle assumes a central role in intermediary metabolism and because of its nearly universal occurrence throughout the living realm, many origin-of-life researchers think this metabolic route may have been one of the first metabolic pathways to arise, predating the emergence of LUCA (the last universal common ancestor). Some origin-of-life researchers think that a primitive version of the Krebs cycle predated the origin of primitive genetic material and primitive cell boundaries. As such, the Krebs cycle operated as an autocatalytic cycle, kick-starting the process of abiogenesis. In other words, for some origin-of-life researchers, understanding the evolutionary pathway that led to the Krebs cycle provides them with a vital clue as they seek to make sense of the mystery of life's origin.

To understand the origin of the Krebs cycle and its role in abiogenesis, renowned origin-of-life investigator Harold Morowitz and a team of collaborators sought to identify the types of prebiotic compounds that might emerge under the conditions of early Earth. They hoped this list of compounds would help them determine if, in principle, the Krebs cycle could emerge spontaneously on early Earth.[23] Morowitz and his team screened the 3.5 million entries in the Beilstein database of organic compounds, applying a set of rules that reflected a realistic set of physical and chemical constraints relevant to the prebiotic conditions of early Earth. Specifically, they focused on relatively low molecular weight compounds consisting of carbon, hydrogen, and oxygen that are water soluble and possessed either aldehyde or ketone functional groups. They excluded compounds with high heats of combustion, chemically unstable compounds, and compounds containing functional groups that would be challenging to generate under plausible prebiotic conditions. Applying this filter, they winnowed the 3.5 million Beilstein entries to 153 compounds, which included all the Krebs cycle intermediates. For Morowitz and his team, this result means that compounds forming the Krebs cycle are emergent features on early Earth that flow naturally out of the properties of carbon chemistry and prebiotic conditions. In fact, they argue that the Krebs cycle intermediates may well constitute a unique set of molecules.

Building upon the work of Morowitz and his colleagues, a team of researchers from France discovered they could generate nine of the eleven compounds that constitute the Krebs cycle intermediates by incubating either pyruvate or glyoxylate in water under an inert atmosphere at 70°C.[24] This set of conditions models mild hydrothermal vents. Conceivably, pyruvate and glyoxylate could have been generated at such hydrothermal vents on the early Earth. These results indicate that the Krebs cycle isn't the product of an evolutionary history, but instead arises out of the dictates of carbon chemistry and the planet's early geochemical conditions. As origin-of-life researchers Eric Smith and Harold Morowitz argue:

> The chart of intermediary metabolism has a universal anabolic core, which should not be understood as merely a result of common ancestry but rather as a solution imposed on life already within the energetically structured environment of the early earth, by details of carbon chemistry and certain transport and transformation functions performed only by biomass.[25]

It appears that these same types of geochemical and physicochemical constraints also account for the design of the glycolytic pathway. Researchers from the UK discovered that the reactions that produce many of the metabolic intermediates in the glycolytic pathway and the pentose phosphate pathway are spontaneously generated under the prebiotic conditions of early Earth's oceans.[26] These researchers incubated pure metabolic intermediates from the glycolytic and the pentose phosphate pathway in water in the presence of salts likely to have existed in the oceans of early Earth, including ferrous iron. They discovered that, when incubated, these individual compounds generated several metabolic intermediates found in these two pathways. In fact, once formed, these metabolic intermediates reacted with one another to form a network of interconverting chemical materials that resembled segments of glycolysis and the pentose phosphate pathway. These researchers conclude that:

> These results indicate that the basic architecture of the modern metabolic network could have originated from the chemical and physical constraints that existed on the prebiotic Earth.... These results therefore support the hypothesis that the topology of extant metabolic network could have originated from the structure of a primitive, metabolism-like, prebiotic chemical interconversion network.[27]

It is highly suspicious that (1) the constraints that arise out of the geochemical and geophysical settings of early Earth, (2) the physicochemical constraints that arise out of carbon chemistry, and (3) the interplay between organic compounds and water conspire to spontaneously generate the compounds that serve as metabolic intermediates for the Krebs cycle and glycolysis (and the other pathways that comprise central carbon metabolism). These physicochemical constraints produce a set of just-right metabolic intermediates that can, in principle, become organized into highly optimal metabolic routes. Without question, the anthropic coincidences abound when it comes to glycolysis and the Krebs cycle. It is absolutely eerie.

But what about the electron transport chain, the last stage in the cell's energy-harvesting pathways?

The Electron Transport Chain

The electron transport chain (ETC) is the metabolic machinery that converts the energy captured in the molecules NADH and $FADH_2$ into ATP molecules. The energy in NADH and $FADH_2$ exists in the form of high-energy electrons. Ultimately, these electrons are transferred to molecular oxygen by the ETC to generate water. Because this process requires molecular oxygen, it is sometimes called oxidative phosphorylation. It is also referred to as cellular respiration.

The ETC consists of three massive protein complexes embedded in the

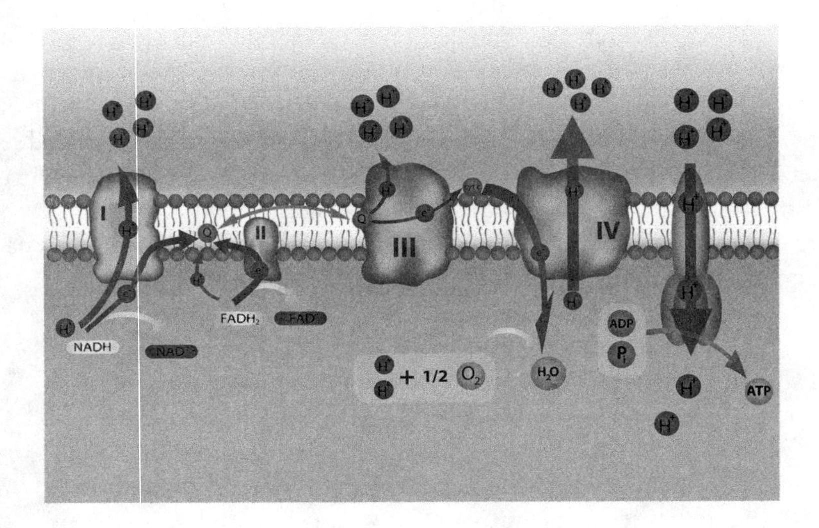

Figure 8.7: The Electron Transport Chain

mitochondrial inner membrane in eukaryotes (and in the plasma membrane of Bacteria and Archaea). The high-energy electrons from NADH and $FADH_2$ are shuffled from one protein complex to the next, with each transfer releasing energy that is used to transport protons from the mitochondrial matrix across the inner membrane, establishing the proton gradient. Oxygen is the final electron acceptor in the ETC.

Because protons are positively charged, the exterior region just outside the inner membrane is positively charged and the interior region is negatively charged. The charge differential created by the proton gradient is analogous to a battery and the inner membrane is like a capacitor. Biochemists call the energy associated with the proton gradient the electrochemical potential.

The coupling of the proton gradient to ATP synthesis occurs through the flow of positively charged protons through the F_0 component of a protein complex called ATP synthase (or F_1-F_0ATPase). This protein complex is embedded in the mitochondrial inner membrane in Eukaryotes and the plasma membrane of Bacteria and Archaea. ATP synthase uses proton flux to convert electrochemical energy into mechanical energy that is used to drive the formation of ATP from ADP and inorganic phosphate.

The process of electron transport begins when NADH transfers two electrons to complex I (also called NADH dehydrogenase or NADH-Q oxidoreductase). Massive in size, complex I has an overall L-shaped structure. One arm lies within the membrane. The other arm extends into the mitochondrial matrix. Electrons from NADH are transferred to a component of complex I called flavin mononucleotide (FMN). From FMN, the electrons are shuffled through a series of iron-sulfur complexes. From there, the electrons are shuttled to a membrane-embedded lipid compound called Q. Once Q accepts electrons it exits complex I, diffusing through the bilayer of the mitochondrial inner membrane. During this process, complex I pumps four protons from the mitochondrial matrix across the inner membrane to the mitochondria's exterior.

The next complex of the ETC is called complex II. (It is also called succinate-Q-reductase complex.) Complex II possesses FAD that is converted to $FADH_2$ by succinate dehydrogenase during the Krebs cycle. Succinate dehydrogenase is part of complex II. The electrons are then transferred from $FADH_2$ to an iron-sulfur complex and then to compound Q.

The next stage of the ETC involves complex III (also called Q-cytochrome c oxidoreductase or cytochrome c reductase). Complex III transfers electrons from Q to cytochrome C. This process transfers two protons across the mitochondrial inner membrane. The last step of the ETC takes place when complex IV (also called cytochrome c oxidase) transfers electrons from cytochrome c to molecular oxygen, producing water. Thanks to this step, four protons are transported out of the matrix across the inner membrane.

The gradient established by the ETC drives the generation of ATP. The enzyme ATP synthase uses the electrochemical potential of the proton gradient to power the formation of ATP from ADP and inorganic phosphate. When everything is said and done, the glycolytic pathway and the Krebs cycle, working in conjunction with the ETC, generate approximately 30 molecules of ATP from each molecule of glucose as it is completely oxidized into carbon dioxide and water.

Electron Transport Chains and the Biochemical Anthropic Principle

As with the case for glycolysis and the Krebs cycle, the use of the ETC to generate proton gradients, which in turn power ATP synthesis, is, in effect, universal among aerobic organisms. In fact, biochemists have come to recognize the widespread importance of proton gradients. Proton gradients are pervasive in living systems. Chloroplasts rely on proton gradients during the process of photosynthesis. Cells use proton gradients to transport material across cell membranes. And proton gradients even power the bacterial flagellum.

In fact, a growing number of origin-of-life researchers hold the view that

LUCA (and maybe even the life-forms that preceded it) made use of proton gradients to harvest energy. For many evolutionary biologists, oxidative phosphorylation (and the use of proton gradients, in general) assumes a position of unique prominence because this process plays a central role in harvesting energy in both prokaryotic and eukaryotic organisms. In other words, understanding the origin of oxidative phosphorylation (and use of proton gradients) is central to understanding the origin of life and the fundamental design of biochemical systems.

Because of its widespread occurrence among all of life's domains and its central importance to energy metabolism (and other biochemical processes), the architecture and operation of the ETC becomes an ideal system to probe for evidence of the biochemical anthropic principle.

The Chemiosmotic Theory

The use of proton gradients to power the process of oxidative phosphorylation was first proposed by biochemist Peter Mitchell. He argued that the ETC generates a proton gradient across the mitochondrial inner membrane and, in turn, exploits that gradient through a coupling process to drive the synthesis of ATP from ADP and inorganic phosphate. He referred to this idea as the chemiosmotic theory.[28]

Mitchell's proposal was initially met with a large measure of skepticism from biochemists. It didn't fit with the orthodoxy characteristic of classical biochemistry. Origin-of-life researcher Leslie Orgel referred to the chemiosmotic theory as one of the most counterintuitive ideas to ever come out of biology, comparing it to the ideas that formed the foundations of quantum mechanics and relativity.[29]

At that time, many biochemists preferred the chemical theory of oxidative phosphorylation over Mitchell's chemiosmotic theory. Researchers thought that the phosphate group added to ADP was transferred from one of the components of the ETC. To support this idea, many biochemists searched frantically for a chemical intermediate with a high-energy phosphate moiety that could power the synthesis of ATP. (The chemical theory was based on a process called substrate-level phosphorylation, exemplified by two reactions that form ATP during glycolysis.) The elusive intermediate was never found and adherents of the chemical theory were forced to abandon their model. Peter Mitchell's idea won the day. In fact, Mitchell was awarded the 1978 Nobel Prize in Chemistry for his contribution to understanding the mechanism of oxidative phosphorylation.

Now that oxidative phosphorylation is understood, biologists and origin-of-life researchers have turned their attention to two questions: How did chemiosmosis originate? Why are proton gradients so central to biochemical operations?

Answers to these questions provide insight that has bearing on whether the anthropic principle applies to the structure and operation of the ETC. Because the use of proton gradients in living systems has historically been viewed as odd and counterintuitive, many origin-of-life researchers and evolutionary biologists are tempted to conclude that chemiosmosis resulted from a historically contingent evolutionary process that co-opted and modified existing systems and designs. The work of origin-of-life researcher Nick Lane reinforced this notion.

Lane and his collaborators conclude that proton gradients must have been integral to the biochemistry of LUCA because proton gradients are a near-universal feature of living systems. If so, then the use of proton gradients must have emerged during the origin-of-life process before LUCA originated. Lane and his team go so far as to propose that the first protocells emerged near hydrothermal vents and made use of naturally occurring proton gradients found in these environments as their energy source.[30]

Once this system was in place, the use of proton gradients was retained in the cell lines that diverged from these early protocellular entities as the ETC evolved from a simple, naturally occurring vent process to the complex process found in both prokaryotic and eukaryotic organisms. In other words, it would seem that the odd, counterintuitive nature of proton gradients reflects the happenstance outworking of chemical evolution that began when the naturally occurring proton gradients were co-opted in the early stages of chemical evolution.

But other work by Lane indicates that, though counterintuitive, the use of proton gradients to harvest the energy required to make ATP makes sense, displaying an exquisite molecular rationale. And if so, it forces a rethink of the explanation for the origin of chemiosmosis, while adding to the case for the biochemical anthropic principle.[31]

The Molecular Logic of Proton Gradients

So, why are chemiosmosis and proton gradients universal features of living systems? Are they an outworking of a historically contingent evolutionary process? Or is there something more at work?

Even though proton gradients seem counterintuitive at first glance, the use of proton gradients to power the production of ATP and other cellular processes reflects an underlying ingenuity and exquisite molecular logic. Research shows that proton gradients allow the cell to efficiently extract as much energy as possible from the breakdown of glucose (and other biochemical foodstuffs). On the other hand, if ATP were produced exclusively by substrate-level phosphorylation, using a high-energy chemical intermediate, then much of the energy liberated from the breakdown of glucose would be lost as heat.[32]

To understand why this is so, consider an analogy. Suppose people in a particular community receive their daily allotment of water in a 10-gallon bucket. The water they receive each day is retrieved from a reservoir with a 12-gallon bucket and then transferred to their bucket. In the process, two gallons of water are lost. Now, suppose the water from the reservoir retrieved with a 12-gallon bucket is dumped into a secondary reservoir that has a tap. The tap allows each 10-gallon bucket to be filled without losing two gallons. Though the procedure is indirect and more complicated, using the secondary reservoir to distribute water is more efficient in the long run. In the first scenario, it takes 60 gallons of water (transferred from the reservoir in five 12-gallon buckets) to fill up five 10-gallon buckets. In the second scenario, the same amount of water transferred from the reservoir can fill six 10-gallon buckets. With each transfer, the additional two gallons accumulate in the reservoir until there is enough water to fill another 10-gallon bucket.

With substrate-level phosphorylation, when the phosphate group is transferred from the high-energy intermediate to ADP to form ATP, excess energy released during the transfer is lost as heat. It takes 7 kcal/mol of energy to add a phosphate group to ADP to form ATP. Let's say that the hypothetical chemical intermediate releases 10 kcal/mol when its high-energy phosphate bond is broken. Three kcal/mol of energy are lost.

On the other hand, using the ETC to build up a proton gradient is like the reservoir in our analogy. It allows that extra three kcal/mol to be stored in the proton gradient. We can think of the F_1-F_0 ATPase as analogous to the tap. It uses 7 kcal/mol of energy, released when protons flow through its channels to drive the formation of ATP from ADP and inorganic phosphate. The unused energy from the proton gradient continues to accumulate until enough energy is available to form another ATP molecule. So, in our hypothetical scenario, if the cell used substrate-level phosphorylation to make ATP, 70 molecules of the high-energy intermediate would yield 70 molecules of ATP with 210 kcal/mol of energy lost as heat. But using the ETC to generate proton gradients yields 100 ATP molecules with no energy lost as heat.

The elegant molecular rationale that undergirds the use of proton gradients to harvest energy and to power certain cellular processes makes it unlikely that this biochemical feature reflects the outcome of a historically contingent process that just happened upon proton gradients. Instead, it points to a set of principles that underlie the structure and function of biochemical systems—principles that appear to have been set in place from the beginning of the universe.

The most obvious and direct way for the first protocells to harvest energy would seemingly involve some type of mechanism that resembled substrate-level phosphorylation, not an indirect and more complicated mechanism that relies on proton gradients. If the origin of chemiosmosis and the use of proton gradients were a historically contingent outcome—predicated on the fact that the first protocells just happened to employ a natural proton gradient—it seems almost eerie to think that evolutionary processes blindly stumbled upon what would later become such an elegant and efficient energy-harvesting process. And a process necessary for advanced life to be possible on Earth.

If not for chemiosmosis, it is unlikely that eukaryotic cells and, hence, complex life such as animals, plants, and fungi, could have ever existed. Substrate-level phosphorylation just isn't efficient enough to support the energy demands of eukaryotic organisms.

In other words, the use of proton gradients points to a set of deep, underlying principles that arise from the very nature of the universe itself and dictate how life must be. The molecular rationale that undergirds the use of proton gradients and their near-universal occurrence in living organisms suggests that proton gradients are an indispensable feature of living organisms. Without the use of proton gradients to harvest energy and drive cellular processes, advanced life would not be possible. Even if life were discovered elsewhere in the universe, it would have to employ proton gradients to harvest energy.

It is remarkable to think that proton gradients, which are a manifestation of the laws of nature, are at the same time precisely the type of system advanced life needs to exist. One way to interpret this "coincidence" is that it serves as evidence for the biochemical anthropic principle.

Does the Anthropic Principle Apply to Energy-Harvesting Pathways?

The goal of this chapter was to determine if the anthropic principle applies to energy-harvesting pathways. As noted, I predicted that we would find anthropic coincidences based on the central importance of these biochemical routes. Specifically, we examined the architecture and operation of glycolysis and the

Krebs cycle (two pathways that also contribute to central carbon metabolism), and the ETC. By analyzing these biochemical systems and surveying the scientific literature, we hoped to learn if these pathways met the criteria for the biochemical anthropic principle. Specifically, we sought to determine if:

- The laws of nature manifest physicochemical constraints that dictate the structural features of these three metabolic processes.
- A molecular rationale exists that explains the structure of these metabolic routes, accounting for their fitness for life.
- These three metabolic pathways appear to be fit for life based on comparisons with close chemical analogs.

We discovered that all three criteria are met for glycolysis and the Krebs cycle. The design of these two pathways displays an exquisite molecular rationale, an impeccable chemical logic. These two pathways also appear to be highly optimal and highly efficient, yielding maximal ATP flux. While biochemists have discovered alternative pathways that could serve the same function, it appears that under the physiological conditions of the cell, none are superior to either of the naturally occurring pathways. The structure of these two pathways also appears to be dictated, at least in part, by constraints that arise out of the laws of nature. Specifically, the metabolites of these two pathways must display low lipophilicity and high water solubility, possess several charged atoms, and have a minimized nonpolar molecular surface area. Of interest is the importance that geometric features, coupled with hydrophobicity, play in dictating metabolite concentrations in the cell. These same types of constraints also dictate the structure of proteins and cell membranes.

The insights from origin-of-life investigators add to the intrigue. These researchers discovered that on top of these physicochemical constraints are the constraints that arise out of:

- the geochemical and geophysical settings of early Earth,
- carbon chemistry, and
- the interplay between organic compounds.

These dictates lead to potential prebiotic reactions on early Earth that spontaneously generate the compounds that serve as metabolic intermediates for the Krebs cycle and glycolysis (and the other pathways that comprise central carbon metabolism).

How fortunate it is that these constraints produced a set of prebiotic molecules that formed just-right metabolic intermediates that comprised highly optimal metabolic routes. Without question, no shortage of anthropic coincidences exists when it comes to glycolysis and the Krebs cycle.

We also discovered that an impeccable molecular rationale undergirds the design of the ETC. At first glance, the use of proton gradients to drive the final steps of oxidative phosphorylation seems counterintuitive. But if not for their use, advanced, complex life would not be possible on Earth. As with the prebiotic chemistry that many origin-of-life researchers believe generated the glycolytic and Krebs cycle intermediates, the origin of proton gradients appears to have emerged under the conditions of early Earth, specifically at a special type of hydrothermal vent system. It would be an incredible stroke of luck if geochemical processes dictated the emergence of proton gradients in living systems only for proton gradients to later become such an elegant and efficient energy-harvesting system necessary for advanced life to be possible on Earth.

Through the course of the last six chapters, I have established that the anthropic principle extends into the arenas of chemistry and biochemistry. In the final chapter, I will explore the scientific and theological implications of this insight. Is there a teleology that undergirds the structure of biochemical systems—the fundamental systems of life on Earth?

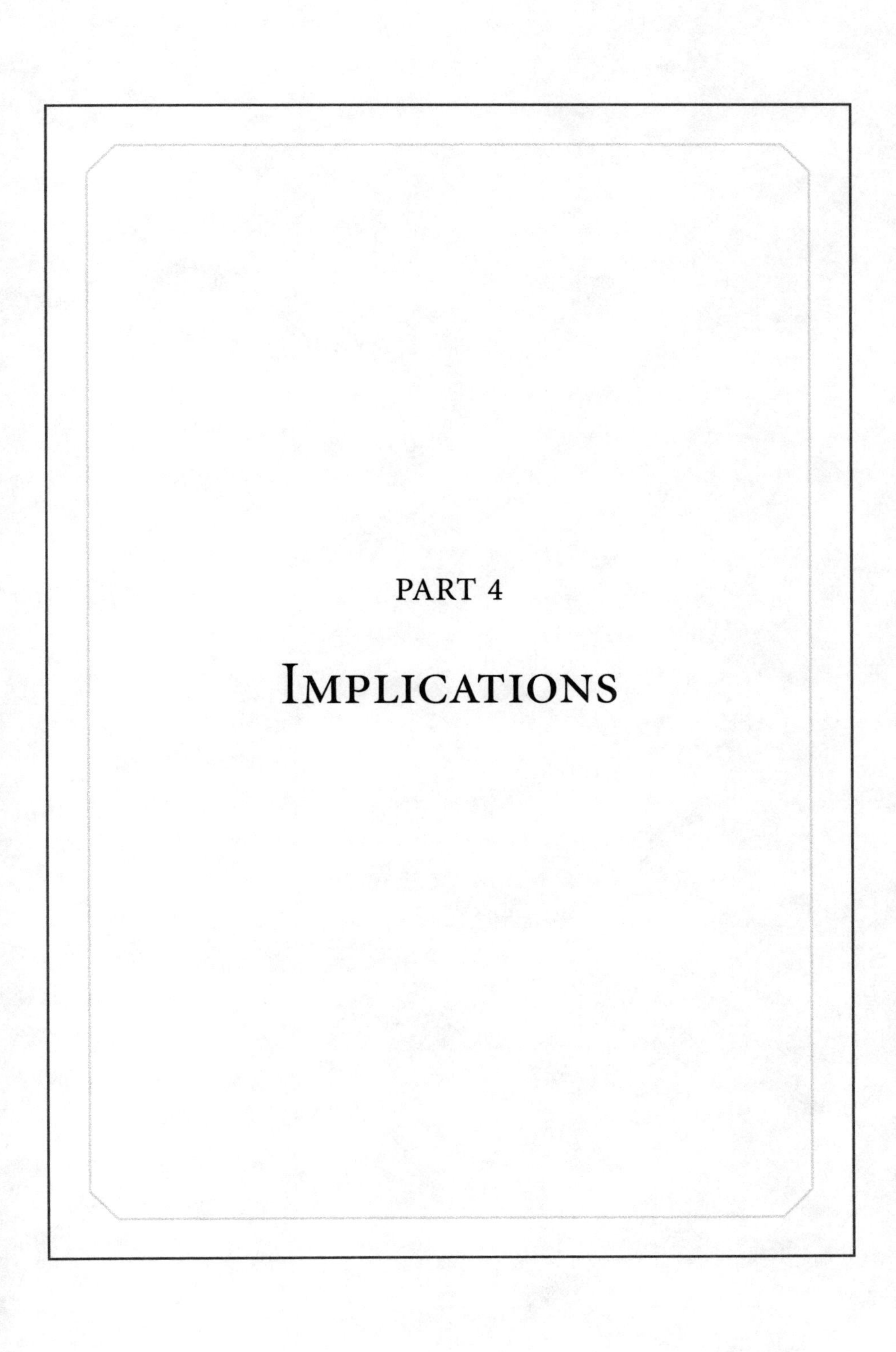

PART 4

Implications

Chapter 9

Implications of the Biochemical Anthropic Principle

I don't think I am alone in feeling that I wouldn't want to relive my high school years. I really struggled as I came of age. I never felt like I fit in. I never felt like I belonged.

Despite my ambivalence about my high school experience, I have some fond memories of special places and magical times spent with my high school compatriots. And for those few instances, I would go back if I could.

I distinctly remember the first time this realization came to me. I was a postdoctoral fellow at the University of Virginia. My wife, Amy, worked nights at the University of Virginia Medical Center in a medical laboratory, while I carried out research in the chemistry department during the day. Shortly after we moved to Charlottesville, Amy gave birth to our second child. Amy took care of the kids during the day, and I watched over our two-year-old and infant daughters in the evening.

One evening, as I tried to come up with a way to entertain the kids, I decided to play a cassette, with the idea that we could dance to the music before bedtime. The cassette happened to be one of my favorite Kansas albums, *Monolith*. I hadn't listened to the album in years. That evening I listened to it with a fresh perspective. While singing along to the track "People of the South Wind," the desire to return to the places and times of my younger years awakened in me.

Kansas released *Monolith* in 1979. The first single released from *Monolith*, "People of the South Wind," is among my favorite Kansas recordings. This track is named for the Kaw, the Native American people who originally settled in what is now the state of Kansas. The great Kaw nation suffered one catastrophe after the other, triggered by the purchase of the Louisiana territory by the United States in 1803.

Throughout these catastrophes—invasion, decimation by smallpox, and loss of their land—the Kaw people held out hope. Part of that hope is embodied in the Kaw legend foretelling a future time when the white people would leave their lands, making it possible for the Kaw nation to return to its past glory. Guitar player Kerry Livgren, the primary songwriter for Kansas, drew upon this legend when he wrote "People of the South Wind." The song expresses a deep longing to return to the days of the songwriter's youth, a time of joy and simplicity. But as he has grown older, those feelings and that innocence have long been replaced with a sense of despondency because of the harsh realities and disappointments of life.

Listening and dancing to "People of the South Wind" with my two daughters that night elicited in me the same feeling Livgren must have had when he wrote that song. I realized that some of the best times of my life took place in high school (despite the struggles I had), college, and graduate school. I can still see the smiling faces of my friends when the times were so good, though the memories of those times are slowly vanishing as I grow older.

Apart from the somber feel of the "People of the South Wind" track, the overall ethos of the *Monolith* album is upbeat and hopeful. This mood stands in sharp contrast to the dark, foreboding sense of hopelessness that defines the songs on the *Point of Know Return* album. Released late in 1977, *Point of Know Return* is Kansas's highest charting album, coming in at number four on the Billboard charts in January of 1978. The album sold over four million copies and is widely recognized for the three singles released from the album, "Point of Know Return," "Dust in the Wind," and "Portrait (He Knew)."

Kerry Livgren was the primary songwriter for both albums. Throughout his tenure with Kansas, Livgren wrote several songs that express spiritual themes and an underlying desire to pursue answers to life's biggest questions. The song "Point of Know Return" is about mustering the courage to take the step into the unknown, fearlessly seeking after spiritual truths.

The difference in tone between *Point of Know Return* and *Monolith* reflects the different stages of Livgren's spiritual journey. It appears that when Livgren wrote the songs for *Point of Know Return*, he questioned whether he or anyone could gain true insight about the meaning of life. The second track on the album "Paradox" is about seeking answers that just can't be found.

Burning with the question in my mind
Seems there's nothing else for me to find

The fourth track on the album, "Portrait (He Knew)," was inspired by Albert Einstein. The song is about someone who discovered a key insight into the nature and meaning of reality, but his ideas weren't understood. Then he died and the understanding he gleaned was lost.

He had a different idea
A glimpse of the master plan

If only he could have been with us
No telling what he might say

The seventh track on the album, "Dust in the Wind," became Kansas's most widely recognized song, though it departs from their signature style. This track is an anthem to atheism and materialism. Inspired in part by biblical passages Genesis 3:19 and Ecclesiastes 3:20, this song is about our mortality and the finality of death, raising questions about whether there is any meaning or purpose to our existence at all.

The next to last song on the album, "Nobody's Home" tells about the horrifying realization that after traversing the point of no return, having boldly searched for the truth no matter where it leads, all that awaits is emptiness—a place where no answers exist and all that was is gone without leaving even a trace behind.

The world that I was sent to reach has got no future now
Across the galaxy to spread the word and no one heard
I came for nothing, I'm alone and nobody's home

Two years later, Kerry Livgren was in a different place in his spiritual journey when he wrote songs for the *Monolith* album, just steps away from converting to Christianity. He no longer questioned if there was meaning to life. He was on the precipice of finding that meaning in the person of Jesus Christ. The second single released from *Monolith*, "Reason to Be," is about the joy in finding your purpose in life.

Someday something will find you
A magical feeling you could not foresee
Your reason to be

Counterbalancing "Nobody's Home" is the fifth track on *Monolith* titled "A Glimpse of Home." Listening to the *Monolith* album that night with my girls led to an epiphany for me. I realized Livgren was singing about finding hope in the person of Christ.

All my life I knew you were waiting, revelation anticipating
All is well, the search is over, let the truth be known
Let it be shown (give me a glimpse of home)

Livgren converted to Christianity during the Kansas Monolith Tour. After his conversion, Livgren recorded three more albums with Kansas before leaving the band. After leaving Kansas in 1983, Livgren formed a new band, AD, and entered the Christian rock market. I learned of AD around the time I converted to Christianity in 1986. When I discovered that Livgren converted to Christianity while with Kansas, I became fascinated with the albums *Point of Know Return*, and *Monolith*, in part because I wanted to gain insight into Livgren's spiritual journey during his time with Kansas. I realized that spiritual themes—even a Christian perspective—could be found in mainstream rock music. Until that point, I naively thought that spirituality and the rock music I listened to growing up were incompatible. I now know that I was wrong.

• • • • • • • • • • • •

Like my initial perception about rock music and spirituality, many people think that science and spiritual ideas are also incompatible. In my view, this perception is due partially to the influence of Charles Darwin's theory of evolution, which in effect stripped teleology (design and purpose) out of biology—and out of science altogether. Prior to publication of *On the Origin of Species* (1859), teleology held a prominent place in the life sciences. With his theory of evolution, not only did Darwin provide an explanation for the origin of species (through natural selection), but he also claimed that this same mechanism, operating over a vast period, could explain the origin of life's major groups and, along with it, account for the biodiversity and biogeographical distribution of life throughout Earth's history. Most significantly, Darwin asserted that natural selection could even account for the appearance of design so pervasive in biological systems. With this maneuver—sometimes called the Darwinian revolution—Darwin expunged teleology from the biological disciplines, replacing the mind of the Watchmaker with the mechanism of natural selection—the blind watchmaker.

In the early 1900s, Harvard physiologist Lawrence J. Henderson attempted to reintroduce design and purpose into the life sciences, and in doing so laid the groundwork for anthropic reasoning. In two works, *The Fitness of the Environment: An Inquiry into the Biological Significance of the Properties of Matter* (1913) and *The Order of Nature: An Essay* (1917), Henderson argued that the chemical environment has been structured so that it displays the necessary properties that make life possible. (See chapter 2.) Henderson saw his idea as a counter perspective to Darwin's theory of evolution. Henderson realized that some environments could never harbor life. For this reason, these environments precluded the evolutionary origin and history of life. Henderson presented evidence that the environment demonstrates a remarkable fitness that makes life possible in the first place. If it wasn't for the fitness of the environment, life couldn't even originate, let alone evolve to become adapted to its surroundings. Henderson thought that the reciprocal interdependence between the environment's fitness and natural selection indicated that there must be a tendency or directional flow to life's evolutionary history that reflects a teleology or purpose to nature.

Henderson's discovery of the environment's fitness resurrected the idea of a teleology, leading him to argue that not only does design and purpose have a place in biology, it has a place in science. In Henderson's scheme, the scientific insight about the fitness of the environment should have pushed science to confront the idea of "design." According to Henderson, we have no choice but to describe the environment's remarkable suitability for life as teleological, because no other concept fits. Henderson's work paved the way for spirituality to regain a place within science. Yet scientists largely ignored these ideas for over 50 years. It seemed as if they were unwilling to allow spiritual considerations back into the arena of science.

Then along came Brandon Carter in the 1970s and the scientific discovery of the cosmological anthropic principle—namely the fine-tuning of the universe's dimensionless constants. Carter sought to make sense of this discovery by presenting an interpretation with an observer-centric tautology. For observers to exist with the capability to discover anthropic coincidences, the universe's physical constants must be fine-tuned and the observers must reside at the just-right location in the universe's space-time. If not, then no observers would exist. In other words, if observers reside in the universe, then the anthropic principle must be true by necessity. Based on the counterfactual studies carried out by astronomers and physicists, we know that if our space-time location situated within the universe's history and values of the physical constants were anything other than what they are, then there wouldn't be any observers present to study the universe and recognize that it wasn't fine-tuned for life.

In 1986, John D. Barrow and Frank J. Tipler published *The Anthropic Cosmological Principle*. In this work, these two physicists addressed theological and philosophical implications of these coincidences, something Carter adeptly sidestepped. Barrow and Tipler recognized that Carter's interpretation was, in effect, true by definition. But they also noted that there are other possible explanations for the anthropic principle. With their so-called strong anthropic principle (SAP), Barrow and Tipler introduced the possibility that an imperative exists such that the universe's physical and cosmological constants *must* assume precise, exacting values so that carbon-based life will exist. (As discussed in chapter 3, it is highly unlikely that non-carbon-based life exists, or is even possible. See pages 86–88). Barrow and Tipler suggested that this imperative arises for one of three possible reasons:

- The universe is the work of an intelligent Agent, influenced by classical design arguments for God's existence.
- The universe is the product of an observer-created reality, influenced by concepts from quantum mechanics.
- An ensemble of universes with different physical and cosmological constants exist by necessity, with our universe being the one that is fine-tuned for life, simply as an outworking of chance.

By introducing a revised version of the SAP, Barrow and Tipler confronted the obvious philosophical implications of the anthropic principle and forced cosmology to straddle science and metaphysics. In the process, they reintroduced teleology into physics and cosmology. In effect, these two physicists accomplished what Henderson failed to achieve.

The discovery of the cosmological anthropic principle prompts the question: Does the anthropic principle manifest in the arenas of chemistry, biochemistry, and biology?

This question carries implications beyond scientific considerations about the nature of chemical systems and life. It has metaphysical implications as well. I think that if a theistic-oriented interpretation of the anthropic principle is valid, then anthropic coincidences would also be observed in chemistry, biochemistry, and biology. It seems to me that if a Creator intentionally designed the universe to be biofriendly then he wouldn't have limited the influence of the universe's physical constants to the processes of star and planetary formation and the production of the chemical elements necessary for life (which is the focus of the cosmological anthropic principle). He would have made every aspect of the creation biofriendly. He would have designed the physical constants to influence chemical and biological systems as well. And I expect that this influence would uniquely manifest in the laws of chemistry and the structure and function of biochemical systems. In other words, *all* creation would display a fitness for purpose—namely humanity's advent and existence.

Does the Anthropic Principle Apply to Chemistry?

In light of this prediction, this book has focused on addressing the question: Does the anthropic principle apply to chemistry and biochemistry? Henderson demonstrated in *Fitness for the Environment* that anthropic coincidences abound in chemistry, but he made his case before the advent of quantum mechanics and a modern understanding of chemical structures and chemical bonding. Using a methodology inspired by Henderson and following in the footsteps of others, I described how water, carbon dioxide, oxygen, and carbon-containing compounds display physicochemical properties that are constrained and dictated by the laws of nature. As a consequence of these constraints, these substances display highly unusual properties that turn out to be ideally suited for life. In fact, if not for these materials and their chemical and physical behaviors, it is hard to imagine how life would exist in the universe. No other known materials have the necessary suite of properties to serve as able substitutes for water, carbon dioxide, oxygen, and organic compounds.

It is highly fortuitous that the laws of nature would produce a chemical environment in the universe that is precisely what is required for life to exist. Equally remarkable is the discovery that the two chemical elements most necessary for life, oxygen and carbon, are also among the most abundant chemical elements in the cosmos. No other elements have the life-support properties that carbon and oxygen display. As it turns out, the triple alpha process which forms these two elements in the cores of stars, relies on a finely tuned mechanism. If even one of several parameters associated with this mechanism deviated much beyond its actual value, carbon and oxygen would exist at relatively low levels in the universe, and life wouldn't be possible. In this respect, the cosmological and chemical anthropic coincidences cojoin. Henderson's insights about the fitness of the environment proved true under the microscope of our modern understanding of the chemical properties of atoms and molecules.

Does the Anthropic Principle Apply to Biochemistry?

So, does the anthropic principle manifest in biochemical systems? In contrast to the theoretical, counterfactual approach used to demonstrate the cosmological anthropic principle, I used an approach that relies on experimental evidence and a comparison of biochemical systems found in nature with conceivable alternatives. Specifically, I sought answers to these questions in the evaluation:

- Does a molecular rationale undergird the structure and function of specific biochemical systems, accounting for their fitness for life?
- Do physicochemical constraints that arise from the laws of nature dictate the structural and functional aspects of biochemical systems, influencing their fitness for life?
- When we compare the structural and functional features of biochemical systems in nature with close chemical and biochemical analogs, do we find evidence that biochemical systems display properties that make them unusually fit for life?

I argued that if these three criteria are met, then we can reasonably conclude that the structural and functional features of biochemical systems display anthropic coincidences.

We focused attention on proteins, nucleic acids (DNA and RNA), and phospholipids (which form the bilayers of cell membranes). We also examined key metabolic processes, such as protein synthesis, DNA replication, and the intermediary metabolic routes involved in energy harvesting and central carbon metabolism (including glycolysis, the Krebs cycle, and the electron transport chain). These biochemical systems were selected because of their central importance, with the idea that if there is a biochemical anthropic principle, it should be reflected in the architecture and operation of these systems.

There is another reason I selected these biochemical systems to probe for the biochemical anthropic principle. Biochemists and origin-of-life researchers have learned that these systems reflect the essential, minimal biochemical processes required for a cellular entity to exist as a bona fide life-form.[1] For life to exist in minimal form, it must be capable of

- DNA replication,
- protein synthesis,
- cell membrane formation and maintenance, and
- energy-harvesting processes.

In other words, the biomolecules and metabolic processes I chose to investigate are the very molecules and systems that constitute minimal life. And if there is a biochemical anthropic principle, we would expect to find evidence for it in these features of the cell's chemistry.

As we soon learned, attempting to identify anthropic coincidences for biochemical systems is no easy task, in part, because of the complexity of biochemistry. We detected a signal for the biochemical anthropic principle—albeit a "noisy" one—in all of the systems examined. We discovered that, for the most part, life's core biochemical systems conform to all three criteria. All the systems examined display an exquisite molecular rationale and logic undergirding their structure and operation. These systems also appear to be highly optimized and, when compared to other conceivable biochemical analogs, they are highly unusual, if not unique. Most of the unusual features of these systems are what make them ideally suited for life.

It also appears that, in large measure, the laws of nature constrain and dictate these systems' physicochemical properties. In the case of proteins, phospholipids, and the metabolic intermediates, molecular geometry and hydrophobicity dictate their structural features. It is remarkable that these two simple features can account for the architecture and activities of much of biochemistry. It is also highly suspicious when we realize that these physicochemical constraints generate biomolecules and metabolic systems with the just-right properties needed for life to exist. How fortunate that the physicochemical constraints that arise out of the laws of nature didn't generate molecular systems lacking the necessary properties needed for life—a reasonable possibility in an undesigned existence.

So, what do we make of this insight? What does it mean for the origin and design of life? What are the implications of this discovery?

The Scientific Implications of the Biochemical Anthropic Principle

If we choose to adopt an approach like Carter's and embodied in Barrow and Tipler's version of the weak anthropic principle (WAP), we could interpret the biochemical anthropic principle in terms of an observer-centric tautology. If so, then of course biochemical systems display a fitness for purpose. Because they do, sapient beings exist who are capable of detecting all the just-right features of biochemical systems necessary for life to exist. If biochemical systems didn't display a type of fitness for purpose, then there wouldn't be any observers in the universe to uncover that fact.

Even if we embrace this interpretation—which must be true by definition—there is still something philosophically disquieting about the existence of the biochemical anthropic principle, particularly as it relates to the intelligibility of biochemical systems. In some respects, it is remarkable to think that biochemical and biological systems are intelligible to the human mind, given the complexity and diversity of living systems. Part of the reason the living realm is accessible to us stems from the universal features of living systems.

All life is made up of cells, either prokaryotic or eukaryotic cells. Effectively, all living systems possess the same biochemical systems. All life relies on proteins made from the same 20 amino acids. All life makes use of DNA as its genetic material. The mechanism of DNA replication is, in effect, universal. The central dogma of molecular biology defines the process by which the information in DNA is expressed in all living systems. The genetic code, which relates the information in the nucleotide sequences of DNA to the information in the amino acid sequences of proteins, is universal. (Nonuniversal codes deviate only slightly from the universal code.)

All life is bounded by cell membranes built from phospholipid bilayers. The energy-harvesting pathways and the metabolic routes that comprise central carbon metabolism are near-universal. So, too, is the use of proton gradients to generate ATP, the universal energy currency of the cell. The universal nature of the cell's chemical systems has profound consequences that make biochemistry possible as a robust scientific discipline. Because of this universality, what we learn as biochemists from studying the biochemistry of, say, the gram-negative bacterium *E. coli* or brewer's yeast (*Saccharomyces cerevisiae*), often applies to virtually every living organism, including human beings.

From an evolutionary perspective, one way to account for the universal nature of biochemistry is that, in effect, the frozen happenstance of a historically contingent evolutionary process resulted in biochemical systems found in the last universal common ancestor (LUCA)—the organism (or community of organisms) that anchors the evolutionary tree of life. As life diverged from LUCA, all living systems retained their biochemical systems. In this scenario, it is rather fortuitous that the instantiation of biochemistry happened with LUCA, rather than later in life's history. If it had happened later, then we could justifiably expect to find disparate and nonuniversal biochemical systems distributed throughout the tree of life. It would also be far more challenging to develop a scientific understanding of life's most basic and fundamental systems.

The existence of the biochemical anthropic principle indicates that biochemical systems cannot be solely the outworking of a historically contingent evolutionary process but appear to be largely shaped by constraints that arise out of the laws of nature. Stated another way, it appears that the laws of nature prescribe the core biochemical systems. This insight implies that the universal nature of biochemistry reflects a deeper reality that has been established by a set of principles. It also holds out the prospect that we can discover these principles. And, when we do, we can fully explain the structure and activity of biomolecules and biochemical systems and move toward a new way of thinking about biochemical systems. If biochemical systems are the result of historically contingent processes, then our investigation should primarily focus on describing and characterizing these systems. In my view, this approach has historically defined much of the research efforts in biochemistry, and still largely does so today. On the other hand, if a set of principles dictates the architecture and operation of life's chemistry, then it presents biochemists the opportunity to extend their work beyond describing and characterizing biochemical systems to actively seeking after these principles. This opportunity extends beyond merely asking the why questions. For if indeed a molecular rationale undergirds the design of biochemical systems, it means that a theoretical framework may well exist that can be discovered and used to yield a fundamental understanding of the nature of biochemical systems. Discovering these principles, assuming they exist, adds to the tractability of the biochemical realm. It grants life scientists greater explanatory power regarding the fundamental nature of life.

The molecular rationale that seems to undergird biochemical systems not only provides the means for biochemists to understand why biochemical systems are the way they are, but can be used to make predictions that can guide biochemical research as well as direct the efforts in synthetic biology. One of the chief goals of synthetic biology is to create artificial, nonnatural life-forms in the lab. Some of this work focuses on creating protocells from the bottom up and other work focuses on reengineering existing life-forms, rendering them unlike anything found in nature. Much of the work involved with the latter efforts centers around introducing new metabolic capabilities into microbes (such as *E. coli* and *S. cerevisiae*), with the hope that these metabolic routes can be used to produce invaluable medicines or other types of products. As synthetic biologists seek to develop metabolic routes from scratch, possessing a set of fundamental principles would be immensely helpful in these efforts.

As a case in point, in the previous chapter I discussed the fact that metabolite concentrations are strongly influenced by four physicochemical features: lipophilicity, solubility, number of charged atoms, and the amount of nonpolar molecular surface area.[2] The investigators who uncovered these insights (which contribute to the case for the biochemical anthropic principle) recognize that this work provides a fundamental understanding of life processes. But it also has practical utility. They write that their findings "can assist in establishing metabolic models that support synthetic biology and metabolic engineering efforts."[3]

The existence of the biochemical anthropic principle also has bearing on astrobiology and the search for life-forms throughout our solar system and beyond. It has become vogue for astrobiologists to suggest that life on Earth—life as we know it—is but one example of life. They argue that other types of life might possibly exist—life as we *don't* know it. But as we discussed in chapter 3, life must be carbon-based and must exist in a liquid water medium. No other chemical element possesses the chemical properties of carbon to make it suitable for life and no other solvent but water displays the wide range of properties needed to support life. Based on the case I built for the biochemical anthropic principle, one could make a reasoned case that carbon-based life in an aqueous medium most likely would:

- make use of proteins assembled from the 20 canonical amino acids;
- employ DNA—comprised of four nucleotides made from adenine, guanine, thymine, and cytosine—as its genetic material;
- use RNA—comprised of adenine, guanine, uracil, and cytosine—as the intermediary to translate the genetic information into the information housed in proteins;
- use phospholipids as components to build cellular boundaries,
- rely on the central dogma to translate its genetic information into the information housed in proteins;
- use the same genetic code as the one found universally throughout the living realm;
- replicate its genetic information using a semiconservative, template-directed mechanism;
- utilize glucose as the primary energy source;
- break down glucose into carbon dioxide and water as the means to harvest energy for the cell using pathways very similar to glycolysis and the Krebs cycle; and
- make use of proton gradients.

It is true that biochemists and origin-of-life researchers have discovered other materials that could be used to build a genetic system. Yet none of these appear to be as optimal as DNA when all their features are considered. In some cases, these alternatives can be excluded from consideration because the building blocks needed to assemble these materials most reasonably would not have been produced by prebiotic reactions on early Earth. It is also true that alternative genetic codes and alternative intermediary metabolic pathways are conceivable. But most are substandard for one reason or the other, lacking the optimality of glycolysis and the Krebs cycle.

In other words, the existence of the biochemical anthropic principle makes it quite likely that terrestrial biochemistry is, in fact, universal biochemistry. Life as we know it on Earth may be and likely is the only way life could be, at least at the biochemical level.

The Metaphysical Implications of the Biochemical Anthropic Principle

In many respects, an interpretation of the biochemical anthropic principle that follows the same line of reasoning as the WAP is self-evident and must be true. Yet this interpretation isn't satisfying to me because it avoids the obvious metaphysical implications of the widespread anthropic coincidences in biochemistry. Generally, scientists are uncomfortable discussing metaphysical implications, in part because of the influence methodological naturalism has on the operation of contemporary science. However, avoiding and sidestepping the metaphysical implications doesn't mean they don't exist, and it doesn't mean we shouldn't explore them.

To their credit, Barrow and Tipler understood this to be so when they presented their version of the SAP. These two physicists recognized that the cosmological anthropic principle reflects an imperative that demands the physics of the universe assume certain features so that carbon-based life will exist. One obvious way to make sense of this imperative is to consider that it arises from a Mind, suggesting that the universe's fitness for purpose evinces a Creator's handiwork. I would assert that if this interpretation has legitimacy, then it follows that all aspects of nature—its physics, chemistry, biochemistry, and biology—should display a fitness for purpose. And, as I have shown, anthropic coincidences abound in chemistry and biochemistry. On the other hand, if we interpret the cosmological anthropic principle in the same vein as Brandon Carter—as an observer-centric tautology—then there is no necessary reason to think that the anthropic principle would extend to all facets of nature. Because of the pervasiveness of anthropic coincidences, I am convinced that the anthropic principle—and the biofriendly nature of the universe—serves as compelling evidence for a Creator. (I will elaborate on two other reasons shortly.)

Of course, if a Creator is involved in the origin and the design of life, this prompts the question: How did the Creator specifically bring life into existence? Can science offer insight into this question? Or must this question be limited to the realm of theology and philosophy?

When confronted with this question, Christian theists generally respond by postulating one of two competing paradigms: theistic evolution or a type of creationism in which God intervenes through direct and personal means to effect the origin and design of life. So, how does the biochemical anthropic principle fit within these two frameworks? Surprisingly, scientific insights help guide us as we attempt to answer this question.

The Biochemical Anthropic Principle and Theistic Evolution

Of course, theistic evolution (or, as more recently referred to in some quarters, evolutionary creation) is the theological idea that the Creator employed evolutionary processes to create. This position is not monolithic. It consists of a spectrum of views about how God made use of evolutionary mechanisms to bring about his creative intents. On one end of the spectrum sits the view that God created a clockwork universe in which he set the laws of nature in motion and from that point on has had no active involvement. The world that we see is the result of an unguided historically contingent process. Oftentimes, people who hold this view believe that God had no hand in guiding the evolutionary process at all and played no role in determining the outworking of evolutionary history. He merely waited until sapient, sentient beings arrived on the scene and then revealed himself to them. These sapient beings didn't necessarily have to be humans. It is merely by happenstance that they were.

On the other end of the spectrum lies the position that God plays an active role guiding and directing the evolutionary history of life. But if so, how did God direct the process? And can we detect his activity? In my experience, most adherents to this version of theistic evolution argue that God's involvement in the evolutionary process is not scientifically detectable. I suspect that this maneuver is an attempt to keep theistic evolution within the realm of theology, so metaphysical ideas can remain separate from science. Adherents to this type of theistic evolution maintain that God has chosen to conceal his involvement in evolutionary history. Perhaps God's activity takes place at the quantum level, hidden from us because of quantum indeterminacy. In other words, most common expressions of theistic evolution agree that we can't identify God's work in evolutionary history because either he wasn't involved at all or he chose to conceal his active role.

Although I once embraced theistic evolution, I no longer find the idea satisfying as a Christian theist. Arguing that God had no role in the evolutionary history of life renders a Creator superfluous. Why posit a Creator at all? Why even think a Creator was involved? Meanwhile, claiming that God was involved with evolutionary history, but concealed that involvement from us is problematic to me for theological reasons. This claim disregards what the Christian Scriptures teach. Both the Old and New Testaments inform us that evidence for God's work as Creator and his attributes is revealed through the record of nature. (For example, see Job 12:7–9, Psalm 19, and Romans 1:20.) If so, then shouldn't we be able to clearly detect God's guiding hand in the evolutionary process—if, indeed, that is how he chose to create? Unfortunately, I don't see any difference between most expressions of theistic evolution and materialistic or naturalistic evolution, once the arguments play out to the end.

On the other hand, the recognition that the anthropic principle extends to the realm of biochemistry makes it possible to conceive of a version of theistic evolution that reveals God's purposeful involvement and *is* scientifically discernable, demonstrating teleology and direction in the evolutionary process—at least with respect to *chemical evolution* and the origin of life (and, hence, the origin of biochemical systems).

As one of the key tenets of the biochemical anthropic principle, the laws of nature preordained the structure and function of biochemical systems. This stands in contradistinction to the commonly held view that evolutionary processes stumbled upon these systems through mechanisms characterized as undirected and historically contingent, operating under the auspices of chance and selection. Accordingly, the structure and function of biochemical systems are the accidental outcome of an evolutionary history. However, if constraints exist that arise out of the laws of nature and dictate the structure and function of biochemical systems, as the biochemical anthropic principle demands, then it becomes reasonable to regard the origin of life (and, consequently, the origin of biochemistry) as a highly probable—if not inevitable—outcome of evolutionary history. In this scheme, the origin of life cannot be considered an outcome that reflects mere happenstance. Instead, it becomes an imperative. Life's genesis becomes a necessity, not a chance outcome, with a predictable endpoint reflected in the universal nature of biochemical systems.

A study by a research team from the Earth-Life Science Institute (ELSI) in Tokyo, Japan, illustrates this point nicely. The ELSI investigators wanted to develop a better understanding of the optimal nature of the universal set of amino acids used to build proteins. (See chapter 4, pages 107–112.) They also wanted to gain insight into the canonical set's evolutionary origin.[4] To accomplish this, they used a library of 1,913 amino acids (including the 20 amino acids in the canonical set) to construct random sets of amino acids. The researchers varied the set sizes from 3 to 20 amino acids and evaluated the performance of the random sets in terms of their capacity to support the folding of protein chains into three-dimensional structures, protein catalytic activity, and protein solubility. They discovered that if a random set of amino acids included even a single amino acid from the canonical set, it dramatically outperformed random sets of the same size without any of the canonical amino acids. Based on these results, the researchers concluded that each of the 20 amino acids used to build proteins stands out, possessing highly unusual properties that make them ideally suited for their biochemical role. The ELSI researchers believe that—from an evolutionary standpoint—these results shed light onto how the canonical set of amino acids emerged. Because of the unique adaptive properties of the canonical amino acids, the researchers speculate that "each time a CAA [canonical amino acid] was discovered and embedded during evolution, it provided an adaptive value unusual among many alternatives, and each selective step may have helped bootstrap the developing set to include still more CAAs."[5]

In other words, the researchers speculate that whenever the evolutionary process stumbled upon a canonical amino acid and incorporated it into nascent biochemical systems, the addition offered such a significant evolutionary advantage that it became instantiated into the biochemistry of the emerging cellular systems. Presumably, as this selection process occurred repeatedly over time, members of the canonical set were added, one by one, to the evolving amino acid set. Eventually this culminated in the full canonical set.

These scientists find further support for this scenario in the following observation: some of the canonical amino acids seemingly play a more important role in optimizing smaller sets of amino acids, some play a more important role in optimizing intermediate-size sets of amino acids, and others play a more prominent role in optimizing larger sets. They argue that this difference may reflect the sequence by which amino acids were added to the evolving set of amino acids as life emerged.

The ELSI researchers speculate that no matter the starting point in the evolutionary process, the pathways will all converge at the canonical set of amino acids because of the amino acids' unusual adaptive properties. In other words, the amino acids that make up the universal set of protein-coding amino acids are not the outworking of a historically contingent evolutionary process but appear to be fundamentally prescribed by the laws of nature. As such, these laws drive life's history to a seemingly predetermined endpoint. This means the constraints of the laws of nature, *not* natural selection operating on historically contingent events, played the more significant role in determining the canonical set of amino acids.

As we saw in chapter 4, the canonical amino acids are an optimal set that displays highly unusual properties that make it just-right for life. It is eerie to think that the reactions of prebiotic chemistry would produce many of the amino acids found in the canonical set. It is doubly eerie to think that the constraints imposed by the laws of nature—which imbue the canonical set with the just-right properties for life—would force the same evolutionary outcome, time and time again. It appears as if the laws of nature preordained the canonical set of amino acids and drove the evolutionary process to the same endpoint. One of the study's authors, Rudrarup Bose, suggests that "life may not be just a set of accidental events. Rather, there may be some universal laws governing the evolution of life."[6] In other words, it appears as if the endpoint of chemical evolution has been designated by the very design of the universe itself—and, by extension, the Mind behind the universe. This means that if we rewound the tape of life and replayed the evolutionary process, it would always wind up at the same place—not because of the influence of natural selection, but because of the constraints the laws of nature place on the composition of the set of R groups that make up the side chains of amino acids.

For those who adopt an evolutionary perspective on life's origin, the biochemical anthropic principle makes it reasonable—for scientific reasons alone—to think that a Mind is responsible for jury-rigging the process to a predetermined endpoint. It looks as if a Mind purposed life to be present in the universe and structured the laws of nature so that, in this case, the uniquely optimal canonical set of amino acids would inevitably emerge. And this jury-rigging is scientifically detectable, bringing metaphysics into the scientific arena. Rejecting the metaphysical implications leaves us with no other conclusion than we have been very, very lucky.

Along these lines, it is provocative to consider the prebiotic chemistry that conceivably occurred on early Earth. Origin-of-life researchers have discovered chemical processes that could have, in principle, generated amino acids (the Strecker reaction), sugars (the formose reaction), fatty acids (the Fischer-Tropsch process), and nucleobases and nucleotides (the Sunderland reaction). As mentioned in chapter 8, the compounds that form the Krebs cycle appear to be emergent features on early Earth. In fact, it looks as if the Krebs cycle intermediates may well constitute a unique set of molecules. It also appears that these same types of geochemical and physicochemical constraints account for the design of the glycolytic pathway.

Again, one could take these results to indicate that the laws of nature have prearranged the design of two key pathways that play the central role in energy-harvesting pathways and central carbon metabolism. It is highly provocative and incredibly fortuitous that the physicochemical constraints that arise out of the laws of nature result in chemical processes that could, in principle, produce organic compounds with the just-right properties required for life. In fact, it is eerie.

So, unlike commonly held views of theistic evolution—in which (1) God let evolutionary processes run on their own once he created the universe, or (2) God's activity is undetectable—the theistic evolutionary model that emerges in light of the biochemical anthropic principle is one that clearly evinces a Creator's handiwork. According to this model, the Creator designed physics, chemistry, and biochemistry in such a way that the origin of life could be understood to be an inevitable outcome—and an outcome that he precisely intended. And, again, these insights can be known to us from scientific investigation. The Creator is no longer concealed. These insights also challenge materialistic or nontheistic conceptions of chemical evolution by returning teleology to biochemistry and origin-of-life research.

Still, as compelling as this version of theistic evolution may be, I am hesitant to fully embrace it, though I know many who will. My hesitancy stems primarily from scientific concerns, not theological ones. Even though it appears as though the laws of nature dictate the production of life's building blocks, we are still a long way off from understanding the origin of life. As astrophysicist Hugh Ross and I argue in *Origins of Life*, every conceivable model for chemical evolution suffers from seemingly intractable problems. Moreover, as I demonstrate in *Creating Life in the Lab*, work in prebiotic chemistry designed to gain insight into the process of chemical evolution and the origin of life through naturalistic processes, has ironically demonstrated the necessary role a Creator must have played, directly, to bring about the appearance of Earth's first life. In other words, evolutionary mechanisms appear insufficient to account for the origin of life (and biochemistry) even if the biochemical anthropic principle is in effect.

Prebiotic Chemistry and the Hand of God

I think few would dispute the statement that the origin of life stands as one of the most enigmatic scientific mysteries of our day. Scientists working on this mystery have gained some clues into the origin-of-life question from the fossil and geochemical records of Earth's oldest rock formations—yet this evidence affords them only a dim glimpse through the glass. (For a detailed discussion of the fossil and geochemical evidence that bears on the origin-of-life problem, see the book *Origins of Life*.)

Because of these limitations, origin-of-life researchers have no choice but to carry out most of their work in laboratory settings where they try to replicate the myriad steps they think contributed to the origin-of-life process. Pioneered in the early 1950s by the late Stanley Miller, prebiotic chemistry has become a scientific subdiscipline in its own right. This is the very work that has identified the chemical routes that could have, in principle, generated life's building blocks on early Earth.

Prebiotic Chemistry

The goals of prebiotic chemistry are threefold.

- *Proof of principle.* The objective of these types of experiments is to determine if a chemical or physical process that could have potentially contributed to one or more steps in the origin-of-life pathway even exists.
- *Mechanistic studies.* Once processes have been identified that could contribute to the emergence of life, researchers study them in detail to understand the mechanisms undergirding these physicochemical transformations.
- *Geochemical relevance.* Perhaps the most important goal of prebiotic studies is to establish the geochemical relevance of the physicochemical processes believed to have played a role in life's start. In other words, how well do the chemical and physical processes identified and studied in the laboratory translate to the likely conditions on early Earth?

Without question, over the last six to seven decades, origin-of-life researchers have been wildly successful with respect to the first two objectives. It is safe to say that prebiotic chemists have demonstrated that, *in principle*, the chemical and physical processes needed to generate life through chemical evolutionary pathways exist. And these pathways can, *in principle*, generate the building-block materials with the just-right properties for life.

But when it comes to the third objective, origin-of-life researchers have experienced frustration—and, arguably, failure. The failure stems from a phenomenon called unwarranted researcher intervention. This problem keeps me from embracing some form of theistic evolution as a way to account for the origin of life and the design of biochemical systems—the biochemical anthropic principle notwithstanding.

Researcher Intervention and Prebiotic Chemistry

In an ideal world, humans would not intervene at all in any prebiotic study, but this ideal isn't possible. Researchers involve themselves in the experimental design out of logistical necessity and to ensure that the results of the study are reproducible and interpretable. If researchers don't set up the experimental apparatus, adjust the starting conditions, add the appropriate reactants, and analyze the product, then, by definition, the experiment would never happen. Utilizing carefully controlled conditions and chemically pure reagents is necessary for reproducibility and to make sense of the results. In fact, this level of control is essential for proof of principle and mechanistic prebiotic studies—and perfectly acceptable.

However, when it comes to establishing geochemical relevance, the controlled conditions of the laboratory become a liability. Here, researcher intervention becomes potentially unwarranted. Needless to say, the conditions of early Earth were uncontrolled and chemically and physically complex. Chemically pristine and physically controlled conditions didn't exist. And, of course, origin-of-life researchers weren't present to oversee the processes and guide them to the desired end. Yet prebiotic simulation studies rarely take the actual conditions of early Earth fully into account in the experimental design. It is rarer for origin-of-life investigators to acknowledge this limitation.

This complication means that many prebiotic studies designed to simulate processes on early Earth seldom accomplish anything of the sort due to excessive researcher intervention. Yet it isn't always clear when examining an experimental design if researcher involvement is legitimate or unwarranted.

I pointed out this concern in my 2011 book *Creating Life in the Lab*. At that time, I felt like a lone voice in the wilderness. However, origin-of-life researchers are beginning to recognize the complications unwarranted researcher involvement causes for prebiotic simulation studies. For example, in 2018, origin-of-life investigator Clemens Richert, from the University of Stuttgart in Germany, acknowledged this very concern in a perspectives piece.[7] As Richert argues, the role of researcher intervention and a clear assessment of geochemical relevance is rarely acknowledged or properly explored in prebiotic simulation studies. Richert's commentary represents an important first step toward encouraging more realistic prebiotic simulation studies and a more cautious approach to interpreting the results of these studies. Hopefully, it will also lead to a more circumspect assessment on the importance of these types of studies when accounting for the various steps in the origin-of-life process.

Researcher Intervention and the Hand of God

Richert neglects to address the demanding and fastidious physicochemical transformations deemed central to chemical evolution by origin-of-life researchers. Even though these reactions reflect the constraints that arise from the laws of nature, they require exacting conditions to matriculate. As I discuss in *Creating Life in the Lab*, mechanistic studies indicate these processes often depend on precise conditions in the laboratory. Put another way, these processes take place only because of human intervention, with the researcher creating the just-right set of conditions in the laboratory for the reaction to successfully proceed. As a corollary, these processes would be unproductive on early Earth. They often require chemically pristine conditions; relatively high concentrations of reactants (that could never be realistically achieved on early Earth); carefully controlled order of additions; carefully regulated temperature, pH, and salinity levels; etc.

As Richert states, "It is not easy to see what replaced the flasks, pipettes and stir bars of a chemistry lab during prebiotic evolution, let alone the hands of the chemist who performed the manipulations. (And yes, most of us are not comfortable with the idea of divine intervention.)"[8]

Even though Richert and his many colleagues do whatever they can to eschew a Creator's role in the origin of life, could it be that abiogenesis (life from nonlife) required divine intervention?

I would argue that this conclusion flows from nearly seven decades of work in prebiotic chemistry and the consistent demonstration of the central role that origin-of-life researchers play in the success of prebiotic simulation studies. It is becoming increasingly evident, for whoever will see, that the hand of the researcher serves as the analog for the hand of God. Or as evolutionary biologist Simon Conway Morris so aptly states:

> Many of the experiments designed to explain one or other [sic] step in the origin of life are either of tenuous relevance to any believable prebiotic setting or involve an experimental rig in which the hand of the researcher becomes for all intents and purposes the hand of God.[9]

That is, intelligent agency becomes an indispensable feature of prebiotic simulation experiments in the lab. By extension, we can expect this reality to be in effect on early Earth. This provides empirical evidence that a Creator must have intervened to bring about the origin of life (and biochemistry).

Unwarranted researcher involvement presents an intractable problem for evolutionary explanations for the origin of life. More importantly, it provides direct empirical evidence that intelligent agency must play an active role in the generation of life from chemicals. For these two reasons, I prefer to interpret the biochemical anthropic principle within the creation model framework. This leads to an important expectation. If a Creator was involved with the origin and design of biochemical systems, then we should expect these systems to display features that indicate they were designed by a Mind.

The Revitalized Watchmaker Argument

As I argue in *The Cell's Design*, insights into the structure and function of biochemical systems provide us with this very evidence of biochemical design features. These features provide the opportunity to present a revitalized Watchmaker argument for a Creator's existence and necessary role in the design of biochemical systems, adding further support to my conviction that a Creator must have played a direct role in creating life.

British natural theologian William Paley (1743–1805) advanced this argument, suggesting that the complex interaction of a watch's precision parts, which serve the purpose of telling time, evince the work of an intelligent designer. By analogy, Paley argued that just as a watch requires a watchmaker, so, too, life requires a Creator, because living systems characteristically display numerous features defined by the precise interplay of complex parts toward accomplishing a purpose. Though Paley's argument hasn't fared well over the last century and a half, advances in biochemistry provide us with the opportunity to give his argument new life. For example, biochemists have made the provocative discovery that a number of protein complexes serve the cell as molecular-scale machines and motors—many of which bear an eerie similarity to machines made by human designers.

To illustrate the revitalized Watchmaker argument, I will describe three examples of protein complexes that resemble machines designed by human engineers: those that play a role in DNA metabolism (replication, transcription, and repair) and two of the proteins involved in the electron transport chain (ETC). I chose these examples because both DNA metabolism and the ETC evince the biochemical anthropic principle.

Turing Machines

At their most basic level of design, all computer operations are based on Turing machines. Named for British mathematician Alan Turing, Turing machines are not physical entities, but abstract, conceptual machines. They are "built" from three components: the input, the output, and the finite control.

The input consists of a stream of data read and altered by the finite control according to a specific set of rules. This process transforms the input into a new stream of data, the output. Both the input and output data streams are made up of a string (or sequence) of characters linked together. The finite control operates bit by bit on each character in the input string to produce the output string. Each finite control is limited to relatively simple transformations. However, complex computations and operations can be achieved by linking together several Turing machines, so that the output string of one Turing machine becomes the input string of another.

Biochemical Turing Machines and DNA Computing

More than a decade ago, computer scientist Leonard Adleman recognized that the proteins responsible for DNA replication, repair, and transcription operate as Turing machines.[10] The similarity between DNA metabolism and computer systems is so stark it inspired the advent of DNA computing, an exciting new arena of biotechnology.

To understand this analogy, let's start with the information stored in the nucleotide sequences of DNA. This information is, essentially, digital information. Whenever a complex biochemical process, such as DNA replication (see chapter 6, pages 214–216), takes place, the cell's machinery (proteins) takes the digital information in DNA as input and alters it in a prescribed manner, producing output strands of digital DNA information. The individual proteins serve as the finite control. While each protein can only perform a limited transformation of the DNA information, complex biochemical operations ensue when these biomolecules work in combination with one another.

In nanotechnology, DNA computers treat the nucleotide sequences as input and output strings. The finite control corresponds to the different chemical, biochemical, and physical processes that can be used to manipulate DNA in the laboratory. These operations alter the inputed DNA sequences, generating altered DNA sequences as outputs. In DNA computing, complex operations can be achieved by linking together simple laboratory operations, with the output of one laboratory operation serving as the input for the next.

DNA computers have several advantages compared to silicon-based computers. They can carry out many operations simultaneously (in parallel) as opposed to single operations, one at a time (serially). Additionally, DNA has the capacity to store an enormous quantity of information. One gram of DNA can house as much information as nearly one trillion CDs. Lastly, researchers believe that DNA computing operates near the theoretical capacity with regard to energy efficiency.

It's mind-boggling to think that the information-based activities of biochemical systems, which routinely take place in the cell, can be used to construct computers in a laboratory setting. The direct correspondence between input and output strings in computer operations and DNA sequences makes it clear: the cell's machinery that manipulates DNA consists of an ensemble of Turing machines. Molecular-level computers have long been the dream of nanotechnologists. By making use of the cell's information systems to build DNA computers, this dream may become a reality. And this hope rests on the astounding similarity between DNA metabolism (carried out by biochemical Turing machines) and a computer system's foundational operations.

The Protein Complexes of the Electron Transport Chain

Two protein complexes in the ETC also display remarkable machine-like features. F_1-F_0ATPase and respiratory complex I are ubiquitous protein complexes that represent two of the most important enzymes in biology because of the central role they play in energy-harvesting reactions.

Located in the inner membrane of mitochondria, F_1-F_0ATPase makes use of a proton gradient across the inner membrane to drive the production of ATP. (See chapter 8, pages 283–285.) In effect, F_1-F_0ATPase is a molecular-scale rotary motor. Displaying a mushroom shape, the F_1 component of the enzyme complex extends above the membrane's surface. The button of the mushroom is literally an engine turbine. The F_1-F_0 ATPase mushroom stalk is literally a rotor. During the catalytic cycle, the rotor interacts with the turbine, causing a mechanical displacement that powers the formation of ATP.

Positively charged hydrogen ions flow through the F_0 component, which is

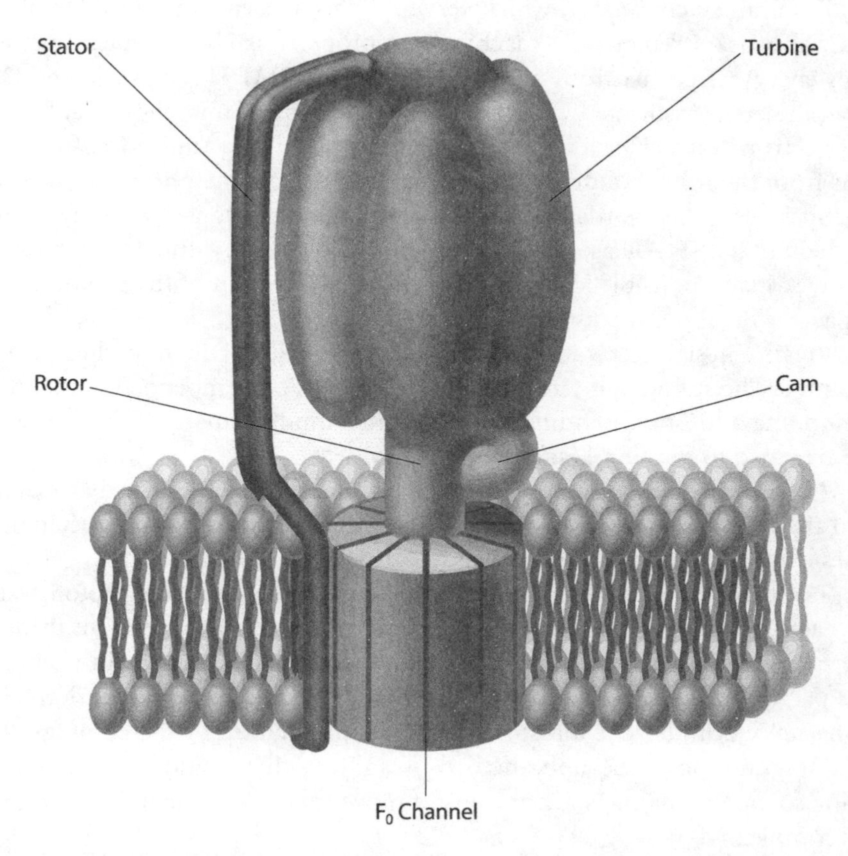

Figure 9.1: F_1-F_0 ATPase

embedded in the cell membrane. This flow drives the rotation of the rotor. Also extending from the membrane surface is a rod-shaped protein structure that functions as a *stator*. This protein rod holds the turbine in a stationary position as the rotor rotates.

The "electrical current" that flows through the channels of the F_0 complex is transformed into mechanical energy, which then drives the rotor's movement. A camshaft that extends at a right angle from the rotor's surface causes displacements of the turbine. These back-and-forth motions are used to produce ATP. A remarkable biochemical machine, this complex far exceeds the best machine produced by human engineers, in terms of its efficiency.

Respiratory complex I, which serves as the first enzyme complex of the ETC, also bears remarkable machine-like properties. The complex transfers high-energy electrons from a compound (called NADH) to coenzyme Q, a small molecule associated with the inner membrane of mitochondria. During the electron-transfer process, respiratory complex I also transports four protons from the mitochondria's interior across the inner membrane to the exterior space. In other words, respiratory complex I helps to generate the proton gradient that F_1-F_0 ATPase uses to generate ATP. By some estimates, respiratory complex I is responsible for establishing about 40 percent of the proton gradient across the inner membrane.

Massive in size, respiratory complex I is comprised of 45 individual protein subunits. The subunits interact to form two arms, one embedded in the inner membrane and one extending into the mitochondrial matrix. The two arms are arranged to form an L-shaped geometry.

The electron-transfer process occurs in the peripheral arm that extends into the mitochondrial matrix. Conversely, the proton-transport mechanism takes place in the membrane-embedded arm.

The machine-like behavior of respiratory complex I drives proton transport across the inner membrane. The process of transferring electrons through the peripheral arm results in conformational changes (changes in shape) in this part of the complex. The conformational changes drive the motion of an alpha-helix cylinder like a piston in the membrane arm of the complex. The pumping motion of the alpha-helix causes three other cylinders to tilt and, in doing so, opens channels for protons to move through the membrane arm of the complex.

Without question, there exists a striking similarity between the machine-like parts of F_1-F_0 ATPase and respiratory complex I and the machines designed and built by human engineers. There is also a startling similarity between the protein machinery that manipulates DNA during processes such as replication, repair, and transcription and Turing machines that serve as the foundation for the operation of computer systems. These similarities form the basis for a revitalized Watchmaker argument. We know from common experience that machines, motors, and computers are the work of human designers. The fact that we have discovered biomolecular motors, machines, and computers in the cell that are identical to the versions designed by humans suggests that these, too, must be the product of a Mind. It is doubly provocative to think that Turing machines are conceptual machines that exist only in human minds, yet inside the cell several actual Turing machines operate on DNA. Moreover, it appears these machine-like systems are part of biochemical systems that seem to be fundamentally dictated by the laws of nature to display the just-right properties needed for life.

The Biochemical Anthropic Principle and Creationism

Good reasons abound to view the biochemical anthropic principle in the context of a creation model approach. The optimal and unusual but just-right features of the chemical environment and the biomolecules and metabolic processes of core biochemical systems reflect both the intentional design of the living realm and a fitness for purpose. The fitness for purpose can be understood as evidence for the Creator's handiwork, instantiated in the creation through his direct involvement.

What about the discovery that the chemical environment and the features of biochemical systems appear to be fundamentally prescribed and constrained by the laws of nature? From a creation model approach, these insights provide further evidence that a Creator designed the universe to be biofriendly, with the universe displaying a fitness for purpose that centers around life's advent. If the Creator intended life to exist in the universe, wouldn't every feature of nature be designed to support that intention?

An analogy will hopefully shed some light on how I see the biochemical anthropic principle fitting into a creation model framework. Imagine a puzzle master who designs and builds jigsaw puzzles. He crafts each piece of the puzzle so all the pieces fit together. If they were off even a little bit, the puzzle could never be put together. Yet, even though each piece of the puzzle is designed to fit together perfectly, it is unlikely that the puzzle will ever assemble simply by shaking all the puzzle pieces together in a box. Instead, someone must play a direct and active role in combining the puzzle pieces. In like manner, in light of the biochemical anthropic principle, we could view life's building blocks, biomolecules, and metabolic systems to have been predetermined by the laws of nature to have the just-right properties needed for life. If they were any different, life could never exist. Still, these individual pieces must be assembled by an intelligent agent to form biochemical systems and the first cell.

The discovery of the biochemical anthropic principle naturally and logically introduces metaphysics into the origin-of-life question and the study of biochemical systems. It strongly indicates that a Mind was involved in the genesis of life—regardless of how one views the nature of Divine action. And this conclusion is music to my ears.

•••••••••••••

I had an epiphany at my friend John Harrelson's funeral, when his good friend described the impact that John had on his life, particularly when he pointed out that Harrelson's music served as the soundtrack to his life. I realized then that over the years I had assembled a soundtrack for my life as well. It features songs that perfectly captured the most important moments of my life—songs that take me back to special times whenever I hear them. Those songs, written by other people, I have appropriated as my own, because they seemed to capture my feelings and experiences—songs so aptly fit for purpose.

I am still assembling the soundtrack to my life. While writing this book, one song that played in my mind is "Michelangelo," the opening track from *Jefferson Street*, the second album released by the Christian blues rock band Three Crosses. Three Crosses isn't just one of my favorite Christian rock bands. It is one of my favorite rock bands of all time. Period.

Formed in New Jersey in 1994, Three Crosses only recorded three albums before they disbanded in 1998. The *Jefferson Street* album is by far their best work. Even though "Michelangelo" was considered a hit on Christian radio stations, the song never really resonated with me—until I began working on this book. Out of nowhere, I began to hum and sing the song. As I did, I found it to fit this time in my life perfectly.

God painted better than
Michelangelo
God painted better than
Pablo Picasso

The design of the universe is like a soundtrack, composed of the just-right features that make life possible. We are still in the process of assembling the universe's soundtrack. And the more scientists probe the workings of the universe—whether the feature under investigation falls in the realm of physics, chemistry, or biochemistry—the more each new insight brings a deep, underlying teleology and an unavoidable sense of design and purpose.

Notes

Introduction

1. Daniel J. Levitin, *This Is Your Brain on Music: The Science of a Human Obsession* (New York: Dutton, 2006).
2. *Encyclopedia of Human Evolution and Prehistory*, ed. Eric Delson et al., 2nd ed. (New York: Garland, 2000), s.vv. "Late Paleolithic," "Later Stone Age."
3. Christopher S. Henshilwood et al., "An Abstract Drawing from the 73,000-Year-Old Levels at Blombos Cave, South Africa," *Nature* 562 (October 4, 2018): 115–18, doi:10.1038/s41586-018-0514-3.
4. Levitin, *Your Brain on Music*, 6.
5. For example, see Iris Berent et al., "Language Universals in Human Brains," *Proceedings of the National Academy of Sciences, USA* 105, no. 14 (April 8, 2008): 5321–25, doi:10.1073/pnas.0801469105; Iris Berent et al., "Language Universals Engage Broca's Area," *PLOS ONE* 9, no. 4 (April 17, 2014): e95155, doi:10.1371/journal.pone.0095155.
6. Catherine Matacic, "Rhythm Might Be Hardwired in Humans," *Science* (December 19, 2016): doi:10.1126/science.aal0531.
7. Virginia Hughes, "Why Does Music Feel So Good?" *National Geographic*, April 11, 2013.
8. Diane Koopman, "The Science Behind Why Music Makes Us Feel So Good," *Lifehack*, lifehack.org/361240/the-science-behind-why-music-makes-feel-good.
9. Christopher Bergland, "Why Do the Songs from Your Past Evoke Such Vivid Memories?" *Psychology Today*, December 11, 2013, psychologytoday.com/us/blog/the-athletes-way/201312/why-do-the-songs-your-past-evoke-such-vivid-memories.
10. Bergland, "Songs from Your Past?"
11. Taylor Pittman, "Why You Like Listening to the Same Song Over and Over Again," *HuffPost Canada*, May 24, 2018, huffingtonpost.ca/entry/why-you-like-listening-same-song_n_5b06c900e4b05f0fc8458fc2.
12. Pittman, "Listening to the Same Song."
13. Brandon Carter, "Large Number Coincidences and the Anthropic Principle in Cosmology," *International Astronomical Union Symposium* 63 (1974): 291–98, doi:10.1017/s0074180900235638.
14. Michael J. Denton, *Nature's Destiny: How the Laws of Biology Reveal Purpose in the Universe* (New York: Free Press, 1998).
15. John D. Barrow et al., ed., *Fitness of the Cosmos for Life: Biochemistry and Fine-Tuning* (New York: Cambridge University Press, 2008).

Chapter 1: The Cosmological Anthropic Principle

1. David Browne, "Baby Hold On: Why Eddie Money Was the Patron Saint of Rock Uncool," *Rolling Stone*, September 13, 2019, rollingstone.com/music/music-features/eddie-money-appreciation-884179/.
2. Browne, "Baby Hold On."
3. A great summary of the history of the anthropic principle can be found in the chapter written by Ernan McMullin called "Tuning Fine-Tuning" for the treatise from John D. Barrow et al., ed., *Fitness of the Cosmos for Life: Biochemistry and Fine Tuning* (New York: Cambridge University Press, 2008), 70–94.
4. An accessible discussion on the effect of altering the universe's dimensions can be found in Geraint F. Lewis and Luke A. Barnes, *A Fortunate Universe: Life in a Finely Tuned Cosmos* (Cambridge: Cambridge University Press, 2016), 221–26.
5. An accessible discussion on the effect of altering the universe's expansion rate can be found in Lewis and Barnes, *A Fortunate Universe*, 221–26, and in Hugh Ross's *The Creator and the Cosmos: How the Latest Scientific Discoveries Reveal God*, 4th ed. (Covina, CA: RTB Press, 2018), 45–57, 237.
6. An excellent discussion of dark energy can be found in Ross's *Creator and the Cosmos*, 45–57.
7. Ross, *Creator and the Cosmos*, 233.
8. Ross, *Creator and the Cosmos*, 233–41.

Chapter 2: The Father of the Anthropic Principle

1. "500 Greatest Songs of All Time," *Rolling Stone*, December 11, 2003, rollingstone.com/music/music-lists/500-greatest-songs-of-all-time-151127/.
2. The material for this section is taken from an essay by Everett Mendelsohn, "Locating 'Fitness' and L. J. Henderson," in John D. Barrow et al., ed., *Fitness of the Cosmos for Life: Biochemistry and Fine Tuning* (New York: Cambridge University Press, 2008), 3–19.
3. Lawrence J. Henderson, *The Fitness of the Environment: An Inquiry into the Biological Significance of the Properties of Matter* (New York: Macmillan, 1913), vi–vii.
4. Henderson, *Fitness of the Environment*, 272.
5. Henderson, *Fitness of the Environment*, viii.
6. Henderson, *Fitness of the Environment*, 269.
7. Henderson, *Fitness of the Environment*, 271.
8. Henderson, *Fitness of the Environment*, 275.
9. Henderson, *Fitness of the Environment*, 279.
10. Henderson, *Fitness of the Environment*, 281.
11. Henderson, *Fitness of the Environment*, 308.
12. Lawrence J. Henderson, *The Order of Nature: An Essay* (Cambridge, MA: Harvard University Press, 1917), 8–9.
13. Henderson, *Order of Nature*, 191–92.
14. Henderson, *Order of Nature*, 201.
15. Mendelsohn, "Locating 'Fitness,'" 3–19.
16. John D. Barrow, "Chemistry and Sensitivity," in Barrow et al., *Fitness of the Cosmos*, 135.
17. Mendelsohn, "Locating 'Fitness,'" 3–19.

Chapter 3: The Chemical Anthropic Principle

1. Owen Gingerich, "Revisiting the Fitness of the Environment," in Barrow et al., *Fitness of the*

Cosmos, 23.

2. Simon Conway Morris and Ard A. Louis, "Is Water an Amniotic Eden or a Corrosive Hell? Emerging Perspectives on the Strangest Fluid in the Universe," in *Water and Life: The Unique Properties of H_2O*, ed. Ruth M. Lynden-Bell et al. (Boca Raton, FL: CRC Press, 2010), 3.
3. For example, see Michael J. Denton, *Nature's Destiny: How the Laws of Biology Reveal Purpose in the Universe* (New York: Free Press, 1998), 19–46. Also, Reasons to Believe hosts a nice seven-part series of articles written by chemists Dr. John Millam, Iain D. Sommerville, and environmental scientist Ken Klos, entitled "Water: Designed for Life," reasons.org/explore/blogs/todays-new-reason-to-believe/read/tnrtb/2013/05/20/water-designed-for-life-part-1-(of-7).
4. Charles Tanford, *The Hydrophobic Effect: Formation of Micelles and Biological Membranes* (New York: John Wiley and Sons, 1973); Charles Tanford, "The Hydrophobic Effect and the Organization of Living Matter," *Science* 200, no. 4345 (June 2, 1978): 1012–18, doi:10.1126/science.653353.
5. Martin F. Chaplin, "Water's Hydrogen Bond Strength," in Lynden-Bell et al., *Water and Life*, 69–86.
6. Morris and Louis, "Water an Amniotic Eden?" 8–9.
7. For a more comprehensive discussion of the beneficial properties of carbon dioxides that exert their influence on a planetary scale, see Denton, *Nature's Destiny*, 90–92.
8. Henderson, *Fitness of the Environment*, 135–36.
9. Ali Naqui, Britton Chance, and Enrique Cadenas, "Reactive Oxygen Intermediates in Biochemistry," *Annual Reviews of Biochemistry* 55 (1986): 137–66, doi:10.1146/annurev.bi.55.070186.001033.

Chapter 4: Proteins

1. For a discussion of the complexity of the simplest cell, see my book *The Cell's Design: How Chemistry Reveals the Creator's Artistry* (Grand Rapids, MI: Baker Books, 2008), 53–67.
2. Michael J. Denton, "Protein-Based Life as an Emergent Property of Matter: The Nature and Biological Fitness of Protein Folds," in John D. Barrow et al., ed., *Fitness of the Cosmos for Life: Biochemistry and Fine Tuning* (New York: Cambridge University Press, 2008), 260
3. Stephen Jay Gould, *Wonderful Life: The Burgess Shale and the Nature of History* (New York: W. W. Norton, 1990).
4. For example, see Toshiki Koga and Hiroshi Naraoka, "A New Family of Extraterrestrial Amino Acids in the Murchison Meteorite," *Science Reports* 7 (April 4, 2017): 636, doi.org/10.1038/s41598-017-00693-9.
5. Arthur L. Weber and Stanley L. Miller, "Reasons for the Occurrence of the Twenty Coded Protein Amino Acids," *Journal of Molecular Evolution* 17, no. 5 (September 1981): 273–84, doi:10.1007/bf01795749; H. James Cleaves II, "The Origin of the Biologically Coded Amino Acids," *Journal of Theoretical Biology* 263, no. 4 (April 21, 2010): 490–98, doi:10.1016/j.jtbi.2009.12.014.
6. Gayle K. Philip and Stephen J. Freeland, "Did Evolution Select a Nonrandom 'Alphabet' of Amino Acids?" *Astrobiology* 11, no. 3 (April 2011): 235–40, doi:10.1089/ast.2010.0567.
7. Philip and Freeland, "Nonrandom 'Alphabet' of Amino Acids?" 235–40.
8. Melissa Ilardo et al., "Adaptive Properties of the Genetically Encoded Amino Acid Alphabet Are Inherited from Its Subsets," *Scientific Reports* 9 (August 28, 2019): 12468, doi:10.1038/s41598-019-47574-x.

9. Matthias Granold et al., "Modern Diversification of the Amino Acid Repertoire Driven by Oxygen," *Proceedings of the National Academy of Sciences, USA* 115, no. 1 (January 2, 2018): 41–46, doi:10.1073/pnas.1717100115.
10. Jayanth R. Banavar and Amos Maritan, "Life on Earth: The Role of Proteins," in Barrow et al., *Fitness of the Cosmos*, 249.
11. Christine A. Orengo and Janet M. Thornton, "Protein Families and Their Evolution—A Structural Perspective," *Annual Review of Biochemistry* 74 (July 7, 2005): 867–900, doi:10.1146/annurev.biochem.74.082803.133029.
12. Denton, "Protein-Based Life," in Barrow et al., *Fitness of the Cosmos*, 256–79.
13. Cyrus Chothia et al., "Protein Folds in the All-β and All-α Classes," *Annual Review of Biophysics and Biomolecular Structure* 26 (June 1997): 597–627, doi:10.1146/annurev.biophys.26.1.597.
14. Chothia et al., "Protein Folds," 597–627.
15. Lijia Yu et al., "Grammar of Protein Domain Architectures," *Proceedings of the National Academy of Sciences, USA* 116, no. 9 (February 26, 2019): 3636–45, doi:10.1073/pnas.1814684116.
16. It should be noted that not all proteins automatically fold into the correct three-dimensional shape after they are produced at the ribosome. Even though physicochemical constraints exist, proteins can often misfold. Correct folding requires the assistance of chaperone proteins.
17. Banavar and Maritan, "Life on Earth," 225–55.
18. Banavar and Maritan, "Life on Earth," 239, 250.
19. Sven Hovmöller, Tuping Zhou, and Tomas Ohlson, "Conformations of Amino Acids in Proteins," *Acta Crystallographica D* 58, no. 5 (May 2002): 768–76, doi:10.1107/s0907444902003359.
20. Peter Y. Chou and Gerald D. Fasman, "Conformational Parameters for Amino Acids in Helical, β-Sheet, and Random Coil Regions Calculated from Proteins," *Biochemistry* 13, no. 2 (January 15, 1974): 211–22, doi.org/10.1021/bi00699a001.
21. Hang Chen, Fei Gu, and Zhengge Huang, "Improved Chou-Fasman Method for Protein Secondary Structure Prediction," *BMC Bioinformatics* 7 (December 12, 2006): S14, doi:10.1186/1471-2105-7-S4-S14.
22. Jeffrey Skolnick and Mu Gao, "Interplay of Physics and Evolution in the Likely Origin of Protein Biochemical Function," *Proceedings of the National Academy of Sciences, USA* 110, no. 23 (June 4, 2013): 9344–49, doi:10.1073/pnas.1300011110.
23. Mu Gao and Jeffrey Skolnick, "A Comprehensive Survey of Small-Molecule Binding Pockets in Proteins," *PLOS Computational Biology* 10 (October 24, 2013): e1003302, doi:10.1371/journal.pcbi.1003302.
24. Skolnick and Gao, "Interplay of Physics," 9344–49.
25. Banavar and Maritan, "Life on Earth," 250.
26. European Molecular Biology Laboratory–European Bioinformatics Institute, "Periodic Table of Protein Complexes," *ScienceDaily*, December 10, 2015, sciencedaily.com/releases/2015/12/151210144539.htm.
27. Sebastian E. Ahnert et al., "Principles of Assembly Reveal a Periodic Table of Protein Complexes," *Science* 350, no. 6266 (December 11, 2015): doi:10.1126/science.aaa2245.
28. Elke Deuerling and Bernd Bukau, "Chaperone-Assisted Folding of Newly Synthesized Proteins in the Cytosol," *Critical Reviews in Biochemistry and Molecular Biology* 39, no. 5–6 (2004): 261–77, doi:10.1080/10409230490892496.

29. Yun-Chi Tang et al., "Structural Features of the GroEL-GroES Nano-Cage Required for Rapid Folding of Encapsulated Protein," *Cell* 125, no. 5 (June 2, 2006): 903–14, doi: 10.1016/j.cell.2006.04.027.

Chapter 5: The Nucleic Acids

1. Francis S. Collins recounts this story in his book *The Language of God: A Scientist Presents Evidence for Belief* (New York: Free Press, 2006), 2–3.
2. Collins, *Language of God*, 3.
3. Georgi Muskhelishvili and Andrew Travers, "Integration of Syntactic and Semantic Properties of the DNA Code Reveals Chromosomes as Thermodynamic Machines Converting Energy into Information," *Cellular and Molecular Life Sciences* 70, no. 23 (December 2013): 4555–67, doi:10.1007/s00018-013-1394-1.
4. F. H. Westheimer, "Why Nature Chose Phosphates," *Science* 235, no. 4793 (March 6, 1987): 1173–78, doi:10.1126/science.2434996.
5. Felisa Wolfe-Simon et al., "A Bacterium That Can Grow by Using Arsenic Instead of Phosphorus," *Science* 332, no. 6034 (June 3, 2011): 1163–66, doi:10.1126/science.1197258.
6. Dan S. Tawfik and Ronald E. Viola, "Arsenate Replacing Phosphate—Alternative Life Chemistries and Ion Promiscuity," *Biochemistry* 50, no. 7 (February 22, 2011): 1128–34, doi: 10.1021/bi200002a.
7. Yu Xu et al., "Structural and Functional Consequences of Phosphate-Arsenate Substitutions in Selected Nucleotides: DNA, RNA, and ATP," *Journal of Physical Chemistry B* 116, no. 16 (April 26, 2012): 4801–11 doi:10.1021/jp300307u.
8. Ryszard Kierzek, Liyan He, and Douglas H. Turner, "Association of 2′–5′ Oligoribonucleotides," *Nucleic Acids Research* 20, no. 7 (April 11, 1992): 1685–90, doi:10.1093/nar/20.7.1685.
9. Gaspar Banfalvi, "Why Ribose Was Selected as the Sugar Component of Nucleic Acids," *DNA and Cell Biology* 25, no. 3 (March 2006): 189–96, doi:10.1089/dna.2006.25.189.
10. Huiqing Zhou et al., "m^{1}A and m^{1}G Disrupt A-RNA Structure through the Intrinsic Instability of Hoogsteen Base Pairs," *Nature Structural and Molecular Biology* 23 (September 2016): 803–10, doi:10.1038/nsmb.3270.
11. For example, see Albert Eschenmoser, "Chemical Etiology of Nucleic Acid Structure," *Science* 284, no. 5423 (June 25, 1999): 2118–24, doi:10.1126/science.284.5423.2118.
12. As a case in point, see Eveline Lescrinier, Matheus Froeyen, and Piet Herdewijn, "Difference in Conformational Diversity between Nucleic Acids with a Six-Membered 'Sugar' Unit and Natural 'Furanose' Nucleic Acids," *Nucleic Acids Research* 31, no. 12 (June 15, 2003): 2975–89, doi:10.1093/nar/gkg407.
13. Stanley L. Miller, "The Endogenous Synthesis of Organic Compounds," in *The Molecular Origins of Life: Assembling Pieces of the Puzzle*, ed. Andri Brack (Cambridge: Cambridge University Press, 1998), 59–85, doi:10.1017/cbo9780511626180.005.
14. Jean-Marc L. Pecourt, Jorge Peon, and Bern Kohler, "Ultrafast Internal Conversion of Electronically Excited RNA and DNA Nucleosides in Water," *Journal of the American Chemical Society* 122, no. 38 (September 1, 2000): 9348–49, doi:10.1021/ja0021520.
15. Dónall A. Mac Dónaill, "A Parity Code Interpretation of Nucleotide Alphabet Composition," *Chemical Communications* 18 (September 21, 2002): 2062–63, doi:10.1039/b205631c; Dónall A. Mac Dónaill, "Why Nature Chose A, C, G and U/T: An Error-Coding Perspective of Nucleotide Alphabet Composition," *Origins of Life and Evolution of the Biosphere* 33 (October 2003): 433–55, doi:10.1023/a:1025715209867.

16. Reiner Veitia and Chris Ottolenghi, "Placing Parallel Stranded DNA in an Evolutionary Context," *Journal of Theoretical Biology* 206, no. 2 (September 21, 2000): 317–22, doi:10.1006/jtbi.2000.2119.
17. For example, see Stephanie A. Havemann et al., "Incorporation of Multiple Sequential Pseudothymidines by DNA Polymerases and Their Impact on DNA Duplex Structure," *Nucleosides, Nucleotides, and Nucleic Acids* 27, no. 3 (March 2008): 261–78, doi:10.1080/15257770701853679 and Shuichi Hoshika et al., "Hachimoji DNA and RNA: A Genetic System with Eight Building Blocks," *Science* 363, no. 6429 (February 22, 2019): 884–87, doi:10.1126/science.aat0971.
18. H. James Cleaves II, Markus Meringer, and Jay Goodwin, "227 Views of RNA: Is RNA Unique in Its Chemical Isomer Space?" *Astrobiology* 15, no. 7 (July 2015): 538–58, doi:10.1089/ast.2014.1213; H. James Cleaves II et al., "One among Millions: The Chemical Space of Nucleic Acid-Like Molecules," *Journal of Chemical Information and Modeling* 59, no. 10 (October 28, 2019): 4266–77, doi:10.1021/acs.jcim.9b00632.
19. Tokyo Institute of Technology, "DNA Is Only One among Millions of Possible Genetic Molecules," *ScienceDaily* (November 11, 2019): sciencedaily.com/releases/2019/11/191111084915.htm.
20. Simon Conway Morris, *Life's Solution: Inevitable Humans in a Lonely Universe* (Cambridge: Cambridge University Press, 2003), 27–31.
21. Conway Morris, *Life's Solution*, 27–31.
22. M. R. Arkin et al., "Rates of DNA-Mediated Electron Transfer between Metallointercalators," *Science* 273, no. 5274 (July 26, 1996): 475–80, doi:10.1126/science.273.5274.475; Peter J. Dandliker, R. Erik Holmlin, and Jacqueline K. Barton, "Oxidative Thymine Dimer Repair in the DNA Helix," *Science* 275, no. 5305 (March 7, 1997): 1465–68, doi:10.1126/science.275.5305.1465.
23. Amie K. Boal et al., "Redox Signaling between DNA Repair Proteins for Efficient Lesion Detection," *Proceedings of the National Academy of Sciences, USA* 106, no. 36 (September 8, 2009): 15237–42, doi:10.1073/pnas.0908059106; Pamela A. Sontz et al., "DNA Charge Transport as a First Step in Coordinating the Detection of Lesions by Repair Proteins," *Proceedings of the National Academy of Sciences, USA* 109, no. 6 (February 7, 2012): 1856–61, doi:10.1073/pnas.1120063109; Michael A. Grodick, Natalie B. Muren, and Jacqueline K. Barton, "DNA Charge Transport within the Cell," *Biochemistry* 54, no. 4 (February 3, 2015): 962–73, doi:10.1021/bi501520w.
24. Elizabeth O'Brien et al., "The [4Fe4S] Cluster of Human DNA Primase Functions as a Redox Switch Using DNA Charge Transport," *Science* 355, no. 6327 (February 24, 2017): eaag1789, doi:10.1126/science.aag1789.
25. Kamaludin Dingle, Steffan Schaper, and Ard A. Louis, "The Structure of the Genotype-Phenotype Map Strongly Constrains the Evolution of Non-Coding RNA," *Interface Focus* 5, no. 6 (December 6, 2015): 20150053, doi:10.1098/rsfs.2015.0053.

Chapter 6: The Synthesis of Proteins and Nucleic Acids

1. Francis Crick, *What Mad Pursuit: A Personal View of Scientific Discovery* (New York: Basic Books, 1988), 109.
2. Ian S. Dunn, "Are Molecular Alphabets Universal Enabling Factors for the Evolution of Complex Life?" *Origins of Life and Evolution of Biospheres* 43, no. 6 (December 2013): 445–64, doi:10.1007/s11084-014-9354-9.
3. Francis Crick, "The Origin of the Genetic Code," *Journal of Molecular Biology* 38, no. 3

(December 28, 1968): 367–79, doi:10.1016/0022-2836(68)90392-6.
4. Crick, "Origin of the Genetic Code," 367–79.
5. Stephen J. Freeland, "Could an Intelligent Alien Predict Earth's Biochemistry?" in John D. Barrow et al., eds., *Fitness of the Cosmos for Life: Biochemistry and Fine-Tuning* (New York: Cambridge University Press, 2008), 300.
6. Hubert P. Yockey, *Information Theory and Molecular Biology* (Cambridge: Cambridge University Press, 1992), 180–83.
7. Malgorzata Wnętrzak et al., "The Optimality of the Standard Genetic Code Assessed by an Eight-Objective Evolutionary Algorithm," *BMC Evolutionary Biology* 18, no. 1 (December 18, 2018): 192, doi:10.1186/s12862-018-1304-0.
8. David Haig and Laurence D. Hurst, "A Quantitative Measure of Error Minimization in the Genetic Code," *Journal of Molecular Evolution* 33 (November 1991): 412–17, doi:10.1007/bf02103132.
9. Gretchen Vogel, "Tracking the History of the Genetic Code," *Science* 281, no. 5375 (July 17, 1998): 329, doi:10.1126/science.281.5375.329; Stephen J. Freeland and Laurence D. Hurst, "The Genetic Code Is One in a Million," *Journal of Molecular Evolution* 47, no. 3 (September 1998): 238–48, doi:10.1007/pl00006381; Stephen J. Freeland, Tao Wu, and Nick Keulmann, "The Case for an Error Minimizing Standard Genetic Code," *Origins of Life and Evolution of the Biosphere* 33 (October 2003): 457–77, doi:10.1023/A:1025771327614; Stephen J. Freeland, Robin D. Knight, and Laura F. Landweber, "Measuring Adaptation within the Genetic Code," *Trends in Biochemical Sciences* 25, no. 2 (February 1, 2000): 44–45, doi:10.1016/s0968-0004(99)01531-5; Stephen J. Freeland and Laurence D. Hurst, "Load Minimization of the Genetic Code: History Does Not Explain the Pattern," *Proceedings of the Royal Society B Biological Sciences* 265, no. 1410 (November 7, 1998): 2111–19, doi:10.1098/rspb.1998.0547; J. Gregory Caporaso, Michael Yarus, and Rob Knight, "Error Minimization and Coding Triplet/Binding Site Associations Are Independent Features of the Canonical Genetic Code," *Journal of Molecular Evolution* 61, no. 5 (November 2005): 597–607, doi:10.1007/s00239-004-0314-2.
10. Dimitri Gilis et al., "Optimality of the Genetic Code with Respect to Protein Stability and Amino-Acid Frequencies," *Genome Biology* 2 (October 24, 2001): doi:10.1186/gb-2001-2-11-research0049.
11. Artem S. Novozhilov, Yuri I. Wolf, and Eugene V. Koonin, "Evolution of the Genetic Code: Partial Optimization of a Random Code for Robustness to Translation Error in a Rugged Fitness Landscape," *Biology Direct* 2 (October 23, 2007): 24, doi:10.1186/1745-6150-2-24; Eugene V. Koonin and Artem S. Novozhilov, "Origin and Evolution of the Genetic Code: The Universal Enigma," *IUBMB Life* 61, no. 2 (February 2009): 99–111, doi:10.1002/iub.146; Stephen J. Freeland et al., "Early Fixation of an Optimal Genetic Code," *Molecular Biology and Evolution* 17, no. 4 (April 2000): 511–18, doi:10.1093/oxfordjournals.molbev.a026331; Wnętrzak et al., "Optimality of the Standard Genetic Code," 192.
12. Freeland et al., "Early Fixation," 511–18.
13. Regine Geyer and Amir Madany Mamlouk, "On the Efficiency of the Genetic Code after Frameshift Mutations," *PeerJ* 6 (May 21, 2018): e4825, doi:10.7717/peerj.4825.
14. Stefan Wichmann and Zachery Ardern, "Optimality in the Standard Genetic Code Is Robust with Respect to Comparison Code Sets," *Biosystems* 185 (November 2019): id. 104023, doi:10.1016/j.biosystems.2019.104023.
15. Wichmann and Ardern, "Optimality."
16. Shalev Itzkovitz and Uri Alon, "The Genetic Code Is Nearly Optimal for Allowing Additional

Information within Protein-Coding Sequences," *Genome Research* 17, no. 4 (April 2007): 405–12, doi:10.1101/gr.5987307.
17. Liat Shenhav and David Zeevi, "Resource Conservation Manifests in the Genetic Code," *Science* 370, no. 6517 (November 6, 2020): 683–87, doi:10.1126/science.aaz9642.
18. Wichmann and Ardern, "Optimality."
19. Freeland et al., "Early Fixation," 511–18.
20. Aline Bender, Parvana Hajieva, and Bernd Moosmann, "Adaptive Antioxidant Methionine Accumulation in Respiratory Chain Complexes Explains the Use of a Deviant Genetic Code in Mitochondria," *Proceedings of the National Academy of Sciences, USA* 105, no. 43 (October 28, 2008): 16496–501, doi:10.1073/pnas.0802779105.
21. Shlomi Reuveni, Måns Ehrenberg, and Johan Paulsson, "Ribosomes Are Optimized for Autocatalytic Production," *Nature* 547 (July 20, 2017): 293–97, doi:10.1038/nature22998.
22. Xinzhu Wei and Jianzhi Zhang, "On the Origin of Compositional Features of Ribosomes," *Genome Biology and Evolution* 10, no. 8 (August 1, 2018): 2010–16, doi:10.1093/gbe/evy169.
23. Zhen Shi et al., "Heterogeneous Ribosomes Preferentially Translate Distinct Subpools of mRNAs Genome-Wide," *Molecular Cell* 67, no. 1 (July 6, 2017): 71–83, doi:10.1016/j.molcel.2017.05.021.
24. Jeffrey A. Hussmann, Hendrick Osadnik, and Carol A. Gross, "Ribosomal Architecture: Constraints Imposed by the Need for Self-Production," *Current Biology* 27, no.16 (August 21, 2017): R798–R800, doi:10.1016/j.cub.2017.06.080.
25. J. D. Watson and Francis Crick, "The Structure of DNA," *Cold Spring Harbor Symposia on Quantitative Biology* 18 (1953): 123–31, doi:10.1101/sqb.1953.018.01.020.
26. Shoko Kimura et al., "Template-Dependent Nucleotide Addition in the Reverse (3′–5′) Direction by Thg1-like Protein," *Science Advances* 2, no. 3 (March 25, 2016): e1501397, doi:10.1126/sciadv.1501397.
27. James E. Graham, Kenneth J. Marians, and Stephen C. Kowalczykowski, "Independent and Stochastic Action of DNA Polymerases in the Replisome," *Cell* 169, no. 7 (June 15, 2017): 1201–13, doi:10.1016/j.cell.2017.05.041.
28. J. William Schopf, "When Did Life Begin?" in *Life's Origin: The Beginnings of Biological Evolution*, ed. J. William Schopf (Berkeley: University of California Press, 2002), 163.
29. Detlef D. Leipe, L. Aravind, and Eugene V. Koonin, "Did DNA Replication Evolve Twice Independently?" *Nucleic Acids Research* 27, no. 17 (September 1, 1999): 3389–401, doi:10.1093/nar/27.17.3389.

Chapter 7: Cell Membranes

1. S. J. Singer and Garth L. Nicolson, "The Fluid Mosaic Model of the Structure of Cell Membranes," *Science* 175, no. 4023 (February 18, 1972): 720–31, doi:10.1126/science.175.4023.720.
2. G. Vereb et al., "Dynamic, Yet Structured: The Cell Membrane Three Decades after the Singer-Nicolson Model," *Proceedings of the National Academy of Sciences, USA* 100, no. 14 (July 8, 2003): 8053–58, doi:10.1073/pnas.1332550100; Donald M. Engelman, "Membranes Are More Mosaic Than Fluid," *Nature* 438 (December 1, 2005): 578–80, doi:10.1038/nature04394.
3. Miles D. Houslay and Keith K. Stanley, *Dynamics of Biological Membranes: Influence on Synthesis, Structure and Function* (New York: John Wiley & Sons, 1982), 152–205.
4. Philippe F. Devaux, "Static and Dynamic Lipid Asymmetry in Cell Membranes," *Biochemistry* 30, no. 5 (February 5, 1991): 1163–73, doi:10.1021/bi00219a001.

5. Houslay and Stanley, *Dynamics of Biological Membranes*, 152–205.
6. Houslay and Stanley, *Dynamics of Biological Membranes*, 98–105.
7. Vereb et al., "Dynamic, Yet Structured," 8053–58; Engelman, "Membranes Are More Mosaic," 578–80.
8. Kai Simons and Elina Ikonen, "Functional Rafts in Cell Membranes," *Nature* 387 (June 5, 1997): 569–72, doi:10.1038/42408; D. A. Brown and E. London, "Functions of Lipid Rafts in Biological Membranes," *Annual Review of Cell and Developmental Biology* 14 (November 1998): 111–36, doi:10.1146/annurev.cellbio.14.1.111.
9. Danilo D. Lasic, "The Mechanism of Vesicle Formation," *Biochemical Journal* 256, no. 1 (November 15, 1988): 1–11, id. 3066342, doi:10.1042/bj2560001.
10. Lasic, "Mechanism of Vesicle Formation," 1–11.
11. For example, see Barry R. Lentz, Tamra J. Carpenter, and Dennis R. Alford, "Spontaneous Fusion of Phosphatidylcholine Small Unilamellar Vesicles in the Fluid Phase," *Biochemistry* 26, no.17 (August 25, 1987): 5389–97, doi:10.1021/bi00391a026.
12. N. L. Gershfeld, "The Critical Unilamellar Lipid State: A Perspective for Membrane Bilayer Assembly," *Biochimica et Biosphysica Acta* 988, no. 3 (December 6, 1989): 335–50, doi:10.1016/0304-4157(89)90009-9.
13. For example, see N. L. Gershfeld et al., "Critical Temperature for Unilamellar Vesicle Formation in Dimyristoylphosphatidylcholine Dispersions from Specific Heat Measurements," *Biophysical Journal* 65, no. 3 (September 1, 1993): 1174–79, doi:10.1016/S0006-3495(93)81157-3.
14. N. L. Gershfeld, "Spontaneous Assembly of a Phospholipid Bilayer as a Critical Phenomenon: Influence of Temperature, Composition, and Physical State," *Journal of Physical Chemistry* 93, no. 13 (June 1, 1989): 5256–61, doi:10.1021/j100350a043.
15. For example, see Lionel Ginsberg, Daniel L. Gilbert, and N. L. Gershfeld, "Membrane Bilayer Assembly in Neural Tissue of Rat and Squid as a Critical Phenomenon: Influence of Temperature and Membrane Proteins," *Journal of Membrane Biology* 119 (January 1991): 65–73, doi:10.1007/BF01868541.
16. For example, see K. E. Tremper and N. L. Gershfeld, "Temperature Dependence of Membrane Lipid Composition in Early Blastula Embryos of *Lytechinus pictus*: Selective Sorting of Phospholipids into Nascent Plasma Membranes," *Journal of Membrane Biology* 171 (September 1999): 47–53, doi:10.1007/s002329900557.
17. For example, see A. J. Jin et al., "A Singular State of Membrane Lipids at Cell Growth Temperatures," *Biochemistry* 38, no. 40 (October 1, 1999): 13275–78, doi:10.1021/bi9912084.
18. For example, see N. L. Gershfeld and M. Murayama, "Thermal Instability of Red Blood Cell Membrane Bilayers: Temperature Dependence of Hemolysis," *Journal of Membrane Biology* 101 (December 1988): 67–72, doi:10.1007/BF01872821.
19. J. N. Israelachvili, S. Marčelja, and R. G. Horn, "Physical Principles of Membrane Organization," *Quarterly Reviews of Biophysics* 13, no. 2 (May 1980): 121–200, doi:10.1017/S0033583500001645.
20. William R. Hargreaves and David W. Deamer, "Liposomes from Ionic, Single-Chain Amphiphiles," *Biochemistry* 17, no. 18 (September 5, 1978): 3759–68, doi:10.1021/bi00611a014.
21. Charles L. Apel, David W. Deamer, and Michael N. Mautner, "Self-Assembled Vesicles of Monocarboxylic Acids and Alcohols: Conditions for Stability and for the Encapsulation of Biopolymers," *Biochimica et Biophysica Acta* 1559, no. 1 (February 10, 2002): 1–9, doi:10.1016/s0005-2736(01)00400-x.

22. David W. Deamer and R. M. Pashley, "Amphiphilic Components of the Murchison Carbonaceous Chondrite: Surface Properties and Membrane Formation," *Origins of Life and Evolution of the Biosphere* 19, no. 1 (1989): 21–38, doi:10.1007/BF01808285; David W. Deamer, "Boundary Structures Are Formed by Organic Components of the Murchison Carbonaceous Chondrite," *Nature* 317 (October 31, 1985), 792–94, doi:10.1038/317792a0.
23. Hargreaves and Deamer, "Liposomes," 3759–68; Apel, Deamer, and Mautner, "Self-Assembled Vesicles," 1–9.
24. Hargreaves and Deamer, "Liposomes," 3759–68; Apel, Deamer and Mautner, "Self-Assembled Vesicles," 1–9.
25. Trishool Namani and David W. Deamer, "Stability of Model Membranes in Extreme Environments," *Origins of Life and Evolution of Biospheres* 38, no. 4 (August 2008): 329–41, doi:10.1007/s11084-008-9131-8.
26. S. E. Maurer et al., "Chemical Evolution of Amphiphiles: Glycerol Monoacyl Derivatives Stabilize Plausible Prebiotic Membranes," *Astrobiology* 9, no. 10 (December 2009): 979–87, doi:10.1089/ast.2009.0384.
27. Bernd R. T. Simoneit, Ahmed I. Rushdi, and David W. Deamer, "Abiotic Formation of Acylglycerols under Simulated Hydrothermal Conditions and Self-Assembly Properties of Such Lipid Products," *Advances in Space Research* 40, no. 11 (2007): 1649–56, doi:10.1016/j.asr.2007.07.034.
28. Sheref S. Mansy et al., "Template-Directed Synthesis of a Genetic Polymer in a Model Protocell," *Nature* 454 (July 3, 2008): 122–25, doi:10.1038/nature07018; David W. Deamer, "How Leaky Were Primitive Cells?" *Nature* 454 (July 3, 2008): 37–38, doi:10.1038/454037a.
29. Gordon Sproul, "Abiogenic Syntheses of Lipoamino Acids and Lipopeptides and Their Prebiotic Significance," *Origins of Life and Evolution of Biospheres* 45, no. 4 (December 2015): 427–37, doi:10.1007/s11084-015-9451-4.
30. MaRosa Infante, Aurora Pinazo, and Joan Seguer, "Non-Conventional Surfactants from Amino Acids and Glycolipids: Structure, Preparation and Properties," *Colloids and Surfaces A: Physicochemical and Engineering Aspects* 123–124 (May 15, 1997): 49–70, doi:10.1016/S0927-7757(96)03793-4; Michael Ambuehl et al., "Configurational Changes Accompanying Vesiculation of Mixed Single-Chain Amphiphiles," *Langmuir* 9, no. 1 (January 1, 1993): 36–38, doi:10.1021/la00025a011.
31. Hiroyuki Fukuda et al., "Bilayer-Forming Ion Pair Amphiphiles from Single-Chain Surfactants," *Journal of the American Chemical Society* 112, no. 4 (February 1, 1990): 1635–37, doi:10.1021/ja00160a057; S. A. Safran et al., "Spontaneous Vesicle Formation by Mixed Surfactants," *Progress in Colloid and Polymer Science* 84 (1991): 3–7, doi:10.1007/BFb0115925.
32. S. M. Gruner et al., "Lipid Polymorphism: The Molecular Basis of Nonbilayer Phases," *Annual Review of Biophysics and Biophysical Chemistry* 14 (June 1985): 211–38, doi:10.1146/annurev.bb.14.060185.001235.
33. S. M. Gruner, "Intrinsic Curvature Hypothesis for Biomembrane Lipid Composition: A Role for Nonbilayer Lipids," *Proceedings of the National Academy of Sciences, USA* 82, no. 11 (June 1, 1985): 3665–69, doi:10.1073/pnas.82.11.3665; P. R. Cullis and B. de Kruijff, "Lipid Polymorphism and the Functional Roles of Lipids in Biological Membranes," *Biochimica et Biophysica Acta* 559, no. 4 (December 20, 1979): 399–420, doi:10.1016/0304-4157(79)90012-1.
34. G. Dennis Sprott, "Archaeal Membrane Lipids and Applications," in *eLS* (Hoboken, NJ: John Wiley & Sons, 2011): doi:10.1002/9780470015902.a0000385.pub3.

35. Parkson Lee-Gau Chong, "Archaebacterial Bipolar Tetraether Lipids: Physico-Chemical and Membrane Properties," *Chemistry and Physics of Lipids* 163, no. 3 (March 2010): 253–65, doi:10.1016/j.chemphyslip.2009.12.006.
36. Daniel Balleza et al., "Ether- versus Ester-Linked Phospholipid Bilayers Containing Either Linear or Branched Apolar Chains," *Biophysical Journal* 107, no. 6 (September 16, 2014): 1364–74, doi:10.1016/j.bpj.2014.07.036; S. Deren Guler et al., "Effects of Ether vs. Ester Linkage on Lipid Bilayer Structure and Water Permeability," *Chemistry and Physics of Lipids* 160, no. 1 (July 2009): 33–44, doi:10.1016/j.chemphyslip.2009.04.003; Antonella Caforio and Arnold J. M. Driessen, "Archaeal Phospholipids: Structural Properties and Biosynthesis," *Biochimica et Biophysica Acta* 1862, no. 11 (November 2017): 1325–39, doi:10.1016/j.bbalip.2016.12.006.
37. Cullis and de Kruijff, "Lipid Polymorphism," 399–400.
38. W. Dowhan, "Molecular Basis for Membrane Phospholipid Diversity: Why Are There So Many Lipids?" *Annual Review of Biochemistry* 66 (1997): 199–232, doi:10.1146/annurev.biochem.66.1.199.
39. Houslay and Stanley, *Dynamics of Biological Membranes*, 51–65.
40. Dowhan, "Membrane Phospholipid Diversity," 199–232.
41. Giuseppe Paradies et al., "Functional Role of Cardiolipin in Mitochondrial Bioenergetics," *Biochimica et Biophysica Acta—Bioenergetics* 1837, no. 4 (April 2014): 408–17, doi:10.1016/j.bbabio.2013.10.006.
42. Anna L. Duncan, Alan J. Robinson, and John E. Walker, "Cardiolipin Binds Selectively but Transiently to Conserved Lysine Residues in the Rotor of Metazoan ATP Synthases," *Proceedings of the National Academy of Sciences, USA* 113, no. 31 (August 2, 2016): 8687–92, doi:10.1073/pnas.1608396113.
43. Dowhan, "Membrane Phospholipid Diversity," 199–232.

Chapter 8: Energy-Harvesting Pathways

1. David Marchese, "The Lion in Johnny Winter: A Tribute to the Guitar Icon," *Rolling Stone*, July 17, 2014, rollingstone.com/music/music-news/the-lion-in-johnny-winter-a-tribute-to-the-guitar-icon-242043.
2. Enrique Meléndez-Hevia et al., "Theoretical Approaches to the Evolutionary Optimization of Glycolysis—Chemical Analysis," *European Journal of Biochemistry* 244, no. 2 (March 1997): 527–43, doi:10.1111/j.1432-1033.1997.t01-1-00527.x.
3. Arren Bar-Even et al., "Rethinking Glycolysis: On the Biochemical Logic of Metabolic Pathways," *Nature Chemical Biology* 8 (June 2012): 509–17, doi:10.1038/nchembio.971.
4. Bar-Even et al., "Rethinking Glycolysis," 509–17.
5. Bar-Even et al., "Rethinking Glycolysis," 509–17.
6. Bar-Even et al., "Rethinking Glycolysis," 509–17.
7. Bar-Even et al., "Rethinking Glycolysis," 509–17.
8. Bar-Even et al., "Rethinking Glycolysis," 509–17.
9. Reinhart Heinrich et al., "Theoretical Approaches to the Evolutionary Optimization of Glycolysis: Thermodynamic and Kinetic Constraints," *European Journal of Biochemistry* 243, nos. 1–2 (January 1997): 191–201, doi:10.1111/j.1432-1033.1997.0191a.x; Oliver Ebenhöh and Reinhart Heinrich, "Evolutionary Optimization of Metabolic Pathways. Theoretical Reconstruction of the Stoichiometry of ATP and NADH Producing Systems," *Bulletin of Mathematical Biology* 63, no. 1 (January 2001): 21–55, doi:10.1006/bulm.2000.0197.
10. Bar-Even et al., "Rethinking Glycolysis," 509–17; Meléndez-Hevia et al., "Evolutionary

Optimization of Glycolysis," 527–43.

11. Bar-Even et al., "Rethinking Glycolysis," 509–17; Melendez-Hevia et al., "Evolutionary Optimization of Glycolysis," 527–43.
12. Meléndez-Hevia et al., "Evolutionary Optimization of Glycolysis," 527–43.
13. For example, see Ebenhöh and Heinrich, "Evolutionary Optimization of Metabolic Pathways," 21–55; Steven J. Court, Bartlomiej Waclaw, and Rosalind J. Allen, "Lower Glycolysis Carries a Higher Flux than Any Biochemically Possible Alternative," *Nature Communications* 6 (September 29, 2015): id. 8427, doi:10.1038/ncomms9427.
14. Ebenhöh and Heinrich, "Evolutionary Optimization of Metabolic Pathways," 21–55.
15. Court, Waclaw, and Allen, "Lower Glycolysis."
16. Meléndez-Hevia et al., "Evolutionary Optimization of Glycolysis," 527–43.
17. Avi Flamholz et al., "Glycolytic Strategy as a Tradeoff between Energy Yield and Protein Cost," *Proceedings of the National Academy of Sciences, USA* 110, no. 24 (June 11, 2013): 10039–44, doi:10.1073/pnas.1215283110.
18. Ebenhöh and Heinrich, "Evolutionary Optimization of Metabolic Pathways," 21–55.
19. Elad Noor et al., "Central Carbon Metabolism as a Minimal Biochemical Walk between Precursors for Biomass and Energy," *Molecular Cell* 39, no. 5 (September 10, 2010): 809–20, doi:10.1016/j.molcel.2010.08.031.
20. Enrique Meléndez-Hevia, Thomas G. Waddell, and Marta Cascante, "The Puzzle of the Krebs Citric Acid Cycle: Assembling the Pieces of Chemically Feasible Reactions, and Opportunism in the Design of Metabolic Pathways during Evolution," *Journal of Molecular Evolution* 43 (September 1996): 293–303, doi:10.1007/bf02338838.
21. Bar-Even et al., "Rethinking Glycolysis," 509–17.
22. Arren Bar-Even et al., "Hydrophobicity and Charge Shape Cellular Metabolite Concentrations," *PLOS Computational Biology* 7, no. 10 (October 6, 2011): e1002166, doi:10.1371/journal.pcbi.1002166.
23. Harold J. Morowitz et al., "The Origin of Intermediary Metabolism," *Proceedings of the National Academy of Sciences, USA* 97, no. 14 (July 5, 2000): 7704–8, doi:10.1073/pnas.110153997.
24. Kamila B. Muchowska, Sreejith J. Varma, and Joseph Moran, "Synthesis and Breakdown of Universal Metabolic Precursors Promoted by Iron," *Nature* 569 (May 2, 2019): 104–7, doi:10.1038/s41586-019-1151-1; Robert Pascal, "A Possible Non-biological Reaction Framework for Metabolic Processes on Early Earth," *Nature* 569 (May 1, 2019): 47–49, doi:10.1038/d41586-019-01322-3.
25. Eric Smith and Harold J. Morowitz, "Universality in Intermediary Metabolism," *Proceedings of the National Academy of Sciences, USA* 101, no. 36 (September 7, 2004): 13168–73, doi:10.1073/pnas.0404922101.
26. Markus A. Keller, Alexandra V. Turchyn, and Markus Ralser, "Non-enzymatic Glycolysis and Pentose Phosphate Pathway-Like Reactions in a Plausible Archean Ocean," *Molecular Systems Biology* 10, no. 4 (April 25, 2014): 725, doi:10.1002/msb.20145228.
27. Keller, Turchyn, and Ralser, "Non-enzymatic Glycolysis," 725.
28. John Prebble, "Peter Mitchell and the Ox Phos Wars," *Trends in Biochemical Sciences* 27, no. 4 (April 1, 2002): 209–12, doi:10.1016/S0968-0004(02)02059-5.
29. Leslie E. Orgel, "Are You Serious, Dr Mitchell?" *Nature* 402, no. 17 (November 4, 1999): doi:10.1038/46903.
30. Nick Lane, John F. Allen, and William Martin, "How Did LUCA Make a Living? Chemiosmosis in the Origin of Life," *BioEssays* 32, no. 4 (March 26, 2010): 271–80, doi:10.1002/

bies.200900131.
31. Nick Lane, "Why Are Cells Powered by Proton Gradients?" *Nature Education* 3, no. 9 (2010): 18.
32. Nick Lane, "Bioenergetic Constraints on the Evolution of Complex Life," *Cold Spring Harbor Perspectives in Biology* 6, no. 5 (May 2014): a015982, doi:10.1101/cshperspect.a015982.

Chapter 9: Implications of the Biochemical Anthropic Principle

1. David Deamer, "A Giant Step towards Artificial Life?" *TRENDS in Biotechnology* 23, no. 7 (July 1, 2005): 336–338, doi:10.1016/j.tibtech.2005.05.008.
2. Arren Bar-Even et al., "Hydrophobicity and Charge Shape Cellular Metabolite Concentrations," *PLOS Computational Biology* 7, no. 10 (October 6, 2011): e1002166, doi:10.1371/journal.pcbi.1002166.
3. Bar-Even et al., "Hydrophobicity and Charge Shape."
4. Melissa Ilardo et al., "Adaptive Properties of the Genetically Encoded Amino Acid Alphabet Are Inherited from Its Subsets," *Scientific Reports* 9 (August 28, 2019): 12468, doi:10.1038/s41598-019-47574-x.
5. Ilardo et al., "Adaptive Properties."
6. Tokyo Institute of Technology, "Scientists Find Biology's Optimal 'Molecular Alphabet' May Be Preordained," *ScienceDaily* (September 10, 2019), sciencedaily.com/releases/2019/09/190910080017.htm.
7. Clemens Richert, "Prebiotic Chemistry and Human Intervention," *Nature Communications* 9 (December 12, 2018): 5177, doi:10.1038/s41467-018-07219-5.
8. Richert, "Prebiotic Chemistry," 5177.
9. Simon Conway Morris, *Life's Solution: Inevitable Humans in a Lonely Universe* (Cambridge: Cambridge University Press, 2003), 41.
10. Leonard M. Adleman, "Computing with DNA," *Scientific American* 279, no. 2 (August 1998): 54–61, doi:10.1038/scientificamerican0898-54.

Index

About the Author

Fazale "Fuz" R. Rana is VP of Research and Apologetics at Reasons to Believe (RTB) and holds a PhD in chemistry with an emphasis in biochemistry from Ohio University.

Fuz conducted postdoctoral work at the Universities of Virginia and Georgia and worked for seven years as a senior scientist in product development for Procter & Gamble. He has published articles in peer-reviewed scientific journals, delivered presentations at international scientific meetings, and addressed the relationship between science and Christianity at churches and universities in the United States and abroad. Since joining RTB in 1999, Fuz has participated in numerous podcasts and videos and authored countless blog articles and several books, including *Origins of Life*, *The Cell's Design*, and *Creating Life in the Lab*.

Fuz and his wife, Amy, live in Southern California.

About Reasons to Believe

We at Reasons to Believe seek to dispel the idea that religious beliefs and scientific studies should be kept separate. Our scholar team, consisting of three PhD scientists and a philosopher-theologian, offers distinctive and fascinating insights on topics ranging from biblical creation to cutting-edge biotechnology.

Our intent is to create a space where ideas and ideologies can be explored fearlessly and where a spirit of curiosity is welcomed. We aim to present research and start a conversation—because people deserve respect and a safe forum for discussing their views.

For more information, visit reasons.org.

For inquiries, contact us via:
818 S. Oak Park Rd.
Covina, CA 91724
(855) REASONS | (855) 732-7667
ministrycare@reasons.org

YouTube